Do Not Trim

£6·99.

**FOCUS
ON
PHYSICS**

# FOCUS ON PHYSICS

**Sylvia Chaplin**
*formerly of Highfield School*

**John Keighley, B.Sc.**
*Marlborough College*

*Illustrated by*
**Lacey Hawkins**

WHEATON
A Division of Pergamon Press

**SECOND EDITION**

A. Wheaton & Company Limited
*A Division of Pergamon Press*
Hennock Road, Exeter EX2 8RP

Pergamon Press Ltd
Headington Hill Hall, Oxford OX3 0BW

Pergamon Press Inc
Maxwell House, Fairview Park, Elmsford, New York 10523

Pergamon of Canada Ltd
75 The East Mall, Toronto, Ontario M8Z 2L9

Pergamon Press (Australia) Pty Ltd
19a Boundary Street, Rushcutters Bay, N.S.W. 2011

Pergamon Press GmbH
6242 Kronberg/Taunus, Pferdstrasse 1, Frankfurt-am-Main,
West Germany

Second Edition Copyright © 1977
Sylvia Chaplin and John Keighley

*134008*

374.53
CHA

First published 1974
Reprinted (with corrections) 1975
Second Edition 1977
Reprinted (with revisions) 1978
*Printed in Great Britain by A. Wheaton & Co. Ltd, Exeter*
ISBN 0 08 021016 3

*0047979*

The course provided in this single volume is designed as an introduction to physics and has been written with the needs of the CSE student especially in mind. It is hoped that the use of colour and the many photographs of industrial and everyday applications of the principles studied will make the book attractive and easily readable.

Physics teaching in schools has changed considerably in the past few years. This book reflects the modern approach. The student is encouraged to think for himself rather than to learn by heart a mass of facts. Questions have therefore been included in the text in order to make the reader consider carefully what he is reading, to help him consolidate his knowledge before he progresses further, and to provide that change of occupation which is so valuable an aid to concentration. To gain the maximum benefit the answers should be thought out and written down before recourse is had to the solutions at the back. However, even if this advice is not always followed the student will at least (we hope!) be thinking about the problem while he is looking up the answer.

SI units have been used throughout. Nearly all topics that find a place in the syllabuses of the various Regional Boards are dealt with in the text and in "Things to work out" at the end of each chapter. The experiments suggested in the "Things to do" section can in most cases easily be carried out at home and with very simple apparatus. One or two that are more difficult have been included, but with a little help from parent or teacher these should give the student who is good with his hands a sense of satisfaction and achievement.

In this new edition we have taken the opportunity to revise some of the text and to add new material on a number of topics which have appeared in CSE examinations since the book was first written. Major additions are the new chapters on motion and energy. We have also added a table (see inside back cover) giving names of units and some idea of their size. Additional questions have also been added at the end of some chapters.

For the most part, we have ensured that the page numbering in this second edition remains the same as the first. We are aware of the difficulties which can confront teachers when the page numbering in new editions of established texts is significantly altered.

We are indebted to teachers who have made comments on the text of the first edition; many of their suggestions are incorporated into this new edition. We shall always be grateful for suggestions for improving the text that arise as a result of using the book in the classroom.

At the first reprint (1978) of the second edition corrections have been implemented and some small alterations made to update and improve the text. In order that students may become familiar with various electrical symbols they are likely to see, we have included some of the alternatives still in common use (and found in 1977 examination papers). The inclusion of these will, it is hoped, offer some practice to pupils in recognizing and using the different symbols they will come across for some time to come.

*Sylvia Chaplin*
*John Keighley*

infra-red rays – reflection of heat radiation – radiant heat from the sun – glass as a transmitter – vacuum flask – heat transfer in space probes and fast aircraft

# LIGHT

**Part One**

*Banks of light used on a film set*
(*from the film* Invitation to the Dance)

# LIGHT TRAVELS IN STRAIGHT LINES

## 1.1 Sources of light

If you have ever played the children's game of Blind Man's Buff you will probably appreciate how the world appears to someone who has really lost his sight. From our birth we learn about our surroundings by using our senses of sight, touch, hearing, taste and smell. To see things we use our eyes. The blind man in Fig. 1.1. is using another pair of eyes — those of a specially trained dog. He is also using his sense of touch to help him find his way down the steps of a bus.

## CHAPTER 1

**Fig. 1.1** *A specially trained guide dog assists a blind man*

However, there are times when those of us who have nothing wrong with our eyes cannot see the things around us. Usually this is when it is dark, when there is no *light energy* for our eyes to receive. Our eyes are sensitive to light and we use them to detect it.

But what is light? Where does it come from? And what are its properties?

Throughout the following chapters we shall try to answer these questions about light, and we shall see how man has used this knowledge to enable him to become even more aware of the world around him.

One of man's first attempts to master his environment was to find some way of overcoming the state of darkness which occurred when his natural source of light was on the other side of the earth. He achieved this by providing himself with artificial sources of light.

*QUESTION 1:* (a) What is the earth's natural source of light? (b) List six artificial sources of light.

Objects that give out light are said to be *luminous*. Some animals, such as the glow-worm, have luminous spots on their bodies which glow in the dark after exposure to light, as do luminous paints (Fig. 1.2). Many

**Fig. 1.2** *Luminous patches help in road safety*

substances will give out light only when heated to a temperature of more than 800°C. Such substances are said to be *incandescent*. Quicklime is one of these incandescent substances: it was once used to provide stage lighting in theatres and we still talk about actors, film stars and famous people being "in the limelight".

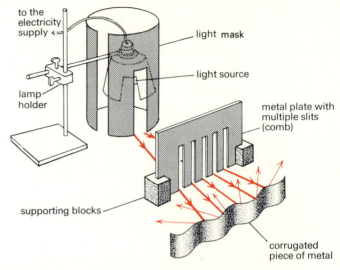

**Fig. 1.3** *Experiment to determine the effect of different materials on light*

## 1.2 What happens when light falls on different materials?

The simple experiment shown in Fig. 1.3 can help us answer this question. A light source is placed behind a metal plate with multiple slits in it (at home you can use a torch placed behind the teeth of a comb) and the light rays are allowed to fall onto the surface of a material positioned as shown. Try this on a succession of different materials: pieces of mirror, plain glass, frosted glass, wood, tin, corrugated metal, tissue paper and newspaper.

You will find that when light falls on a material, one of the following, or a combination of the following, takes place:

a. The light bounces back. This *reflection* of light takes place either (i) regularly (as in the case of the piece of mirror, which has a flat surface), or (ii) irregularly (as in the case of the corrugated metal, which has an uneven surface).

b. The light passes through. When this *transmission* of light takes place the material is said to be *transparent*. The transmission occurs either (i) regularly (as in the case of plain glass), or (ii) irregularly, when it is said to be *diffused* (as in the case of frosted glass). Pearl bulbs and shades are used in our homes to give diffused lighting and to cut down the hard glare from unshielded bulbs, which is tiring and harmful to the eyes.

c. The light does not pass through but is *absorbed*. A material that does not let light through is said to be *opaque*.

A combination of these actions is seen in the materials that allow some light to pass through but not all. Such materials are said to be *translucent*; in the experiment the tissue paper and newspaper were seen to be translucent, to varying degrees.

## 1.3 Shadow

The artificial lighting equipment used on film sets can be arranged and adjusted by the film cameraman to give the required "mood" or atmosphere to the film scene. Fig. 1.4, with its dramatic contrasts in brightness and shadow shows how lights can be used to create a tense and frightening atmosphere.

*QUESTION 2*: How would you produce an effect similar to this, at home or on the school stage?

Shadows (areas of darkness) occur when light is

**Fig. 1.4** *Shadows used to create an atmosphere of fear (from the film* Notorious)

directed onto an opaque object; this is because light does not bend around the object and illuminate the other side. The formation of shadows shows us that light travels in straight lines.

The appearance of a shadow can be changed by using a different size of light source. Look at Fig. 1.5(a). The card with the hole punched in it, placed in front of a 100W household bulb, provides, in effect, what is called a *point source* of light. When you do this experiment

**Fig. 1.5** *(a) Apparatus to investigate the shadow produced by a point source*

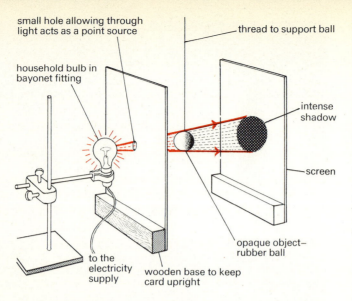

small hole allowing through light acts as a point source

thread to support ball

household bulb in bayonet fitting

intense shadow

screen

to the electricity supply

wooden base to keep card upright

opaque object— rubber ball

**Fig. 1.5** *(b) Apparatus to investigate the shadow produced by an extended source*

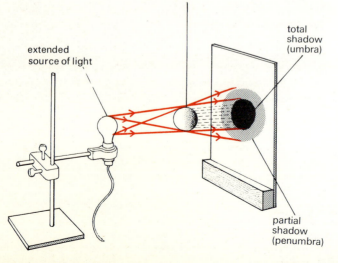

extended source of light

total shadow (umbra)

partial shadow (penumbra)

observe first the shadow of the ball or other opaque object lit by this point source, then remove the card and observe the new shadow. The change in the appearance of the shadow is illustrated in Fig. 1.5(b).

A sharp, uniformly dark shadow is produced by a point source of light. This is because there is a complete absence of light behind the object when the light source is so small. When the light source is extended (when it becomes in effect many point sources) the shadow becomes blurred, larger, and lighter at the edges. The reason for this is that some light from this extended source now reaches the area behind the object. However, there is not enough of this light to illuminate this area completely and it forms a region of partial shadow called the *penumbra*. The central region of total shadow, complete darkness, is called the *umbra*.

Look again at Fig. 1.4 and notice the variety of shadow effects. In photography and filming the cameraman uses many lights to prevent the formation of hard dark shadows, unless of course he needs them for a special purpose.

### 1.4 Eclipses

An eclipse of the sun occurs when the moon comes between the sun and the earth (see Fig. 1.6), cutting off the light rays and forming a region of darkness on the earth's surface. People in the umbra region of the moon's shadow will view a total eclipse of the sun. People in the penumbra region of the moon's shadow will view a partial eclipse. However, as the moon moves in its orbit, so its shadow travels over the surface of the earth, and this has frequently been observed by airmen flying in the region of an eclipse. The shadow travels at approximately 160 kilometres per hour. Any one town or city can expect to see only one total eclipse every 400 years, although taking the world as a whole there are, on average, five eclipses of the sun every two years.

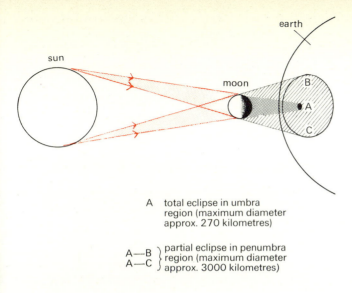

A total eclipse in umbra region (maximum diameter approx. 270 kilometres)

A—B ⎫ partial eclipse in penumbra
A—C ⎭ region (maximum diameter approx. 3000 kilometres)

**Fig. 1.6** *Eclipse of the sun (not drawn to scale)*

*QUESTION 3:* The moon goes round the earth once every 28 days. Why do we not have 12 eclipses of the sun in every year?

When an eclipse does occur it is a spectacular event and people travel to the most remote places to view it. Scientists, particularly astronomers, are especially interested because during an eclipse they can carry out observations which are normally very difficult, for instance the study of the sun's *corona*.

Figure 1.7 shows a typical eclipse where the disc of the moon appears to "eat" its way across the disc of the sun until it covers it. There is now left in the sky a black disc surrounded by the spectacular flame-like prominences

that make up the corona. This can extend to heights of about 960 000 kilometres. At the same time, darkness descends on the earth (though it is not as dark as night with a full moon), animals and birds become quiet, stars are seen, the temperature drops rapidly and the eclipse wind begins to blow. It is no wonder that such events caused terror and were regarded as bad omens in the past. This stage can last up to a maximum of $7\frac{1}{2}$ minutes, after which sunlight is gradually restored.

*QUESTION 4:* What are the positions of the sun, moon and earth when an eclipse of the moon occurs?

### 1.5 The pinhole camera

A day seldom goes by without our seeing an *image* of ourselves in a mirror, in some shiny surface or on a photograph. An image is a likeness of an object. In the frontispiece to Part 1 you can see not only the lighting equipment needed to make a film, but also the cameras that record the images on the film. We could not try to

**Fig. 1.7** *Total eclipse of the sun, showing the corona* (Photo: Science Museum, London)

**Fig. 1.8** *Construction of a pinhole camera*

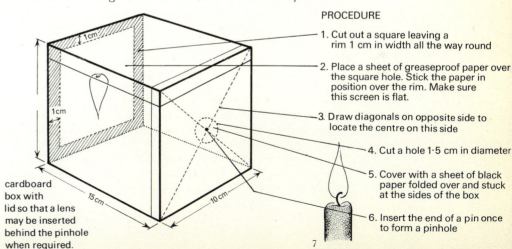

PROCEDURE

1. Cut out a square leaving a rim 1 cm in width all the way round

2. Place a sheet of greaseproof paper over the square hole. Stick the paper in position over the rim. Make sure this screen is flat.

3. Draw diagonals on opposite side to locate the centre on this side

4. Cut a hole 1·5 cm in diameter

5. Cover with a sheet of black paper folded over and stuck at the sides of the box

6. Insert the end of a pin once to form a pinhole

cardboard box with lid so that a lens may be inserted behind the pinhole when required.

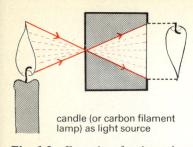

candle (or carbon filament lamp) as light source

**Fig. 1.9** *Formation of an image in a pinhole camera*

construct one of these complicated cameras, yet with the simplest of materials we can make what was probably the first form of camera (see Fig. 1.8 and "Things to do" section, page 10). This is the pinhole camera, which produces an upside-down image because light travels in straight lines (see Fig. 1.9). As the camera is moved nearer to the object the size of the image is increased. The distance to the object and the length of the camera are the only factors that affect the size of the image.

*QUESTION 5:* If half-a-dozen pinholes, a few millimetres apart, are made in the front of the camera, what would you expect to see on the screen?

The image viewed on the screen of a pinhole camera is not very bright, as such a small amount of light passes through the pinhole. A brighter image can be obtained by enlarging the pinhole, but this is unsatisfactory because the image is then blurred.

## 1.6 Waves

So far we have seen that light travels in straight lines and we have used the term *light rays*. Light is energy, but what form does it take as it travels through materials or space?

To help us solve this question we must turn briefly to a study of waves and wave motion. We are all familiar with the waves on the sea-shore and the energy they possess as they knock us over. To study the properties of waves more closely in the laboratory we use a shallow ripple tank (Fig. 1.10).

Produce simple circular ripples in the tank by dipping your finger or a pencil into the water, or by dropping water drops from a medicine dropper. Position a piece of cork in the water. Move your finger up and down more vigorously. Does this cause any changes in the waves? What happens to the cork?

8

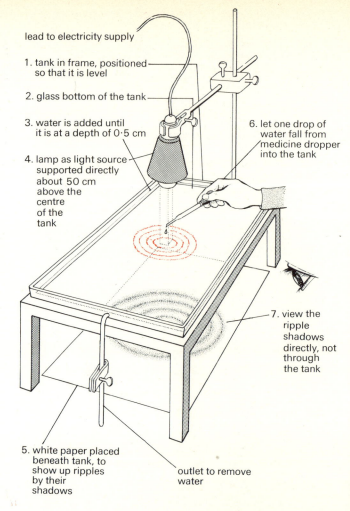

lead to electricity supply

1. tank in frame, positioned so that it is level

2. glass bottom of the tank

3. water is added until it is at a depth of 0·5 cm

4. lamp as light source supported directly about 50 cm above the centre of the tank

5. white paper placed beneath tank, to show up ripples by their shadows

6. let one drop of water fall from medicine dropper into the tank

7. view the ripple shadows directly, not through the tank

outlet to remove water

**Fig. 1.10** *Ripple tank used for the study of waves and wave motion*

The up-and-down movement in the water creates waves, which are a series of crests and troughs. The distance between two adjacent crests or two adjacent troughs is called the *wave-length*. The waves cause a cork on the water to bob up and down. When more energy is used to disturb the water, then the greater movement of the cork indicates that the waves have greater energy. The cork does not move along with the wave. This shows that the water does not move along with the wave motion.

Waves are energy carriers.

The energy is carried as a series of pulses, and these cause the particles of the material through which the pulses are travelling to vibrate. Straight lines drawn perpendicular to the wave front show the direction of travel of the waves. In connection with light such lines showing direction of travel are called *rays* (Fig. 1.11).

## 1.7 Reflection of waves

We have all seen sea waves pounding against a sea wall. Let us introduce such a barrier into the ripple tank and study the effect that it has on the waves.

Produce straight waves by placing your fingers flat on the upper surface of a glass rod placed at one end of the tank.

Slowly roll the rod slightly forwards and backwards (Fig. 1.12). Observe what happens when the wave hits the barrier. Repeat the procedure if necessary. Move the barrier so that the waves strike it at different angles and observe the effect.

You will find that when the waves strike the barrier (we call these the *incident* waves) they bounce back; in other words, they are *reflected* (Fig. 1.13).

*QUESTION 6:* Using Fig. 1.13 and your own observations, can you state the relationship between the angle at which the waves strike the barrier and the angle at which they are reflected?

## 1.8 Refraction of waves

Obtain a rectangular sheet of glass and position four pennies in the water so that they support the glass horizontally at its four corners. The water covering the glass must be as shallow as possible and certainly not more than 1mm deep. At the opposite end of the tank suspend a piece of wood horizontally from two rubber bands so that its edge is just below the water level. When the wood is made to vibrate up and down it produces plane waves similar to those produced by the glass rod. Observe the effect on the waves as they arrive at the glass surface. Move the glass plate so that the waves arrive at various angles to the surface. The observations you will have made are illustrated in Fig. 1.14. The waves are slowed down as they enter shallow water. If the waves strike the boundary of the shallow area at an angle, then the change of speed causes them to change direction. We say that the waves can be *refracted* (bent).

These experiments have shown us that

a. waves travel in straight lines,
b. they can be reflected,
c. they can be refracted.

Now that we have discovered some of the properties of waves, we shall continue with our investigation into the properties of light and compare the two sets of results.

Does light travel through space in waves? We have already seen that light travels in straight lines, but is it also reflected and refracted? If so, it may well be that light is some kind of wave motion. We shall investigate some properties of light in the next few chapters.

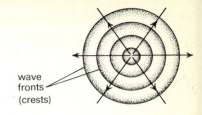

**Fig. 1.11** *Lines drawn perpendicular to the wave front show direction of travel*

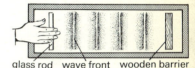

glass rod  wave front  wooden barrier

**Fig. 1.12** *Plane waves moving towards a barrier*

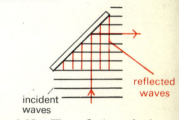

reflected waves

incident waves

**Fig. 1.13** *The reflection of plane waves at a barrier*

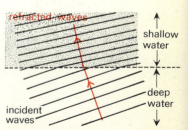

refracted waves

shallow water

deep water

incident waves

**Fig. 1.14** *Waves are slowed down and refracted as they pass through shallow water*

**THINGS TO DO**    A.    Make a pinhole camera with a movable screen (Fig. 1.15).

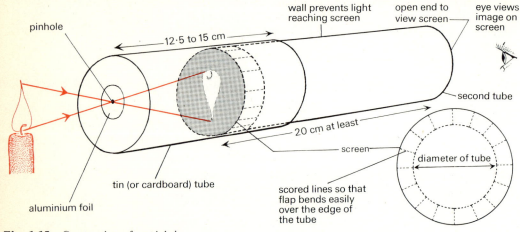

pinhole

12·5 to 15 cm

wall prevents light reaching screen

open end to view screen

eye views image on screen

second tube

20 cm at least

screen

diameter of tube

tin (or cardboard) tube

aluminium foil

scored lines so that flap bends easily over the edge of the tube

**Fig. 1.15** *Construction of a pinhole camera with a movable screen*

Obtain a long round tin. A long fruit tin (425g) is ideal. Remove the top. Paint the inside of the tin with matt black paint. Punch a hole about 1cm in diameter, in the exact centre of the bottom of the tin. Stick across this hole a perfectly flat piece of aluminium foil (a milk-bottle top will do). Push the tip of a sewing needle *once* through the exact centre of the foil, in order to make a very fine hole.

Select a cardboard postal tube which will just slide inside the tin and move freely. (You can make the tube yourself, if you wish, from newspaper. Wind the newspaper around a cylindrical object of the correct diameter. Glue in between each layer of paper with flour paste. This will form a very strong tube when dry.) Cut off at least 20cm

length of tube and paint the inside black. At one end construct a screen by cutting out a circle of greaseproof paper, wider than the tube, and then sticking this in place so it is perfectly flat and taut over the open end of the tube. This should not hinder the movement of the tube inside the tin. Push the screen end of the tube into the tin, direct the pinhole towards a brightly illuminated object, and look at the screen of the camera. (Do not place your eye too close to the screen. About 25cm away is a good distance.) Adjust the position of the screen and thereby its distance from the pinhole by moving the tube in and out of the tin. Observe the effect that changing the screen distance has on the image. Finally keep the screen stationary, but, using a pin, gradually enlarge the pinhole. What effect does this have on the image?

B.    Collect samples of as many different materials as you can, in addition to those listed on page 4, and classify them as opaque, transparent or translucent.

C.    Set up a shadow stick.

Take a stick about 1m long and fix its lower end in the ground so that it is firm and upright. Observe the length of the stick's shadow every hour throughout the day, and repeat this once every week if possible until a full year's record has been obtained. Explain how a shadow stick can be used by a traveller to find direction.

D.    Draw and describe a sundial; it can be one in your own neighbourhood or one that you have seen in a book.

E.    Collect pictures and photographs for an illustrated file on the history of artificial lighting.

F.    Draw a diagram showing how you would design the lighting of a room to give the best form of illumination with no glare. Arrange the lighting so that it can be used for any activity from reading to television viewing.

1. Show by means of a diagram how an eclipse of the sun occurs. What is the difference between a partial and total eclipse?

2. Explain the following terms: opaque, transparent, translucent, luminous, non-luminous, shadow, image, penumbra, umbra.

3. Explain, using a diagram, how a pinhole camera works.

   What happens to the picture if (a) a large number of pinholes are made in the front of the camera, (b) one large hole is made in the front of the camera?

   What is the effect on the size of the image if (a) the camera is moved towards the object, (b) the length of the camera is increased?

   Is it possible to use a pinhole camera, provided with film, to photograph a fast-moving object?

4. Describe an experiment to demonstrate that light travels in straight lines. State three everyday experiences which support this theory.

5. A householder replaces a 100W clear glass electric light bulb with a 100W bulb which has a white coating on the glass (a pearl bulb). What changes will take place in lighting effect and shadow formation in the room?

6. In order to read a book with ease and without any harmful tiring effect on your eyes, how would you position (a) an unshaded electric light bulb, (b) a shaded electric light bulb? Give reasons.

7. Shadows formed by objects exposed to sunlight on the moon are intensively dark compared to those formed by the same objects on the earth. Explain this.

8. The main light in a room is at a height of 240cm. What length of shadow is formed by a dog 90cm high standing 120cm away from the point immediately below the light?

**CHAPTER 2**

## 2.1 Mirrors and the reflection of light

Figure 2.1 shows a film being shot out of doors on a sunny day. The cameraman needs as much natural light as possible directed onto the set. Can you think what the metal sheets supported on stands are being used for?

How do plane (i.e. flat) mirrors—for this is what the metal sheets are — reflect light? The experiment illustrated in Fig. 2.2 will help us to answer this question.

# REFLECTION OF LIGHT

**Fig. 2.1** *All set for the shooting of a film out of doors* (Sergeant York)

A plane mirror is set upright, with a ray of light directed onto it from a light source with a single slit. The ray is thrown back, or reflected, from the mirror. The position of the mirror and rays may be plotted by placing a sheet of paper beneath the apparatus, as shown. Thus, it is easy to draw in the *normal* (the line at right angles to the

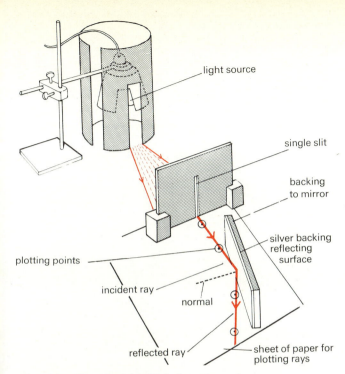

**Fig. 2.2** *The reflection of light by a plane mirror*

light source

single slit

backing to mirror

silver backing reflecting surface

plotting points

incident ray

normal

reflected ray

sheet of paper for plotting rays

mirror surface at the point where the rays strike it). The angle between each ray and the normal can then be measured in turn.

These angles are found to be equal. This is true no matter at what angle the incident ray is directed onto the mirror. Using the terms for the rays and angles set out in Fig. 2.3 we can now state the laws of reflection.

1. The angle of incidence is equal to the angle of reflection.
2. The incident ray, the reflected ray and the normal are all in the same plane.

Therefore, we can use mirrors, or flat polished surfaces, as reflectors to bend light and send it in the required direction. This is done by placing the mirror so that it receives the incident light at the angle required to reflect it correctly.

*QUESTION 1:* (a) At what angle would you direct light onto a mirror in order to get it to bend through 90°? (b) Describe two ways in which you could position a second mirror so that it reflects the light coming from the mirror in (a) through another 90°? (c) What instrument have you constructed in each case?

We have seen that the film cameraman used reflectors to illuminate the dark side of the object to be photographed; in many other ways plane reflectors play an important part in everyday life. For instance, the laws of reflection explain the use of car mirrors, warning mirrors at blind corners and mirrors in our homes. Reflecting number plates on motor cars are an additional safety

**Fig. 2.4** *Reflecting and non-reflecting number plates: white at the front; yellow at the rear*

factor on our roads (Fig. 2.4). Reflectors are also used as signalling devices: every astronaut carries a heliograph in his survival kit (Fig. 2.5).

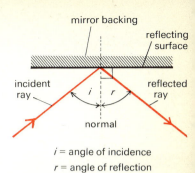

mirror backing

reflecting surface

incident ray

reflected ray

$i$

$r$

normal

$i$ = angle of incidence
$r$ = angle of reflection

**Fig. 2.3** *Terms associated with the reflection of light*

**Fig. 2.5** *An astronaut uses his heliograph to send a message*

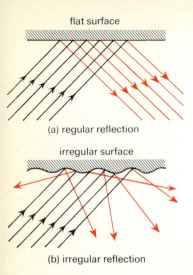

flat surface

(a) regular reflection

irregular surface

(b) irregular reflection

**Fig. 2.6** *Different surfaces reflect light differently*

## 2.2 Regular and irregular reflection

It is obvious to us that flat polished surfaces such as mirrors are reflecting surfaces because we are used to seeing the results of these reflections. We call them *images*. Such images are not to be seen when reflection occurs at rough unpolished surfaces.

*QUESTION 2:* Why is it impossible for you to see yourself in the cover of this book?

Imagine that a smooth and a rough surface are magnified greatly. You can then represent these two surfaces by a flat strip of metal and a corrugated strip of metal. When parallel rays of light, coming from a light source with a lens and multiple slit, are shone onto each of these surfaces the differences can be seen (Fig. 2.6). The flat strip gives *regular* reflection, and the corrugated strip *irregular* or diffuse reflection which scatters the light in all directions. In fact, we see most of the objects around us by diffuse reflection.

**Fig. 2.7** *A baby searches for his image*

## 2.3 Where is the image formed in a plane mirror?

All the photographs you have looked at so far in this book are permanent images. They are the result of light from an object falling onto a light-sensitive paper. Such images formed on a screen are known as *real* images.

Look at Fig. 2.7. The baby is trying to touch the image of himself that he sees in the mirror, but in fact all he is touching is the glass of the mirror surface. The image here is known as a *virtual* image.

Where is the image of the baby formed?

We can find the answer to this question either by means of an experiment using the apparatus in Fig. 2.8 or by applying the laws of reflection to a construction diagram as shown in Fig. 2.9.

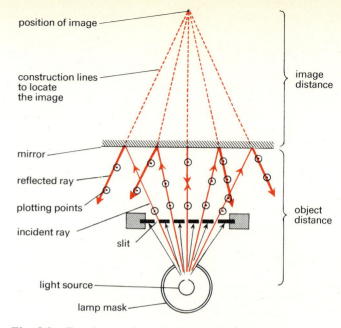

position of image

construction lines to locate the image

image distance

mirror

reflected ray

plotting points

incident ray

object distance

slit

light source

lamp mask

**Fig. 2.8** *Experiment to locate the image in a mirror*

In the experiment a cone of divergent rays from a light source and multiple slit is directed onto the upright plane mirror. The positions of the light source, the rays and the reflecting surface of the mirror are plotted on the paper placed beneath the apparatus. On removing the apparatus the rays may be drawn in and the reflected rays traced back to the point behind the mirror from which they appear to come.

As the baby explores his surroundings he learns to use his eyes and his hands to seek out objects. He relates these two experiences and forms some idea of distance. In this way our brains get accustomed to using the direction from which rays enter the eye so as to judge the

14

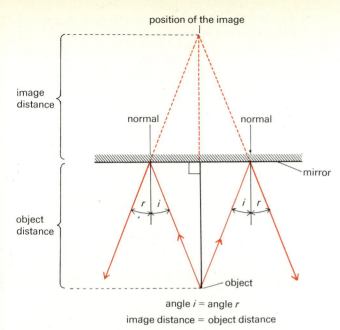

position of the image

image distance

normal    normal

mirror

object distance

r | i      i | r

object

angle *i* = angle *r*

image distance = object distance

**Fig. 2.9** *Constructional method of locating the apparent position of an image in a mirror. The image is where the reflected rays appear to come from*

distance of an object. We carry out this judgement when our eyes receive the reflected rays of light from a mirror. Thus, we judge the distance of the image in the mirror and its position as being where the rays of light appear to come from. The image appears to be at a distance behind the mirror equal to the distance of the object in front, and on an extension of a straight line from the object to the mirror and at right angles to it.

A *virtual image* appears at a point where a cone of rays entering the eye *appears* to come from.

If you have ever seen yourself on a television screen while actually being televised, perhaps at an exhibition or at school, then you probably found one thing very curious. Your image always raises its right hand when you raise your right. When you next have an opportunity to see your television image try and comb your hair using the television screen as a mirror. It is very difficult. This is because we are used to seeing the *lateral inversion* of images in mirrors. When you raise your right hand, your mirror image appears to raise its left. How often, in films, the detective holds a piece of blotting paper in front of a mirror to re-invert the blotted name or message which is a vital clue!

The properties of a plane-mirror image are:
1. It appears to be at the same perpendicular distance behind the mirror as the object is in front.
2. It is a virtual image.
3. It is laterally inverted.
4. It is the same size as the object.

## 2.4   Images formed by a number of mirrors
Three images are formed if an object is placed between two mirrors with their edges at right angles. The number of images increases as the angle between the mirrors is decreased.

When the mirrors are parallel to one another a very large number of images can be seen. Will they cease after a certain number of images have been formed? Figure 2.10 shows how a film director used mirrors to produce a number of images.

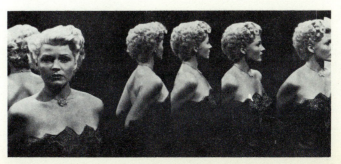

**Fig. 2.10** *Multiple images in a system of mirrors (from the film* The Lady from Shanghai)

A.    Construct a periscope (Fig. 2.11).

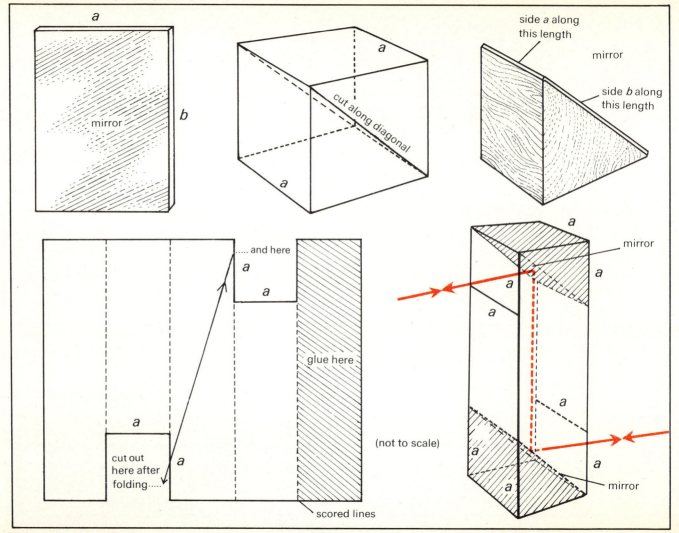

**Fig. 2.11**  *Construction of a periscope*

Take two handbag mirrors (plane). Measure sides *a* and *b* (these are likely to be about 5cm × 8cm). Select a cube of wood of measurement *a*, and then cut it diagonally in half to provide bases that will hold each mirror at an angle of 45°.

Cut out a piece of stiff card approximately 40cm long and five times the width of *a*. Score the card with four vertical lines that are *a* cm apart. Paint the card matt black. Fold it so that the two outermost sections overlap and glue these together to form a square tube. Cut equal sections out of the top and bottom of opposite sides of the box. Glue each mirror and its base onto the sides of the box, so that the mirrors face cut-out areas and are parallel to one another. Figure 2.12 shows clearly how the image of the yacht is conveyed to the eye of the boy below the harbour wall.

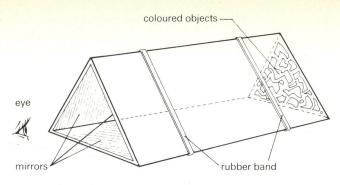

**Fig. 2.13** *Construction of a kaleidoscope*

coloured paper cut-outs. Look into the kaleidoscope by viewing over the top of one mirror into the angle made by the other two. Either move the coloured pieces around or move the kaleidoscope itself in order to view the patterns that are formed.

C.  Produce your own "ghost".

The usual stage set-up for "Pepper's Ghost" is illustrated in Fig. 2.14. The ideal situation at home

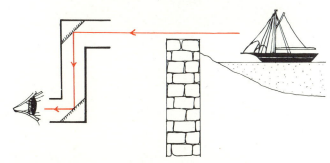

**Fig. 2.12** *The reflection of light within a periscope*

B.  Construct a simple kaleidoscope (Fig. 2.13).

Select three mirrors identical in size. Put their longer sides together so that the three form a three-sided tube with the mirrored faces innermost. Bind the mirrors together with rubber bands. Place one open end of this combination of mirrors (the kaleidoscope) over a selection of small

**Fig. 2.14** *A ghost on a stage*

**Fig. 2.15**

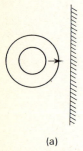

(a)

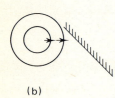

(b)

is a full-length window which can open outwards either into a darkened room or to the outside of the house during the hours of darkness. The curtains will then act as a screen or drapes on either side of the glass.

D.  Collect together material for an illustrated file on the history and everyday applications of plane mirrors. Find out what form the first mirrors took, and how modern mirrors are made.

E.  Stick into a file a full range of ten different shades and textures of wallpaper. Compare their reflecting powers. Make a drawing of a living room that has only one small window as a source of light. Which paper or papers in your range would be suitable for this room, and why?

F.  Run water into a bath until the water is about one centimetre deep. Use a block of wood or some other object to make a plane barrier. Dip the tip of your finger in the water and observe the spherical waves which spread out. Observe the waves reflected from the barrier. Where do they appear to come from? Does this tie up with any of the statements on page 15? Carry out investigations to enable you to complete the diagrams in Fig. 2.15. The diagrams show two successive positions (0.1s apart) of a wave crest travelling towards a reflect-ing barrier. Add lines or curves to the diagrams to show where the crests will be at the end of each of the next four intervals of 0.1s. (If you can use one of the school's ripple tanks this will be easier than using a bath.)

G.  Design, build and use two enoscopes for measuring the speed of a car. The principle is shown in Fig. 2.16. Try and improve on the design.

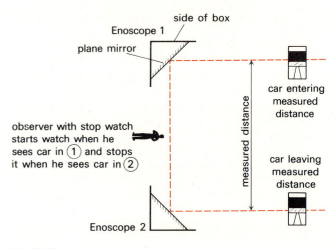

**Fig. 2.16**

1. A vehicle which is often seen moving through the streets has the following written in large letters just above the vehicle's radiator:

# AMBULANCE
*(mirror-reversed)*

What is the vehicle? Why is the name of the vehicle written out in this particular manner? Draw a diagram to illustrate your explanation, using one of the letters.

2. A young boy wishes to draw several objects life-size on a piece of paper. He is not a very good artist, but a friend tells him that all he needs is a sheet of plane glass. With this glass he can actually copy the object as it stands before him on the table. Draw a diagram of the set-up and explain why he is able to do this.

3. Copy the diagrams in Fig. 2.17 and draw in the reflected rays. In (b) and (c) what do you notice about the incident and reflected rays? Can you see how this idea is used in bicycle reflectors?

4. Draw a diagram to show how you would position two mirrors in order to see behind your left ear.

5. Flat plate glass is often troublesome when used in shop windows. This is because the shopper often sees reflections of himself and his surroundings rather than the goods displayed. Suggest a way of overcoming this difficulty so that no reflections are seen by the shopper. Give diagrams if appropriate.

6. A ray of light strikes a mirror at an angle of 30° (Fig. 2.18(i)). (a) Redraw the diagram and draw in the reflected ray. (b) In Fig. 2.18(ii) the mirror has been rotated through 15°. The incident ray has not moved. What is the angle *i*? (c) Redraw Fig. 2.18(ii) and draw in the reflected ray. (d) Through what angle has the reflected ray turned when the mirror was turned through 15°?

7. Why is it advisable to wear dark glasses when skiing at a winter resort in Switzerland?

8. Objects situated within small dark depressions on the surface of the moon are not visible to astronauts. They therefore use their shiny space-suits to view these objects. How do they do this, with the sun as their light source?

9. Explain, with the help of a diagram, why it is not possible for you to see a clear image of yourself in a page of this book.

10. A shopkeeper is stocktaking in his store. He discovers that the ladder used to reach the highest and deepest shelf (an arm's length above his head) is broken. Describe, with a diagram, how he could use two mirrors to view what is tucked away at the back of the highest shelves.

11. Describe an experiment used to demonstrate the laws of reflection.

12. A film producer wishes to give the impression of a hall made up of pairs of pillars stretching endlessly into the distance behind the actors. Explain (with a diagram) how he can achieve this effect by using only one pair of pillars and two very large mirrors.

13. Explain what is meant by the terms *absorption*, *transmission* and *irregular reflection*, with reference to light rays incident upon different materials. Which of these three effects is responsible for our being able to view non-luminous objects?

14. A boy whose eye-level is 120cm from the ground is walking along the street at night. Directly in front of him, at a distance of 240cm is a puddle in which he can just see the reflection of light from a small gap in between the curtains at a house window. The reflected ray makes an angle of 45° with the horizontal. The height of the gap is 360cm. How far away is the house from the boy?

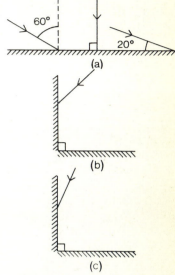

**Fig. 2.17**

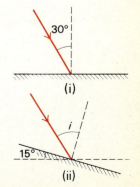

**Fig. 2.18**

**CHAPTER 3** **3.1 Distortion**

You have probably seen many times the distorting effect of patterned (reeded) glass shown in Fig. 3.1.

*QUESTION 1:* Can you suggest why it is happening?

**Fig. 3.1** *The distorting effect of reeded glass*

# REFRACTION OF LIGHT

In an earlier chapter it was stated that we see objects because light from these objects is received by our eyes. If the objects appear distorted when viewed through transparent materials then obviously these transparent materials must be affecting the rays of light in some way. Ordinary window glass does not do this unless there is a fault in it, in which case a waviness is seen. Irregularities of this kind can be prevented by using the flotation process in the manufacture of the glass: the glass is floated on mercury.

**Fig. 3.2** *A staircase wall constructed from glass blocks*

## 3.2 Glass blocks

Figure 3.2 shows the staircase area of a modern house where glass in the form of blocks has been used in the structure. These blocks are manufactured with their faces cut in such a way that they let through a large amount of light (75 per cent of the amount let through by normal glass) and at the same time spread it so that it illuminates what would normally be the darker areas of the room. How does the manufacturer cut such blocks? He needs to know exactly what happens to light as it travels through the glass.

You can find out the principles on which he works by using a glass block yourself, in the following experiment.

## 3.3 Experiment to determine what happens to light as it travels through a glass block

A single ray of light from a light source with a single slit is directed at an angle onto the side of the glass block (Fig. 3.3.). The position of block and rays are plotted on a piece of paper placed beneath the apparatus. It is seen clearly that the light is bent at the point where it enters and leaves the block. This bending of light as it enters or leaves a transparent substance is called *refraction*.

On removing the apparatus, the normals can be drawn in on the paper at both surfaces of the block, and the direction of the bending of the light becomes more obvious.

*QUESTION 2:* In which direction is the light bent (away from or towards the normal) when the light (a) enters the glass, and (b) leaves the glass? (Use Fig. 3.3. to help you.)

When the experiment is repeated several times using different angles of incidence, and with different materials (water for example), then it can be seen that light on entering the substance is bent towards the normal and

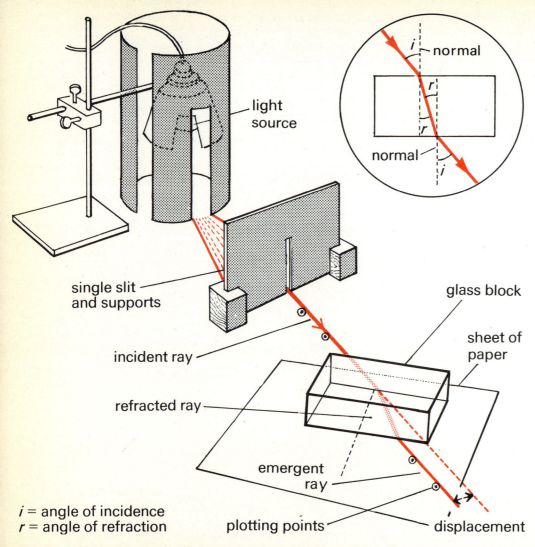

light source

i — normal

r

normal

r

i

single slit and supports

incident ray

refracted ray

glass block

sheet of paper

emergent ray

plotting points

displacement

i = angle of incidence
r = angle of refraction

**Fig. 3.3** *Experiment to determine the passage of light through a glass block*

on leaving the substance is bent away from the normal. These findings are expressed in the law of refraction:

When a ray of light enters a material of greater optical density at an angle, it is bent towards the normal. When a ray of light enters a material of lesser optical density at an angle, it is bent away from the normal. The incident ray, the normal and the refracted ray are all in the same plane.

### 3.4 Refractive index

You could use the apparatus shown in Fig. 3.3 in another way. The glass block could be replaced by different blocks of materials which transmit light. If the angle of incidence is kept the same for each block, do you think the angle of refraction will always be the same?

Experiment shows that the angle of refraction depends on the material from which the block is made. A material which gives a large angle of refraction is refracting the light more than a material which gives a small angle of refraction (when the angle of incidence is the same). Materials which refract light by a large amount are said to have a high *refractive index*. Materials which do not refract the light much are said to have a low refractive index. We shall discuss how the refractive index is calculated on page 28. Its value for water is about 1.3, for glass 1.5 (average value for various kinds of glass) and for diamond 2.4. If light falls on glass and diamond, the angle of incidence being the same in each case, then the light falling on the diamond will be refracted more than the light falling on the glass.

If a material has a high refractive index, it reflects a greater amount of light than a material of low refractive index. For example, diamond has a refractive index of 2.4, and it reflects one-fifth of the light falling on it perpendicularly (normal incidence). If the light is incident at an angle to the normal (grazing incidence), then the percentage reflected is increased. Diamond reflects five times the amount of light that is reflected from glass.

*QUESTION 3:* (a) What percentage of light at normal incidence is reflected back from glass? (b) Which has the higher refractive index, glass or diamond?

## 3.5   Back to the glass blocks

Look again at the staircase area in Fig. 3.2 and then at Fig. 3.4. You will see now how the manufacturer has applied his understanding of the laws of refraction to the cutting and moulding of these glass blocks so that they spread light where it is required within a house.

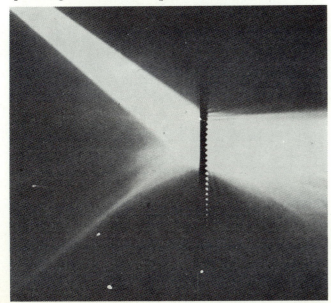

**Fig. 3.4**   *Light passing through a glass block with prismatic cut face*

## 3.6   Refraction in water: the disappearing coin

Put a small coin at the bottom of a sink, then fill the sink with water. Position your head so that you can just see the coin over the edge of the sink. Get someone to pull out the plug while you keep your head still. Your coin will shortly disappear from view.

Figure 3.5 shows how refraction accounts for the disappearance of the coin. In (a) when there is water in the sink, the rays of light are refracted away from the normal at the water surface. Thus the eye sees the coin in its apparent, not its real, position. In (b) there is no water in the sink and the rays of light are not refracted; therefore, the coin can no longer be seen.

If you have ever dived to pick up objects from the bottom of a swimming bath, you will realize that there is a difference between the apparent depth of water and its real depth. When you look down through water the bottom seems much nearer than it actually is. Refraction of light accounts for this too.

*QUESTION 4:* What happens to rays of light that strike a glass/air surface at an angle of incidence of 90°?

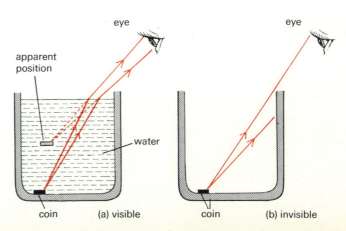

**Fig. 3.5**   *The disappearing coin*

### 3.7 Fish-eye view

A lens often used by cameramen, particularly in advertising, is a fish-eye lens. Look at Fig. 3.6(a), which is a photograph of a street taken through this type of lens. Compare this photograph with Fig. 3.6(b), which is of the same street, but taken through a standard camera lens.

*QUESTION 5:* What has the fish-eye lens been able to achieve that is not possible with the standard lens?

The fish-eye lens is so named because it gives a fish-eye view. A fish sees everything above it in the distorted manner shown in the photograph.

*QUESTION 6:* By referring to the diagram in Fig. 3.7 can you explain why the fish has such a wide-angled view?

**Fig. 3.6**(*a*) *Street scene photographed with a fish-eye lens*

**Fig. 3.6**(*b*) *Same street photographed with a standard camera lens*

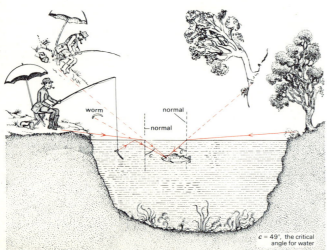

**Fig. 3.7**

24

### 3.8 Total internal reflection

Roll up a newspaper into a tube. How can you read the print on the inside of the tube? A similar sort of problem is met every day by engineers and doctors. They need to inspect the internal or inaccessible area of pipes, body tissues and machinery. Nowadays all of this is possible, and probably some of you will have seen the exciting films that have been produced showing journeys through the breathing and digestive systems of man. The journeys have been made possible by some startling and ingenious inventions. First of all, light must be introduced into the inaccessible parts to illuminate the object, and then a means of transmitting an image back to the viewer is required.

Obtain a Perspex rod, approximately 1.5cm in diameter. Place a small light source at one end and observe the passage of light through the rod. Gently curve the rod (warm it gently but do not apply direct heat) and repeat the experiment. You will find that light can be transmitted or "piped" along the rod, with very little loss in illumination as long as the curve is not too great.

The following experiment helps us to investigate the principles involved.

### 3.9 Experiment to find out about total internal reflection

A semi-circular glass block is used and a single ray is directed through its curved surface (Fig. 3.8).

The light source with its single slit is moved round, so that the angle of incidence is increased, but it is ensured that the incident ray is still striking the glass-to-air surface at the same point A. If angles of incidence of 15°, 30°, 42° and 60° are chosen and the rays plotted as in previous experiments then the following results are obtained: (a) as the angle of incidence is increased, so the angle of refraction is increased and the refracted ray

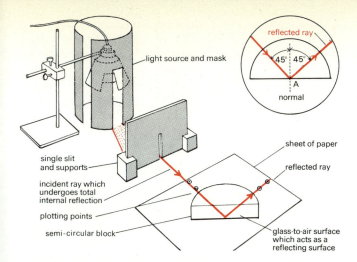

**Fig. 3.8**

moves farther away from the normal; (b) when the angle of incidence of 42° is reached the refracted ray runs along the boundary and the angle of refraction equals 90° (at this point the angle of incidence is called the *critical angle*); (c) with angles of incidence greater than the critical angle the ray never leaves the glass but is reflected back into it.

*QUESTION 7:* Why is it necessary to use a semi-circular block in the experiment?

Total internal reflection occurs when the following conditions are satisfied:
(a) the ray is travelling from an optically dense to an optically less dense medium, and (b) the angle of incidence is greater than the critical angle (see Fig. 3.9).

In the case of glass the critical angle is 42° and in the case of water it is 49°.

We can now see that within the Perspex rod the rays of light undergo a series of total internal reflections, that is, the light is transmitted along the length of the rod. With a total internal reflecting surface, such as glass to air, there is very little loss of light intensity by absorption. In fact 99.9 per cent of the light is reflected from such a surface, compared with silver, which reflects 95˙ per cent, or aluminium, which reflects about 85 per cent of the light incident on it.

*QUESTION 8:* Can you think of another advantage of a glass-to-air reflecting surface over a silvered mirror surface?

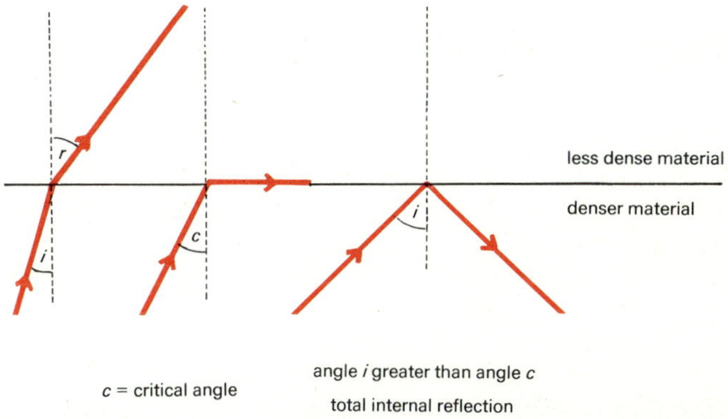

26

less dense material

denser material

*c* = critical angle

angle *i* greater than angle *c*

total internal reflection

**Fig. 3.9**

### 3.10 Some applications of total internal reflection
1. Fibroscope. To produce a flexible rod able to see round corners and into dark and inaccessible places, a bundle of fine glass fibres is used in place of a single rod.

**Fig. 3.10** *Light transmitted through a fibroscope*

2. Reflecting prisms. Right-angled prisms (90°, 45°, 45°) are used in optical instruments in the following ways: (a) to reflect the light through 90°, thereby acting as untarnishable reflecting surfaces which produce a single image and not a multiple one as seen in mirrors — such prisms are used in periscopes for tanks and submarines (Fig. 3.13); (b) to reverse the direction of rays of light, and used for this purpose in prism binoculars (see page 66).

3. Diamond cutting. Diamonds are admired above all other stones, and mainly for their brilliance, that is to say, for their reflection of light. The cutting of a diamond increases this brilliance. The facets are so cut that most of the light is incident at a greater angle than the critical angle of 22.5°. Thus, most of the light is totally internally reflected and returned towards the observer, a simple physical fact which helps to make an uncommon material even more precious and outstanding.

A typical fibroscope consists of a source of light, a number of optical fibres to transmit the light to whatever is being viewed (Fig. 3.10) and a number of other optical fibres to transmit back the image. Lenses assist in the focusing and the viewing of the image. For illumination the fibres may be arranged at random, giving greater flexibility in the light pipe, but for viewing they must be arranged regularly. Each fibre then records and transmits the various regions of light and dark so that the whole image is built up and viewed like a jig-saw (see Fig. 3.11). Obviously, to do this accurately a large number of fibres is required. For example, in a bundle only 3mm square there could be 100 000 fibres, each about 0.01 mm in diameter. Such an instrument would easily cope with viewing the inside of a newspaper (see Fig. 3.12) or the very complicated inspection of an aircraft engine.

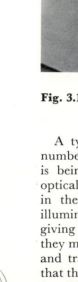

total internal
reflection along
a single fibre

optical fibres
(not to scale)

illuminated object
(a letter)

letter A as viewed
at the end of the
light pipe

**Fig. 3.11** *How a fibroscope works*

### 3.11 Mirages

Refraction in the atmosphere can give rise to strange effects. Light passing through the atmosphere is refracted whenever different layers of air are at different temperatures and hence different densities. Your eye cannot imagine that light rays entering it have been bent so what you see is an image rather than the object itself. Refraction in the atmosphere can produce so much distortion that the images seen bear little or no resemblance to the object that gave rise to them and fantastic sights can occur.

One result of this refraction can be the formation of a mirage (Fig. 3.14). The diagram illustrates how a puddle may appear in the middle of a hot tarmac road on a dry sunny day. The hot air near the road surface is less dense than the air higher up. The light is bent away from the normal (A in Fig. 3.14) as it travels into the less dense air and the resulting path of the light is curved. The eye sees an image of the sky at B which looks like a shimmering puddle in the road. An explanation of the change of direction at X is that the total internal reflection occurs there. However, with this explanation it is not easy to see why the ray should be a continuous curve, and some people find it easier to consider the wavefront PQ (Fig. 3.14). The end Q of the wavefront which is in the less dense air travels faster than the end P (compare Fig. 1.14, page 9). The gradual change in direction gives rise to a curved path.

If the air near the ground is colder than the air higher up, then the light will be bent in the opposite direction and this can give rise to the flying saucer effect.

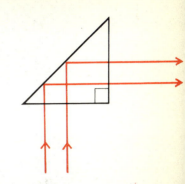

**Fig. 3.13** *Total internal reflection in right-angled prisms*

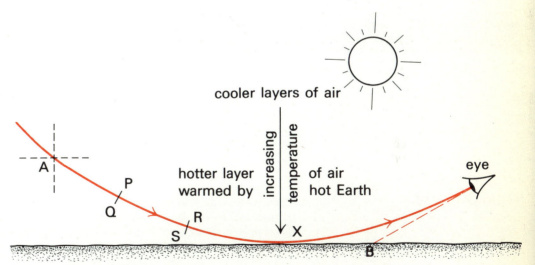

**Fig. 3.14** *The formation of a mirage. The refractive index of the air decreases as the earth is approached. The path of light is a curve. The light entering the eye appears to be coming from B.*

**Fig. 3.12** *A fibroscope being used to view a rolled-up newspaper*

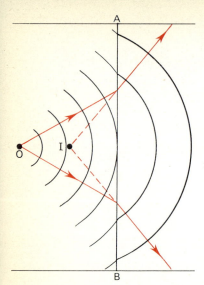

**Fig. 3.15** *Waves and rays leaving a point O. The waves travel faster in the medium to the right of AB*

### 3.12 Refraction of waves

The refraction of waves may be investigated using a ripple tank as discussed on page 9.

Spherical waves may be formed by dipping your finger in the water. Figure 3.15 shows spherical waves originating from a point O in shallow water and moving into deep water to the right of the barrier AB. The waves travel faster in the deeper water, hence the crests are further apart. They appear to come from the point I. The red lines represent *rays of light* leaving O. The rays are perpendicular to the wavefront. After refraction the rays appear to come from I.

### 3.13 Calculation of refractive index

In the experiment illustrated in Fig. 3.3 on page 22 it is found that the angle of refraction changes as the angle of incidence is changed. The table below shows a series of readings obtained when the angle of incidence was varied. In each case the fraction $\dfrac{\sin i \text{ (air)}}{\sin r \text{ (glass)}}$ has been calculated and it is found to be a constant (about 1.5). The constant is known as the refractive index of the material.

| $i$ | $r$ | sine $i$ | sine $r$ | $\dfrac{\text{sine } i}{\text{sine } r}$ |
|---|---|---|---|---|
| 15° | 9.75° | 0.2558 | 0.1691 | 1.53 |
| 37° | 23.5° | 0.6017 | 0.3987 | 1.51 |
| 44.25° | 27° | 0.6977 | 0.4540 | 1.53 |
| 45° | 28° | 0.7071 | 0.4695 | 1.50 |
| 56° | 33.25° | 0.8290 | 0.5483 | 1.51 |

Snell's law states that when a ray of light passes from one medium to another then the ratio of the sine of the angle of incidence to the sine of the angle of refraction is a constant.

Stated mathematically, Snell's law may be written:

$$\frac{\sin i}{\sin r} = \text{constant.}$$

## THINGS TO DO

A.   Turn water into a light rod.

Find a glass jar with a screw lid. Make a fairly large hole in the centre of the lid, and a smaller hole near the rim of the lid, so that when the jar is almost full of water and turned sideways, air enters by the small hole, and the water flows from the central hole. First fill the jar and hold it vertically. Place a torch with its front against the bottom of the glass jar. Switch the torch on and wrap a thick piece of card around both the torch and jar so that it acts as a cylinder to prevent light leaving the system. Thus, all the light from the torch passes upwards through the jar. Now turn the system sideways so that the small hole is above the central hole. Observe the stream of water that leaves the central hole and let it fall on a mirror held at different levels beneath the jar. As the water falls you will see the light is trapped in the water.

B.  Measure the refractive index of water.

   Place a pin (A) at the bottom of a glass of water. View the pin through the surface of the water. At the same time move a second pin (B), held horizontally, up and down along one side of the glass in order to locate the image. You will have located the image when the second pin (B) and the image of the pin A remain in the same straight line when you move your head to the left and then to the right. Measure the real depth of the object, that is, the distance from the surface of the water to the bottom of the glass. Measure the apparent depth of the object, that is, the distance from the surface of the water to the position of pin B. Now, using the formula below, work out the refractive index of water.

$$\text{Refractive index of water} = \frac{\text{real depth}}{\text{apparent depth}}$$

C.  Observe a pencil through a medicine bottle which is completely filled with water.

D.  Construct an illustrated catalogue showing all the applications of total internal reflection in our everyday lives, for example: diamond cutting, vehicle reflectors, household lighting.

1.  Using a diagram explain why a diver, picking up something from the bottom of a swimming pool, has to dive deeper than seems necessary when the object is viewed from the surface of the water.

2.  Each of the diagrams in Fig. 3.16 shows a ray of light arriving at a boundary between two different media. Sketch the diagrams (the angles given need not be measured accurately) showing the path of the ray after it has crossed the boundary.

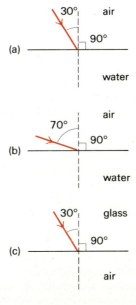

**Fig. 3.16**

3.  Fig. 3.17 shows a ray of light arriving at a glass block. Redraw the diagram and show the path of the ray as it passes through the block. Does all the light arriving at the glass block pass into the block?

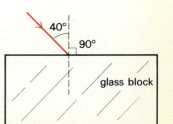

**Fig. 3.17**

29

fly

air

water

**Fig. 3.18**

4. Why do cracks in plate glass appear to be silvered in appearance when viewed at an angle?

5. A very thick glass mirror produces several images of one object placed in front of it. Explain, using a diagram, why this occurs and which image will be most prominent.

6. Fig. 3.18 shows two rays leaving a fly which is in the air above some water. Show the path of the rays after they enter the water. Mark the apparent position of the fly as seen by a fish.

7. Using a diagram explain clearly what is meant by (a) angle of incidence, (b) angle of refraction. Under what conditions is a ray bent towards the normal when it crosses a boundary between two different media?

8. Why does a pencil which has its bottom half immersed at an angle in water appear to be bent? Draw a diagram to explain your answer (refer to Fig. 3.5, page 23, if you need some help).

9. The refractive index of water is 1.33. Explain what is meant by this statement.

10. A boy immerses his finger in water contained in a glass vessel. He is very surprised when he views the under surface of the water at certain angles through the glass because he cannot see the portion of his finger above the water. Explain this effect, and describe the appearance of the under surface of the water.

11. Explain: (a) why it is important to clean spectacle lenses, windscreens, etc., with soft, non-abrasive (non-scratching) materials; (b) how you would use a 45°-angle prism to turn light through an angle of 90°; (c) how you would use two such prisms in a periscope, and what advantage a periscope with prisms has over one made of mirrors.

12. What is a mirage? Explain how the refraction of light accounts for natural phenomena such as this.

13. State Snell's Law.

A ray of light strikes the surface of a glass block, 50cm thick. The angle of incidence is 0°. What will happen to the light? A second ray of light strikes the same glass block at an angle of 30°. The angle of refraction is 18°. What is the displacement of the emergent ray in cm? (Use a construction diagram.)

14. In Fig. 3.19 (a), (b) and (c) show successive positions (0.1s apart) of wave crests travelling towards a refracting surface. AB between two different substances. Redraw the diagrams and show the crests at the end of each of the next four intervals of 0.1s. Assume that in each case the wave travels faster after it has crossed AB. In (c) the crests originate from O.

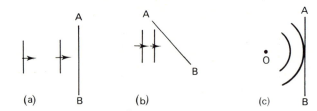

(a)  (b)  (c)

**Fig. 3.19**

15. When recording the position of stars and planets astronomers must make a slight adjustment to the position as viewed through the telescope. Explain why this adjustment is necessary.

# CURVED MIRRORS AND LENSES

## 4.1 Curved mirrors

Figure 4.1 shows the grotesque effect that is produced by a fun-fair mirror. Plane mirrors cannot produce such an effect, but try looking at yourself in any other form of polished surface around the home. You will no doubt quickly find that you are able to produce an effect similar to that of the fun-fair mirror (Fig. 4.1) by looking at yourself in any curved and polished surface. The two surfaces of a spoon are ideal for this. Note that one surface curves inwards and the other outwards.

**Fig. 4.1** *Image in a fun-fair mirror (from the film* The Lady from Shanghai)

*QUESTION 1:* (a) Look at yourself in each side of the spoon in turn, keeping the spoon the same distance away from you each time. Are the images identical? (b) Now hold a pen or pencil with a coloured top in front of the spoon. Look at its image on both sides of the spoon, as before, but this time vary the distance between it and the spoon. Does the image formed change in any way?

**Fig. 4.2** *Apparatus used to determine the action of curved mirrors on light*

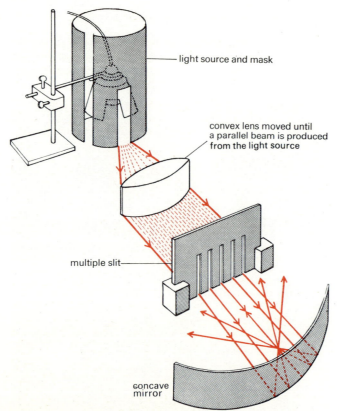

light source and mask

convex lens moved until a parallel beam is produced from the light source

multiple slit

concave mirror

## 4.2 Concave and convex mirrors

*Concave* is the word used to describe a reflecting surface that curves inwards (like a cave), and *convex* is used to describe a reflecting surface that curves outwards. Most curved reflecting surfaces are part of a sphere or cylinder. For instance, a tin can with its bottom and top removed, and cut in half downwards through its length, will provide you with cylindrical concave and cylindrical convex mirrors. Half the surface of a silver ball used in Christmas tree decorations will provide you with a spherical convex mirror. The shape of the mirror is important when we consider what effect such mirrors have on rays of light.

## 4.3 What is the effect of concave and convex mirrors on light?

We can determine this by the following experiment (Fig. 4.2). A light source with a lens and multiple slit is set up to produce parallel rays of light. These are directed onto a cylindrical concave and convex mirror, each in turn.

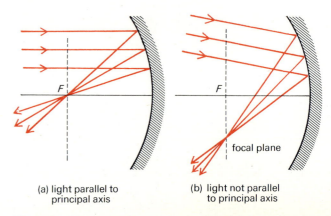

*F*

*F*

focal plane

(a) light parallel to principal axis

(b) light not parallel to principal axis

**Fig. 4.3** *The action of a concave mirror on light*

Figure 4.3(a) shows the result of the experiment using a concave mirror. You can see that each individual ray obeys the laws of reflection and the total effect of the concave mirror is that it converges the rays of light. When the rays of light reaching a concave mirror are parallel they are converged to a point which is called the *principal focus*, or to a common point on the focal plane as shown in Fig. 4.3(b). The distance along the principal axis from the centre of the mirror to the principal focus is called the *focal length* of the mirror.

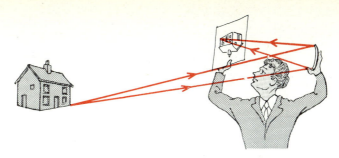

**Fig. 4.5** *Image of a distant object formed by a concave mirror*

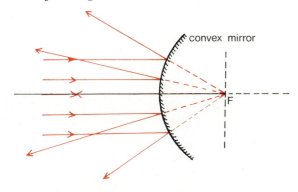

**Fig. 4.4** *The action of a convex mirror on light*

Figure 4.4 shows the result of the experiment using a convex mirror. You can see that a convex mirror diverges the rays of light. A convex mirror has a *virtual* focus.

*QUESTION 2:* Figure 4.5 shows a boy with a concave mirror and a piece of paper which is to act as a screen. He moves the screen backwards and forwards in front of the mirror until he obtains upon the screen a clear image of a distant object. He knows that the rays of light from a point on a distant object are considered to be parallel when they arrive at the mirror. A friend measures the distance between the screen and the mirror. (a) What is the boy finding out about the concave mirror? (b) Could he repeat this experiment with a convex mirror and achieve a similar result?

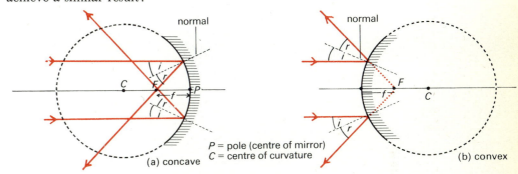

P = pole (centre of mirror)
C = centre of curvature

(a) concave     (b) convex

**Fig. 4.6** *Construction diagrams showing the action of curved mirrors*

In Fig. 4.6, (a) and (b) are construction diagrams of the results shown in Fig. 4.3(a) and Fig. 4.4. The curvature of the mirror has in each case been extended to give the full circle of which the mirror is a part. The centre of this circle is the *centre of curvature* of the mirror.

*QUESTION 3:* Measure with a ruler the construction diagrams (a) and (b) in Fig. 4.6, and then state the relationship between the focal length and the radius of curvature (the distance from the centre of the mirror to the centre of curvature).

33

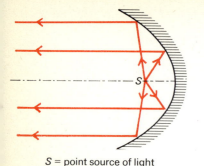

S = point source of light

**Fig. 4.7** *Reflection of light from a parabolic-shaped mirror*

### 4.4 Use of concave mirrors to produce parallel beams of light

Light is reversible. Thus if a point source of light is positioned at the principal focus of a curved mirror, parallel rays of light will be obtained. This is the reason for the use of concave polished surfaces in flash guns, headlights, foglights, searchlights and torches. If, however, the concave mirror has a wide aperture (that is, if the width of the mirror is large) the resulting beam is slightly convergent. This defect is called *spherical aberration*. It may be overcome by making the curved mirror parabolic in shape (Fig. 4.7); this is done in car headlamp reflectors.

In car headlamps two quartz iodide light sources and two reflectors are used within the structure of the headlamp (Fig. 4.8).

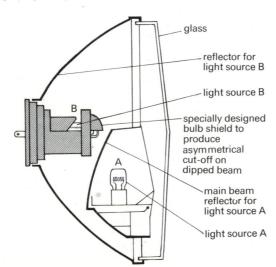

glass

reflector for light source B

light source B

specially designed bulb shield to produce asymmetrical cut-off on dipped beam

main beam reflector for light source A

light source A

**Fig. 4.8** *Car headlamps use two light sources and two reflectors (S.E.V. Marchal dipping quartz iodide optique)*

*QUESTION 4:* What is the effect on the beam of light coming from the headlamp if the motorist switches over to light source B from light source A?

The largest known reflectors are those used in reflecting telescopes, the principles of which are discussed in Chapter 6.

**Fig. 4.9** *A convex mirror in use (from the film* The Servant)

### 4.5 Experimental and constructional methods for accurately locating and sizing images formed by curved mirrors

Figure 4.9 shows how a film director has created the

effect he requires by using the image formed in a convex mirror surface.

*QUESTION 5:* Can you suggest the reason why such a mirror was used in preference to a plane mirror surface? (Assist yourself in answering by using a plane mirror or any plane polished surface and a convex polished surface in a similar situation.)

Earlier in the chapter we considered the images produced in the two surfaces of a spoon and found that the distance of the object from the mirror surface affected the nature and size of the image. Now that the terms associated with curved mirrors are known, we can work out how different types of image are formed. This can be done by an experimental method for a concave mirror, using the apparatus shown in Fig. 4.10, or by construction diagrams for both concave and convex mirrors.

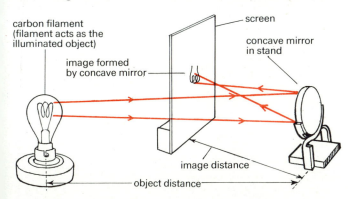

**Fig. 4.10** *Experiment to study the images formed by a concave mirror*

In the experiment the illuminated object is placed at a certain distance (the object distance) from the concave mirror, and the screen is moved along until a clear image

of the object is obtained upon it. The size of the image in relation to the object can be seen, and its distance (the image distance) from the mirror can be measured. The experiment is repeated with the object positioned at different distances from the mirror.

In a construction diagram the paths of two rays of known direction are used to find the position, size and nature of the image when the object is at different distances (Fig. 4.11). You will observe that in these constructional diagrams the reflecting surface of the mirror is represented by a straight line. If this were not done the results would be inaccurate, because on this scale of diagram the small section of the mirror carrying out the reflecting is considered to be straight.

**Fig. 4.11** *Constructional method of locating images formed by curved mirrors*

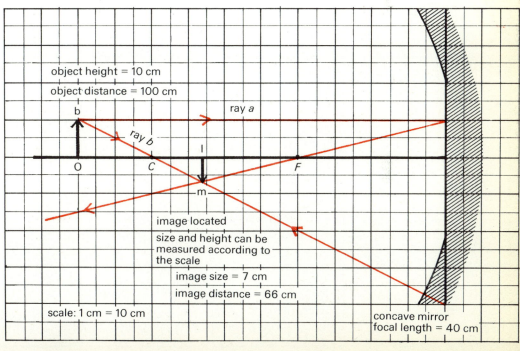

object height = 10 cm

object distance = 100 cm

ray a

ray b

O    C    I    F

m

image located

size and height can be measured according to the scale

image size = 7 cm

image distance = 66 cm

scale: 1 cm = 10 cm

concave mirror focal length = 40 cm

# TABLE 1

F = principal focus    Ob = the object    Im = the image

| | | OBJECT DISTANCE | IMAGE POSITION | NATURE OF IMAGE | CONSTRUCTIONAL DIAGRAM | USES |
|---|---|---|---|---|---|---|
| CONCAVE MIRROR | 1 | at infinity | at focus | real inverted smaller | | reflecting telescope in astronomy |
| | 2 | between infinity and centre of curvature | between focus and centre of curvature | real inverted smaller | | |
| | 3 | at centre of curvature | at centre of curvature | real inverted same size | | in a projector to direct light forwards from the lamp |
| | 4 | between centre of curvature and focus | between centre of curvature and infinity | real inverted enlarged | | floodlight |
| | 5 | at focus | at infinity | parallel beam (if aperture of mirror small) | | car headlight searchlight photographic lights |
| | 6 | between focus and pole of mirror | behind the mirror | virtual erect enlarged | | make-up mirror (X) |
| CONVEX MIRROR | 7 | at any position stated above | behind the mirror | virtual erect smaller | | Viewing mirror on the top of the staircase of a double-decker bus (used by the conductor) Wall mirror in the home Car mirror (Y) |

Ray *a* is a ray parallel to the principal axis of the mirror, which after reflection at the mirror surface passes through the focus (F).

Ray *b* is a ray which passes through the centre of curvature of the mirror. Therefore it strikes the mirror surface at an angle of 90° and is reflected along its own path.

The results of this practical work and diagram construction are summarized in Table 1 opposite. The last column in this table shows some of the uses of curved mirrors.

*QUESTION 6:* Can you suggest two other uses for a concave mirror which could be entered in the space marked (X) in the last column, and one other use for a convex mirror to go in the space marked (Y)?

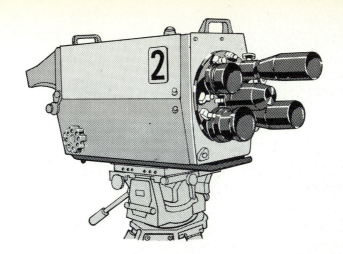

**Fig. 4.12**  *A television camera*

## 4.6  Lenses

"Through the eye of a camera." This is a common expression, and we often see the world around us in this manner — when we look at television, go to the cinema or look at photographs.

*QUESTION 7:* What do you think the word "eye" refers to in the camera?

Perhaps a visitor from outer space might mistake the television camera shown in Fig. 4.12 for a four-eyed monster!

In our everyday life, lenses are used a great deal. But what is a lens, and what effect does it have on the light rays that travel through it? If you run your fingers very gently over the surfaces of a variety of lenses found in the home, you will find that these surfaces are curved either inwards or outwards. This need be true for only one surface. The same terms, *concave* and *convex*, are used for

lenses as for mirrors. Figure 4.13 shows a range of such lenses. In the same way as mirrors, lenses also may either be spherical or cylindrical. Now let us look at the action of these lenses.

## 4.7  The formation of images by lenses

Simple lenses found in spectacles, cameras and magnifiers may be used in this experiment.

A spherical convex lens is held so that it is able to receive light from a distant object, such as a tree. A piece of paper is then positioned on the other side of the lens. The paper is moved backwards and forwards in relation to the lens until a clear image of the object is obtained upon it, as in Fig. 4.14(a).

A convex lens produces a real, inverted image of the distant object. The image is formed whether the screen is present or not. In fact, the image can be viewed without a screen, although this method of viewing may

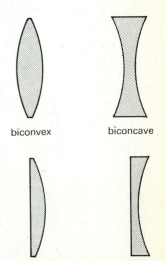

biconvex        biconcave

plano-convex     plano-concave

**Fig. 4.13**  *A range of lenses*

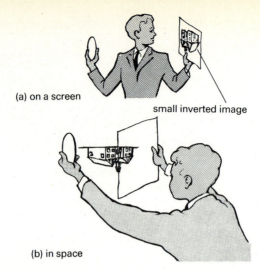

(a) on a screen

small inverted image

(b) in space

**Fig. 4.14**  *Image of a distant object formed by a convex lens*

take you a little while to perfect. At first, use a thin piece of paper as the screen so that you can see the image as you view from behind the paper. Then move the paper so that the image is just on the edge, with half of the image on the paper and the other half visible to you in space, as in Fig. 4.14(b). Concentrate on focusing your eyes on this image in space, then remove the paper so that you view the whole image in this manner. You may find this technique difficult to master. This is not surprising because your eye is not accustomed to focusing on objects in mid-air, but with a little practice you will soon be able to do it.

If you repeat this experiment with a concave lens you will find that it is impossible to form a real image of the distant object.

From these observations it is clear that the convex lens is refracting the rays of light to a point to form a real image. Thus the convex lens must be converging the rays of light.

## 4.8  The action of convex and concave lenses on light

To show this refraction of light by a lens, an experiment can be carried out using apparatus identical to that used in the experiment with curved mirrors (page 32).

The parallel rays from the source of light, lens and multiple slit are directed onto a cylindrical lens which is set upright on the flat surface along which the rays are travelling. The results of such an experiment can be seen in Fig. 4.15 (a) and (b).

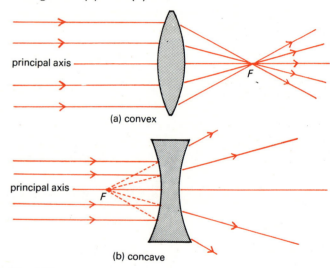

principal axis

*F*

(a) convex

principal axis

*F*

(b) concave

**Fig. 4.15**

It is clear that

a convex lens converges light,
a concave lens diverges light.

When the rays of light reaching a convex lens are parallel to the principal axis of the lens, they pass through the lens and are converged to its *principal focus*.

If they are parallel to one another but not to the

principal axis then they are converged to some other point in the *focal plane*.

The distance from the centre of the lens, along the principal axis to the principal focus is the *focal length* of the lens.

In the case of a concave lens the light rays are diverged and appear to come from a virtual focus.

*QUESTION 8:* If you were to repeat the experiment in section 4.7 (Fig. 4.14) and were to get someone to measure the distance from the convex lens to the screen, what information would you then obtain?

Refraction occurs at both surfaces of the lens, but in diagrams a sum total of this refraction is shown as taking place at a centre line drawn through the lens. This can be seen in Fig. 4.16 which shows the terms used in connection with lenses.

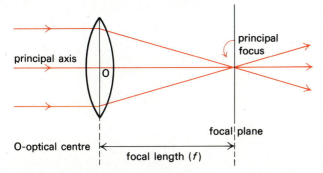

**Fig. 4.16** *Terms associated with lenses*

*QUESTION 9:* (a) Which is the stronger lens (i.e. which one refracts the light more), one with a long focal length or one with a short focal length? (b) How could you determine whether or not a lens is strong merely by looking at it or feeling its surface?

## 4.9 What types of image can be produced by lenses?

We have already seen that a convex lens produces a real, inverted and smaller image of a distant object. If a spherical convex lens is used to form on a screen an image of an illuminated object (for example, the filament in a straight filament bulb), the image is seen to be inverted and may be larger in size. When the lens is brought nearer to the object, the image size increases even more and the screen has to be moved away from the lens in order to focus the image clearly.

It is clear, then, that the distance of the object (object distance) from the convex lens affects the size and position of the image (image distance). The nature of the image changes when the object distance is less than the focal length of the lens: in this case no image can be received on the screen.

No such changes in the positioning of an object in front of a concave lens will result in the formation of an image on the screen. However, if the object is viewed through the lens an upright image can be seen, which is always smaller than the object but which increases in size as the object is brought nearer to the concave lens.

## 4.10 Experimental and constructional methods for accurately locating and sizing images formed by lenses

The general observations discussed above can be considered in more detail by both experimental and constructional methods.

In the experimental method a convex lens is placed between a screen and a carbon filament lamp (the filament acts as an illuminated object). The screen is moved until a clear image of the filament is focused on it. Measurements of image size and position are taken for different object distances. Such an experimental method cannot be conducted with a concave lens.

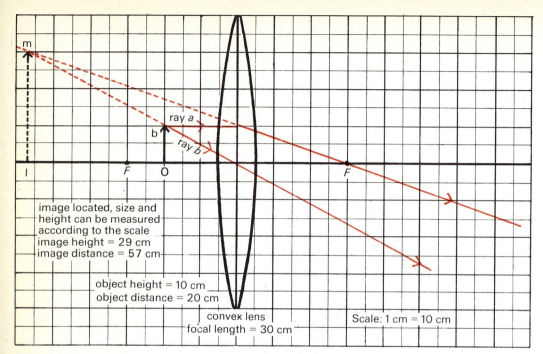

m

ray a

b

ray b

I    F    O    F

image located, size and
height can be measured
according to the scale
image height = 29 cm
image distance = 57 cm

object height = 10 cm
object distance = 20 cm

convex lens
focal length = 30 cm

Scale: 1 cm = 10 cm

**Fig. 4.17** *Constructional method of
locating images formed by lenses*

The results of the above practical work and constructional diagrams are summarized in Table 2 (opposite).

*QUESTION 10:* Both experimental and constructional methods show that when an object is positioned at a shorter distance from a convex lens than its focal length, then a virtual erect and enlarged image is formed. This may be viewed through the lens (see Fig. 4.18). What do we call a convex lens when it is used in this manner? (See Table 2.)

In the constructional diagrams (which can be used for concave and convex lenses alike) the paths of two rays of known direction are used to locate the image (Fig. 4.17).

Ray *a* is a ray parallel to the principal axis of the lens and, after refraction, passes through the principal focus of the lens. Ray *b* passes through the optical centre of the lens without being refracted. (The central section of a lens acts as a very thin parallel glass block and for all practical purposes the ray is considered to pass straight through without deviation.)

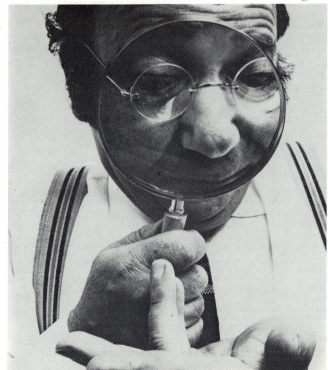

**Fig. 4.18** *Virtual, erect and enlarged image viewed through a convex lens*

TABLE 2

F = principal focus      Ob = the object      Im = the image

| | OBJECT DISTANCE | IMAGE POSITION | NATURE OF IMAGE | CONSTRUCTIONAL DIAGRAM | USES |
|---|---|---|---|---|---|
| **CONVEX LENS** | **1** at infinity | at F | real inverted smaller | | objective lens in a telescope |
| | | | | | a burning glass |
| | **2** between infinity and 2F | between F and 2F | real inverted smaller | | camera lens eye lens |
| | **3** at 2F | at 2F | real inverted same size | | copying camera |
| | **4** between 2F and F | between 2F and infinity | real inverted enlarged | | enlarger objective of a microscope cinema projector |
| | **5** at focus (F) | at infinity | parallel beam | | spotlights searchlights |
| | **6** between focus and centre of lens | same side of lens as the object, and at a greater distance | virtual erect enlarged | | a magnifying glass eyepieces of microscopes and telescopes |
| **CONCAVE LENS** | **7** at any position stated above | same side of lens as the object and nearer to the lens | virtual erect smaller | | eyepiece lens used in a Galilean telescope |

41

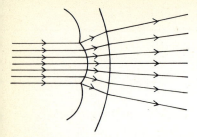

Fig. 4.19 *Cross-sectional diagram of a headlamp glass, showing how the light beam is controlled horizontally*

## 4.11 The everyday uses of lenses

The use of lenses will be discussed fully in Chapter 6, which deals with the construction of optical instruments. However, a study of Table 2 at this point will indicate many ways in which lenses are used. In (2), for example, a diminished image is produced and this you are familar with in camera work. In (5) a parallel beam of light is produced and this effect is used in spotlights. Lens shapes are also cut on surfaces of glass in order to throw light in particular directions or to form beams of particular shapes. The penetrating beam from a lighthouse is produced in this way, and the same process carried out in the making of motor car headlamps (Fig. 4.19) ensures an adequate and safe illumination of the road.

## 4.12 The reflection and refraction of plane waves at curved surfaces

Experiments with the ripple tank (page 8), in which convex and concave barriers are placed, will show us how waves are reflected at curved surfaces (Figs. 4.20 and 4.21). Refraction of waves by different areas of shallow

water, such as a "lens-shaped" area, can also be observed (Fig. 4.22).

The results show that a concave reflector and a "lens-shaped" area both focus the waves.

We have obtained similar effects in our experiment with rays of light. On page 9 we suggested that we might compare the properties of light and waves. Our experiments have shown that light and waves can travel in straight lines and be reflected and refracted both at plane and curved surfaces. Certainly, light seems to possess the properties of waves.

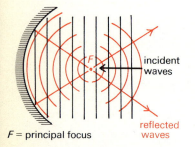

*F* = principal focus

Fig. 4.20 *The reflection of plane waves at a concave surface*

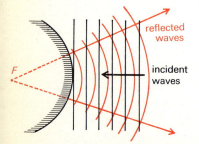

Fig. 4.21 *The reflection of plane waves at a convex surface*

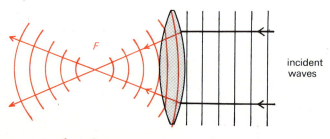

Fig. 4.22 *The refraction of plane waves by a "lens shape"*

42

A. Make a magnifying glass.

Select a piece of card about 100mm × 50mm. Towards one end of the card cut out a circular hole approximately 10mm in diameter. Stick over this a transparent layer of cellophane paper. Then, using a medicine dropper, allow a drop of water to settle on the transparent paper. The surfaces of the drop of water remains curved and the water will act as a magnifying glass when it is held over any object or surface.

B. Make Perspex-and-water lenses (Fig. 4.23).

Construct two flat bases of plasticine, approximately 70mm × 40mm and 10mm deep. Cut rectangular strips — four pieces measuring 50 × 10mm, and two pieces measuring 15 × 10mm — from a flexible piece of Perspex. (Perspex may be bought at any shop that stocks materials for model making.)

To construct a convex lens, take two of the 50 × 10mm pieces and hinge them together with waterproof sticky tape as shown in Fig. 4.23. If the tape is opaque it must not cover very much of the curved lens surface. Hold the lens between thumbs and forefingers at the hinged edges, and push inwards so that the curvature of the surface becomes greater. When the lens is of the required shape keep it to this shape by pushing the bottom edge into the plasticine base. Pour water into the lens shape.

To construct a concave lens use two pieces 50 × 10mm and two 15 × 10mm, hinged together as illustrated. The curvature of both types of lens can be adjusted easily and a full range of lenses of different focal lengths constructed.

C. Construct either an illustrated project or a display chart on the history of lenses and their uses.

D. List, with illustrations, the everyday applications of convex and concave reflecting surfaces.

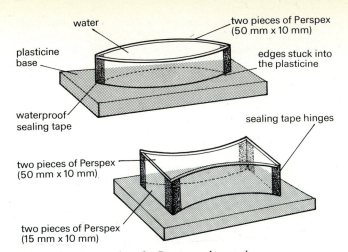

water

two pieces of Perspex (50 mm x 10 mm)

plasticine base

edges stuck into the plasticine

waterproof sealing tape

sealing tape hinges

two pieces of Perspex (50 mm x 10 mm)

two pieces of Perspex (15 mm x 10 mm)

**Fig. 4.23**  *Construction of a Perspex-and-water lens*

# THINGS TO WORK OUT

1. Explain what is meant by the following: focal length of a lens, convex lens, diverging lens, radius of curvature, concave mirror, parabolic mirror.

2. What is the difference between a virtual image and a real image?

3. Draw diagrams to show how a convex lens may be used to form a virtual image and a real image. In which case is the lens being used as a magnifying glass?

4. What type of lens may be used as a "burning" lens? Draw a ray diagram to show its action.

5. What shape of polished metal surface is generally used as a reflector behind a source of light? Draw a diagram to show the action of such a reflector in a searchlight.

6. A convex mirror is used as a car wing-mirror, whereas a concave mirror is used by a dentist when he wishes to look at your teeth. Explain, in turn, why each mirror is so well suited to its specific task and why each would not be satisfactory if used for the other job.

7. A concave mirror has a radius of curvature of 20cm. An object 10cm high is placed at a distance of 40cm from the mirror, By using a construction diagram, find the position, size and nature of the image.

8. An object 5cm high is placed at a distance of 10cm from a concave mirror which has a focal length of 15cm. By a construction diagram find the size, nature and position of the image.

9. A concave mirror and convex mirror have been framed in boxes so that their outlines cannot be detected. A movable source of light producing a parallel beam is placed in front of each mirror in turn. You observe the effect of each mirror on the parallel beam. How do you decide, on the basis of your observations, which mirror is convex and which is concave?

10. Describe what is meant by the term *focal length* with reference to lenses. Describe how you would find the approximate value of the focal length of a convex lens, using only a sheet of paper as a screen.

   Having found the focal length of the convex lens, describe how you would use this knowledge to create the following beams of light: (a) converging, (b) parallel and (c) diverging.

11. Under what conditions is an inverted image formed by (a) a convex lens, (b) a concave lens? Under what conditions is a virtual image formed by (a) and (b)?

12. An object 2cm high is placed at a distance of 20cm from a convex lens. An image is produced on the opposite side of the lens, and the image is exactly the same size as the object. Use a construction diagram to find the focal length of the lens.

13. Figure 4.24 shows a lens of focal length 10cm. An object is placed 20cm from the lens. (a) Redraw the diagram and show the paths of the rays after they have passed through the lens. (b) Is the image magnified or diminished, real or virtual, inverted or erect?

14. Draw a ray diagram in each case to find the position and size of the images of an object 2cm high placed at distances of 3cm, 5cm and 8cm respectively from a convex lens of 6cm focal length.

15. Figure 4.25 shows light from an illuminated object passing through a converging lens and reflected at a plane mirror. The object and image coincide at O. What is the focal length of the lens? Explain your answer.

16. Draw diagrams showing spherical waves originating from a point O and being (i) reflected from a concave surface, (ii) refracted by a convex 'lens shape'. Mark the position of the image in each case.

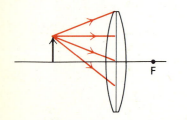

**Fig. 4.24**

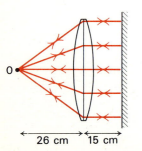

26 cm    15 cm

**Fig. 4.25**

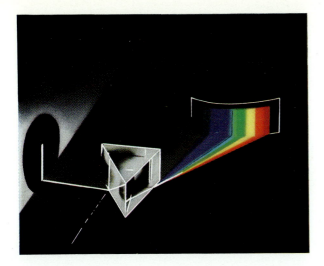

**Fig. 5.1** *Experiment to determine what happens when light passes through a prism*

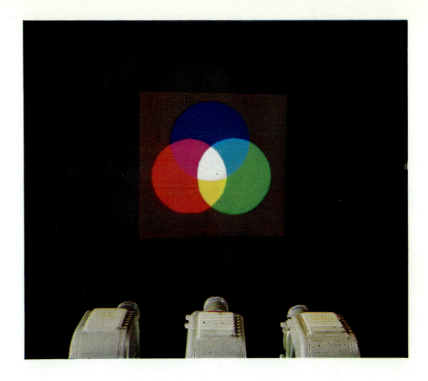

**Fig. 5.4** *The mixing of coloured lights*

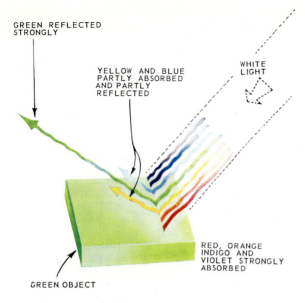

GREEN REFLECTED STRONGLY

YELLOW AND BLUE PARTLY ABSORBED AND PARTLY REFLECTED

WHITE LIGHT

RED, ORANGE INDIGO AND VIOLET STRONGLY ABSORBED

GREEN OBJECT

**Fig. 5.7**  *How we see a colour object*

1

4

2

3

**Fig. 5.8** *Colour printing. From left to right:*
*1 yellow print, 2 magenta print, 3 yellow + magenta,*
*4 cyan print, 5 yellow + magenta + cyan, 6 black*
*print (tone), 7 final print (yellow + magenta + cyan*
*+ black)*

5

6

7

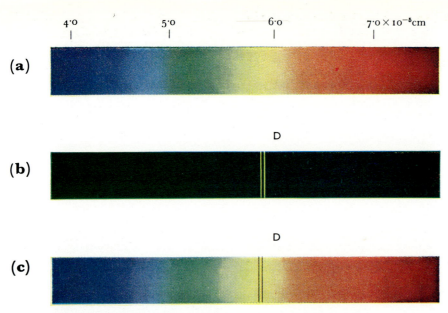

**Fig. 5.11** (a) *Continuous spectrum from a carbon arc* (b) *Sodium vapour emission line spectrum* (c) *Sodium absorption spectrum*

# COLOUR

## 5.1 A colourful world

We are living in what might be called the psychedelic age. The development and use of exciting colours in industry, pop art, architecture and clothing are outstanding features of the twentieth century.

Colour can affect the way we feel. "Pop" stars perform against backgrounds of flashing, intensely coloured lights which are intended to create a feeling of excitement. Bright colours in our clothing and possessions in the home make us feel gay and cheerful. It would be rather dull to have a world of black and white, but have you ever wondered why we have colours, or how we are able to see them?

*QUESTION 1:* What is meant by the "fire" of a diamond? What is it that is so attractive about cut-glass vases and chandeliers?

## 5.2. How are colours obtained from light?

We can obtain colours from light by using a prism — a piece of glass with a triangular section (Fig. 5.1, facing page 44). If a ray of light from a light source is directed onto the side of a triangular prism we obtain a band of colours in which seven colours are prominent: red, orange, yellow, green, blue, indigo and violet, in that order. This band of colours is known as the *spectrum* of white light. This phenomenon of *dispersion* of light, as it is called, was first observed by Isaac Newton in 1660 when he placed a prism in the path of sunlight streaming through a hole in his window blind. He came to the conclusion that white light is made up of light of different colours.

Each of the colours seen in the spectrum has a different wave-length (see page 343); red light has the longest, violet light the shortest. Because of these different wave-lengths each colour is refracted at a slightly different angle (red at the smallest angle) and in this way the colours are separated out. This dispersion occurs when light passes through uneven panes of glass and through lenses, but is not seen in the case of a rectangular glass block because any dispersion at the air-to-glass surface is cancelled out by the refraction at the glass-to-air surface.

### 5.3 Sky colour

Light travelling from the sun through the earth's atmosphere is scattered by dust particles and water droplets. As the blue light is scattered much more than the red this gives the sky a blue appearance as our eyes receive these rays on the earth. When the sun is low in the sky (sunrise and sunset) there is a greater scattering of blue light over the greater distance travelled through the atmosphere, thus light coming from the sun has little blue in it and the sun appears reddish in colour.

*QUESTION 2:* Why does an astronaut see a black sky in outer space?

### 5.4 A pure spectrum

If you look closely at the spectrum in Fig. 5.1, (fac-

ing page 44) you will see that the prominent colours are merged and in some cases are not distinct from one another. This is an impure spectrum, arising because the individual colours are not focused to the same spot on the screen. In order to obtain a pure spectrum the apparatus illustrated in Fig. 5.2 must be used. The light rays entering the prism must be parallel so that the individual colours after refraction are parallel (for example, all red rays should be parallel to one another). These parallel rays must then be focused to the same point on the screen.

*QUESTION 3:* In order to produce a pure spectrum, what must be the distances marked $x$ and $y$ in Fig. 5.2?

### 5.5 Re-combining the spectral colours

*QUESTION 4:* If you were given a second prism to use in re-combining the spectral colours to form white light, how would you position it in the apparatus shown in Fig. 5.1. (Think about fitting the two prisms together to make a glass block.)

The re-combining of spectral colours to form white light may also be carried out by spinning Newton's disc

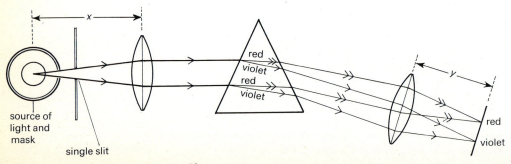

**Fig. 5.2** *Apparatus needed to produce a pure spectrum*

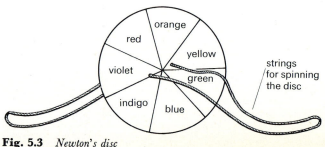

**Fig. 5.3** *Newton's disc*

(Fig. 5.3). A memorable demonstration of this apparatus is said to have been achieved by Faraday, who attached such a disc to a bicycle wheel. The bicycle was ridden into the lecture room by an attractive young lady!

## 5.6   Coloured lights

In the cinema, theatre and even in your school hall coloured lights are used to give "atmosphere" to the stage set. These coloured lights are obtained by shining white light through filters made from transparent pieces of coloured glass or gelatine.

Study the mixing of coloured lights with filters, using torches or lights on the school stage as light sources. Put red, green and blue filters over three such light sources and direct these coloured lights onto a white screen so that they overlap as shown in Fig. 5.4 (facing page 44).

When red, blue and green lights are mixed together they give white light. They are therefore called *primary* colours. A mixing of two of the primaries produces a *secondary* colour.

If red light and blue light are mixed on a white screen the secondary colour is magenta.

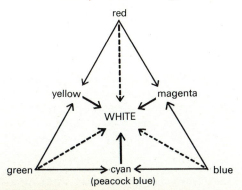

**Fig. 5.5**  *Colour triangle for mixing of lights*

If red light and green light are mixed on a white screen the secondary colour is yellow.

If blue light and green light are mixed on a white screen the secondary colour is cyan (peacock blue).

This colour-mixing of lights involves the adding together of lights of different wavelengths and is called *additive mixing*.

If any of the two colours mentioned above give white light when they are mixed together, then they are called *complementary* colours. Figure 5.5 shows the relationship of all of these colours to one another.

## 5.7   The colour of transparent and opaque objects

a.  *Transparent objects in the form of filters*

A light source or torch may be used as in the previous experiment. Two or three filters of both primary and secondary colours are placed in different combinations in front of the source of light. The colour of the light leaving these different combinations of filters is observed on the screen. Alternatively the filters may be looked at against the light from a window.

When white light is passed through a filter, only colours in the filter's own colour region of the spectrum are allowed to pass through. The other spectral colours are absorbed. For example, a green filter will allow through predominantly green light (and possibly a little yellow and blue).

The colour that leaves a combination of filters is the one that has been allowed through by all of them. If all the spectrum colours are absorbed then no light will pass, and black will be seen. Because this form of colour mixing deals with the removal of certain wavelengths of light it is called *subtractive combination*. Some examples of this are seen in Fig. 5.6.

In Fig. 5.6(a) magenta and red filters are placed together. Red is a primary colour and will absorb all

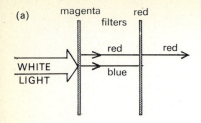

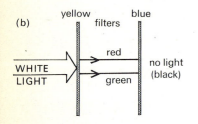

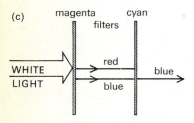

**Fig. 5.6** *Subtractive combination of colours*

other light except red light, which it allows to pass through. Magenta is a secondary colour made up of red and blue light. Thus, it will allow through red and blue light, absorbing all other colours. However, no blue light can pass through the red filter. Thus, the only colour to pass through both filters and emerge from the combination is red light.

In (b) yellow and blue filters are placed together. Yellow is a secondary colour made up of green and red light. Thus, it will allow only these two colours to pass through it. However, the blue filter will absorb these colours because it allows through only blue light. Thus, no light leaves the combination and black is seen.

In (c) cyan and magenta filters are placed together. Both of these colours are secondary colours and the only colour common to both is blue. Thus, blue light is seen.

*QUESTION 5:* What would you see on looking through a combination of cyan and yellow filters?

*b. Opaque objects*
Solid objects appear to have a particular colour because they reflect only that colour. In the following experiment coloured solid objects were placed in a darkened room and a series of lights of different colours was played upon them. (Alternatively, the objects could be viewed through coloured filters in daylight.) A series of results for a yellow object is given in the following table.

| Colour of light (or filter) | Apparent colour of object |
|---|---|
| white | yellow |
| yellow | yellow |
| blue | black |
| red | red |
| green | green |
| cyan | green |
| magenta | red |

In daylight, or under a white light, the yellow pigment in the object absorbs all colours of the spectrum except those that are in its region in the spectrum. These colours it reflects. The predominant colour that is scattered is yellow, plus a small amount of orange and green. Also, when the colour of the object is a secondary colour it reflects its primary colours. Thus the yellow object reflects both red and green light. When none of these colours is present in the light incident upon the object, then there is nothing to reflect, and the object appears black. (For white light on a green object see Fig. 5.7, facing page 44.)

*QUESTION 6:* A green ball has yellow spots. How would it appear under (a) red light, (b) yellow light, (c) cyan light?

### 5.8 Uses of the mixing of colours
*a. Colour printing*
The subject to be reproduced is photographed three times, using successively a red, a green and a blue filter. In each case, the photographic plate receives and records only the colour transmitted by the filter used. Printing blocks are then made from the film, and each is etched so that the portions eaten away by the acid are the areas which were exposed to the light coming through the filters.

In printing (Fig. 5.8, facing page 44) the ink on each block is of a colour complementary to the colour of the filter used in making the block. Cyan ink is used on the block prepared with the red filter, magenta ink is used on the green block and yellow ink on the blue block.

*b. Photography*
A photographer finds that an ordinary black and white film gives a false impression of a scene, because some

colours naturally appear brighter than others to the human eye. For example, yellow always appears approximately ten times as bright as blue and three times as bright as red. These differences can never be represented in a black and white photograph because the film emulsion cannot have the correct sensitivity to each. But if filters are used in front of the camera lens then the different degrees of brightness can be made more life-like. The filter controls the amount of light reaching the screen in two ways: (i) it admits more light from white objects or objects that are the same colour as the filter, and (ii) it absorbs and reduces the light from objects that do not possess the same colour as the filter.

QUESTION 7: A photographer is to take a landscape photograph in black and white and wishes to produce well-defined clouds in the sky (that is, white clouds on a black sky). He has two filters, blue and yellow. Which one should be used and why?

c. *Theatre and art*
Lights of different colours, directed on actors and scenery, are used to create the required atmosphere on the stage. A demon, for example, would be spotlighted in green or red. Objects painted on the scenery can even be made invisible under one light, and suddenly appear when a different light is used. See if you can paint such a picture.

In some art colleges students in search of new ideas experiment with the effects of coloured moving lights on people and things. Spectacular effects can be obtained by passing light through coloured dyes dropped into a small trough of water and using this in a slide projector.

Make-up under such changing lights can produce startling effects. How many girls have been dismayed at the grotesque effect of their day-time make-up under sodium lighting! Obviously, models and film make-up technicians must pay attention to the type of lighting used on the set. Under the strong lighting used in television heavy make-up is always essential if people are to look normal on the screen.

d. *Paints and dyes*
When an artist mixes paints, the colour he sees is the one that is scattered, the rest are absorbed by the pigments he has put together (Fig. 5.9).

QUESTION 8: The machine in Fig. 5.10 is very useful to the interior decorator who wants an exact colour match. Can you say what it is used for?

It is very difficult to match colours in paints and dyes exactly. Girls know that it is unwise to buy knitting wool from two different batches of dyeing: the dye colour always varies slightly in spite of the fact that it is supposed to be the same. The variation in colour is very noticeable when the garment is knitted up, if not before. Differences in colour in dye and paint are the result of impurities in material, slight differences in quantities used, and other human errors of one sort or another.

e. *Colour television*
In a shadow mask television receiver there are three electron guns (see page 334). The screen is composed of phosphor dots in groups of three, uniformly distributed. Each dot in the group glows in a different colour—red, blue or green—when the electrons reach it. The guns are controlled by primary colour signals. Thus, when a red colour signal is received by the correct gun it releases electrons which are directed onto the correct phosphor dots by the shadow mask. Similarly electrons from the other two guns activate the blue and green phosphor dots. When all the dots glow, they are too small to be seen individually and a colour picture is viewed.

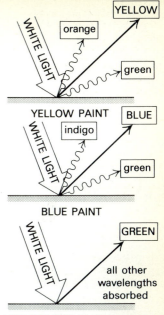

**Fig. 5.9** *The mixing of yellow and blue paint: the only colour not absorbed is green*

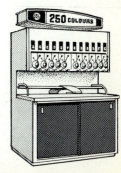

**Fig. 5.10** *A machine useful to a decorator*

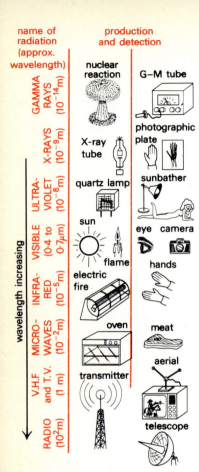

| name of radiation (approx. wavelength) | production and detection |
|---|---|

GAMMA RAYS ($10^{-14}$ m) — nuclear reaction / G-M tube

X-RAYS ($10^{-9}$ m) — X-ray tube / photographic plate

ULTRA-VIOLET ($10^{-8}$ m) — quartz lamp / sunbather

VISIBLE (0·4 to 0·7 μm) — sun / eye, camera

INFRA-RED ($10^{-5}$ m) — flame, electric fire / hands

MICRO-WAVES ($10^{-2}$ m) — oven / meat

V.H.F. and T.V. (1 m) — / aerial

RADIO ($10^2$ m) — transmitter / telescope

wavelength increasing

**Fig. 5.12** *The electromagnetic spectrum. The visible region is a very tiny fraction of the whole. The figures in brackets give an order of magnitude for the wavelength, but there is considerable overlapping of the different classes e.g. V.H.F. and television waves are radio waves of short wavelength.*

## 5.9   How are we able to see colours?

The eye is something like the television screen. It is thought that the retina of the eye possesses three groups of nerve endings each of which is sensitive to one of the three primary colours. When these groups of nerve endings are stimulated to varying degrees the brain is able to build up its colour picture.

## 5.10   Other spectra

On page 46 we discussed the formation of a pure spectrum.

The apparatus illustrated in Fig. 5.2 is the basis of an instrument called a *spectrometer*. This is used to identify elements by the spectra they produce. In the case of incandescent solids and gases at high temperatures and pressure the spectrum is a *continuous emission* spectrum (Fig. 5.11a, facing page 45). In the case of incandescent gases at low pressures, a *band emission* or *line emission* spectrum is produced. Such emission spectra are formed by the hot material releasing some of the energy it had previously absorbed. The released energy takes the form of light of specific wave-lengths.

The spectrum of a sodium lamp (yellow street light) is an example of a line spectrum, as in Fig. 5.11(b). If white light is shone through sodium vapour before it enters the spectrometer an *absorption spectrum* is seen. This is because the matter, in this case the sodium vapour, absorbs certain wave-lengths, and this absorption creates dark lines in the continuous spectrum. Such an absorption spectrum can be seen in Fig. 5.11(c).

*QUESTION 9:* Look at Fig. 5.11, (b) and (c). What do you observe about the sodium line spectrum and the dark lines in the absorption spectrum? What do you think the sodium vapour has done?

A spectrum like Fig. 5.11(c) enables you to state without a doubt that sodium is present in the source. This is how scientists have used the spectrometer to identify elements existing as vapours on other planets. By observing the sun's spectrum, helium was discovered on the sun even before it was found to be present on the earth. Absorption spectra are used for identification purposes in pathology and in forensic (i.e. concerned with the law courts) science laboratories.

## 5.11   Electromagnetic spectrum

The emission of light by a source is caused by the acceleration of electrons. The resulting wave motion consists of electric and magnetic vibrations and the waves are called *electromagnetic waves*. The colour we see depends on the wave-length of the electromagnetic wave. There are many electromagnetic waves which the eye cannot see, such as radio waves, infra-red waves, ultra-violet rays, X-rays and gamma-rays. The difference between electromagnetic radiations and light is basically a difference in wavelength but all are caused by changes in the acceleration of electrons (see page 371). The electromagnetic spectrum is shown in Fig. 5.12

In this spectrum the electromagnetic waves are arranged according to their wavelengths measured in metres. You will see that visible light is only a very small part of the electromagnetic spectrum.

## 5.12   Uses of electromagnetic waves

You make use of electromagnetic waves when you sunbathe: a good suntan is produced by careful and gradual exposure to ultra-violet rays coming from the sun. Special lamps that emit infra-red rays are often used medically in the treatment of rheumatic and muscular complaints.

Another use is illustrated in Fig. 5.13, where we see food being cooked in a canteen by a new method called

**Fig. 5.13** *A microwave cooker in use in a canteen*

Laser sources are being put to practical use in many ways, and there is no doubt that they will play an increasingly important part in the development of communications (Fig. 5.14).

**Fig. 5.14** *Television signal transmitted via a laser signal*

microwave cooking. The advantages of this kind of cooking are (a) that it is very quick (by the time you have read this paragraph it has cooked a meal), and (b) the food is cooked from the centre outwards. Thus you can, in actual fact, cook a hot pudding in the centre of a block of ice cream which will itself remain very cold.

In 1960 Dr Charles Townes, an American, showed how atomic electromagnetic radiation in the form of light waves could be manipulated in a laser beam. The light from this new kind of light source is thousands of times stronger than normal light. It can produce a temperature of millions of degrees, and can cut through metal and vaporize diamond. A beam sent from the earth and reflected at the moon is still strong enough to be detected on its return to earth.

A laser beam is made up of light waves of identical wavelengths which are exactly "in step" with one another (*coherent*). It is produced by stimulating the atoms in either a simple rod of a synthetic ruby crystal or in a helium-neon mixture. The atoms take in energy and radiate it out as light in an intense parallel beam: hence the name laser (Light Amplification by Stimulated Emission of Radiation).

Harmless electromagnetic waves are those with a long wavelength and low energy (e.g. radio waves). Penetrating and dangerous electromagnetic waves are those with a very short wavelength and a high energy (e.g. high-energy $\gamma$-rays, see page 375).

*QUESTION 10:* Name two forms of harmless electromagnetic waves, and two forms of electromagnetic waves dangerous to man.

**THINGS TO DO**

A. Produce your own rainbow.

    a. Out of doors. Use a garden hose with a fine spray nozzle attachment. Position yourself with your back to the sun, and direct the spray upwards into the air in front of you. The rainbow will appear in the spray.

    b. Indoors. Fill a washing-up bowl with water. Take a handbag mirror and position it under the water so that the bottom edge is on the bottom of the bowl and the top edge some distance up the side of the bowl. Thus, the reflecting surface of the mirror is at a steep angle to the water surface. Let sunlight pass through the water and fall onto the mirror. Look at the ceiling to observe the rainbow.

B. Use a strong magnifying lens to view colour pictures and see if you can detect the printing technique.

C. Paint a series of "magic pictures", that is pictures which have a section that is invisible when viewed in a dark room under one primary-coloured light but reappears when viewed under a different primary-coloured light. You could do a "treasure cave", for example: paint the mouth of the cave brown, and the treasure trove inside the cave in red and yellow colours. Paint all the surrounding areas black. Under a blue light all the interior of the cave will appear black, but when a red light is switched on the treasure trove will appear like magic.

D. Make an illustrated *Who's Who* file on the work of Sir Isaac Newton in the field of optics.

E. Do a small illustrated project showing how colours are part of our everyday lives. The project may include such topics as:

a. Colour decoration in house interiors. Many paint firms give advice in their booklets on what colour to select to give the right mood or "shape" to a room.

b. The use of colour by famous artists: how they create the effect they require, for example how they use complementary colours.

c. The effect of artificial lighting on colours in make-up and dress fabrics.

d. The use of colour in signs (e.g. road-traffic signs) and in language (e.g. "red with rage").

F. Do a small illustrated project showing the use of colour in nature, including such topics as camouflage, cross-fertilization in plants, courtship and communication.

G. Find out all you can about lasers.

1. Describe how you would use a single prism to produce a spectrum of light. Explain why such a spectrum is formed. How may the colours be recombined using a second prism?

2. Explain (a) how a rainbow is formed in the sky, (b) why the sky appears to be blue.

3. Explain what differences you would see in a spectrum produced by a prism if (a) a red filter, (b) a yellow filter and (c) a green filter were placed between the prism and the screen.

   If the spectrum from a prism were allowed to fall on a blue surface what change would take place in its appearance?

4. In a dress shop where there is fluorescent lighting a customer insists on taking dresses into the daylight at the doorway. Is her action reasonable? Explain.

5. How are the coloured pictures in magazines produced?

6. Draw Union Jacks and colour them as they would appear in: (a) daylight, (b) red light, (c) yellow light, (d) green light.

7. What paints would you mix to form green paint? Use diagrams to explain the result. Explain why mixing the same coloured lights would not produce the same result.

8. What is meant by the following terms: dispersion of light, infra-red light, ultra-violet light, complementary colours, Newton's disc?

9. People sometimes talk about "viewing the world through rose-coloured spectacles". Explain with diagrams what you would see if you viewed a green field in which there were red, yellow and white flowers through (a) red spectacles, (b) blue spectacles, (c) turquoise spectacles.

# OPTICAL INSTRUMENTS

**CHAPTER 6**

## 6.1  Camera on the moon

The camera played an important part in the *Apollo 11* moon landing. As Neil Armstrong climbed down from the lunar module a television camera relayed back to earth pictures of this first step onto lunar soil — "one small step for a man, but one giant leap for mankind".

**Fig. 6.1**  *Neil Armstrong sets foot on lunar soil 20 July 1969*

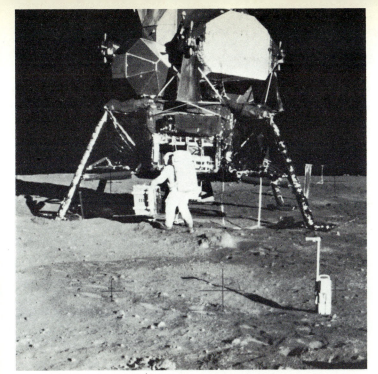

**Fig. 6.2** *Lunar module* Eagle *1969: in the foreground is the stereo camera (with walking-stick attachment) which was used for detailed photographs of the moon's surface*

In Fig. 6.1 we see Armstrong's first footprint. Figure 6.2 is a photograph taken by Armstrong with a 70mm camera. It shows Edwin Aldrin unloading equipment, and in the foreground is the 35mm stereo camera used for close-up photography of the moon's surface.

The camera, here on earth, on the moon and in explorations into space, has become an indispensable instrument, but it is only one of the many optical instruments used by man to increase his depth and breadth of vision.

## 6.2  Pinhole to box camera

A form of the pinhole camera (see Chapter 1), was invented by Baptista Porta at the turn of the sixteenth century. He called his invention the *camera obscura*. But in fact the production of images by using a light-proof compartment with a very tiny hole for the entry of light had been known for some time before this. With the camera obscura arrangement, images were produced on the walls and ceilings of rooms (Fig. 6.3).

**Fig. 6.3**  *Camera obscura*

A pinhole camera would look rather out of place among a selection of modern cameras. However, it can be used to produce excellent photographs although there has to be a long and often inconvenient exposure time of 3 to 4 minutes. The image becomes sharper if the hole is made smaller, and becomes bigger if the distance of the screen from the hole is increased. To convert a pinhole camera to a box camera position a convex lens behind the pinhole, which should be enlarged to pass more light (see "Things to do", page 68). The enlarged pinhole becomes the *aperture* of the camera.

The larger the aperture, the brighter the image. Thus, on a dull day the aperture of the camera should be large. But the size of the aperture also affects the focusing of the image. In fact, it affects what is termed the *depth of field*, that is the distance between the nearest object in focus and the furthest object in focus.

### 6.3 Aperture and depth of field

If you look closely at photographs taken with a pinhole camera, or view the image on the screen, you will observe that objects at any distance from the camera are in focus. Use a simple box camera (Fig. 6.31 page 68 shows a camera with an adjustable screen) to determine how the size of aperture affects the depth of field. Place in front of the camera, in turn, pieces of card with (a) a small aperture, (b) a medium size aperture, and (c) a large aperture. Vary the screen distance and notice the effect on the image with each aperture in place.

In the pinhole camera focusing presents no problem because, with a small aperture and no lens, the depth of field is virtually from 0 to ∞ (infinity). In any camera with a lens the depth of field depends upon the size of aperture.

When the aperture is large there is a small depth of field.

### 6.4 The human eye and the camera

A careful dissection of a bullock's eye reveals all the main structures and these are similar to the structures of the human eye. Likewise, a careful examination of a camera with its back removed reveals its main components. By operating the adjustable parts on the outside of the camera you will see how they work. All these details of eye and camera are fully described in the red pages 57–9.

56

*QUESTION 1:* What is the difference between the yellow spot and the blind spot (see Fig. 6.5) in the eye?

### 6.6 Stereoscopic (binocular) vision and 3D films

Can you make a hole in your hand without any pain? Try the following experiment.

Take a piece of file paper and roll it lengthwise to form a tube approximately 2.5cm in diameter. Hold the tube in your right hand and put it to your right eye. Close the left eye and bring up the left hand, as illustrated in

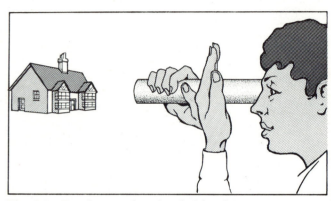

**Fig. 6.4** *Experiment to show the principles of stereoscopic vision*

Fig. 6.4, so that it is next to the tube and positioned halfway along it. Now open your left eye. You should be able to see through a "hole" in your hand! Why is this? 

Try another experiment. Close your right eye and hold up your finger so that it appears to cover an object some distance in front of you. Without moving your finger, close your left eye and open your right eye. The object now comes into view and your finger appears to have moved over to the left. Close your right eye and open your left eye and the object again disappears behind your finger.

What is the explanation?

# THE EYE

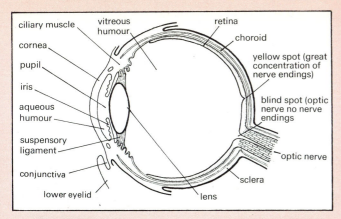

**Fig. 6.5** *Diagram of the human eye*

## STRUCTURE

The tough outer sclerotic coat forms a light tight ball, except for a transparent window at the front called the *cornea*. A black pigment in the *choroid* coat prevents reflection of light inside the eye.

## ENTRY OF LIGHT

(a) The *eyelid* can come across the eye to prevent the entry of light.

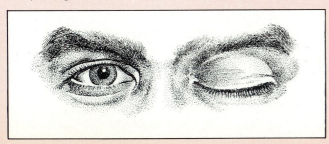

**Fig. 6.6** *The eyelid prevents light from entering the eye*

# THE CAMERA

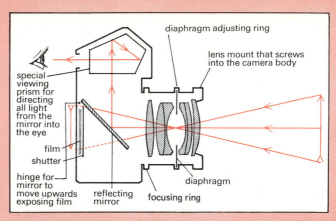

**Fig. 6.7** *Diagram of a reflex camera*

## STRUCTURE

The body case of the camera forms a light tight box. A coating of black matt paint prevents reflection of light inside the camera.

## ENTRY OF LIGHT

(a) The *shutter*. A leaf shutter consists of a series of over-lapping metal sections, which, when the shutter release is pressed, move apart and then spring back after allowing the light to enter the camera. The time for which the shutter is open, (exposure time for the film) may be adjusted to the required speed which is indicated on a knob on the outside of the camera. The shutter speeds are given in fractions of a second, e.g. $\frac{1}{1000}$, $\frac{1}{500}$, $\frac{1}{250}$, $\frac{1}{125}$, $\frac{1}{60}$, $\frac{1}{30}$, $\frac{1}{15}$, $\frac{1}{8}$, $\frac{1}{4}$, $\frac{1}{2}$, 1 (see Fig. 6.8). To photograph a fast-moving object we would select a fast speed. Why?

(b) The *diaphragm* and *aperture*. A series of overlapping metal strips acts as the diaphragm (Figs. 6.11 and 6.12). The movement of the diaphragm, and thus the size of the aperture in the centre, is controlled by the diaphragm

**Fig. 6.8** *The knob with numbers on controls the shutter speed*

(b) The *iris diaphragm* is the coloured area of the eye when viewed from the front. This is a circle of muscular tissue, with a hole in the centre called the *pupil*. The muscles by contracting and relaxing are able automatically to increase the size of the pupil in poor light (Fig. 6.9), and decrease its size in bright light (Fig. 6.10). Which person is looking at the bright light?

**Fig. 6.9** *Dilated pupil allows more light to enter the eye*

**Fig. 6.10** *Contracted pupil allows less light to enter the eye*

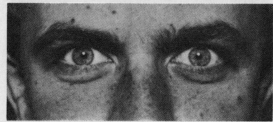

## FORMATION OF THE IMAGE
The lens shape can be altered by the contracting and relaxing of the *ciliary* muscles (*power of accommodation*, Fig. 6.13). When the object is near (less than 6 metres away) the muscles contract, the lens bulges, so that the image is formed on the retina. If the object is nearer than 25 centimetres, it cannot easily be brought into focus. This is called the *least distance of distinct vision*. When the object is more than 6 metres away, the lens shape is made thinner by the relaxing of the muscles. The two humours (jelly-like liquids filling the eyeball) also refract the light and help to produce a focused image on the retina.

58

adjusting ring (Fig. 6.15). The diameter of the aperture is expressed as fractions of the focal length of the lens. Therefore the diaphragm ring is marked in *f*-stops, e.g. *f*/2.8 to *f*/22. The smaller the *f*-number the larger the aperture. On a dull day one would select a large aperture with a slow shutter speed, e.g. *f*/2.8 with $\frac{1}{60}$s. When we go up an *f*-number we halve the aperture, hence only half as much light enters the camera.

**Fig. 6.11** *Aperture opened up to allow more light to enter the camera*  **Fig. 6.12** *Aperture closed down to allow less light into the camera*

## FORMATION OF THE IMAGE
In taking a picture the focusing ring (Fig. 6.15) is revolved so that it gives the required reading of the distance of the object from the camera. The rotation of the ring causes a movement of the lens within its screw mounting. Thus the distance of the lens from the film is adjusted. The movement of the lens enables near or distant objects to be clearly focused on the film (Fig. 6.16). When focusing on a near object the lens is moved away from the film.

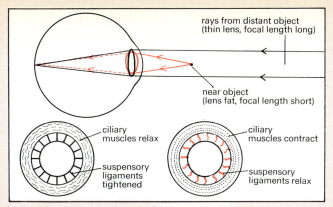

**Fig. 6.13**  *The eye's power of accommodation*

## THE SCREEN FOR THE IMAGE

The *retina* is the nerve tissue layer of the eye. It has many nerve endings called rods and cones embedded in it. These are stimulated by the breakdown of a chemical called *visual purple*, which takes place when light falls on the retina (Fig. 6.14). The stimulated *nerve endings* send an impulse to the brain, which "interprets the messages". At the present time it is thought that the retina itself may also assist in the interpretation of what is seen. The interpretation also corrects the inversion of the image received on the retina.

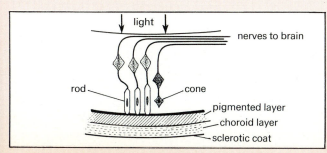

**Fig. 6.14**  *Diagram of the retina of the eye*

Figs 6.8, 6.11, 6.12 and 6.15 are photographs of a Minolta camera, specially taken for this book by Japanese Cameras Ltd.

focusing ring
(set for distance of
object from camera)

diaphragm ring
(changes aperture)

**Fig. 6.15**  *Diaphragm adjusting ring and focusing ring*

## THE SCREEN FOR THE IMAGE

This is a light sensitive film at the back of the camera. A black and white film consists of salts of silver nitrate (or silver bromide) on a gelatine base. These silver salts undergo a chemical change on exposure to light, and have to be chemically developed in order that we may obtain first a negative and then a positive picture.

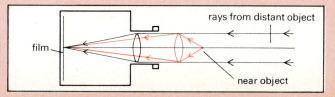

**Fig. 6.16**  *The focusing action of a camera*

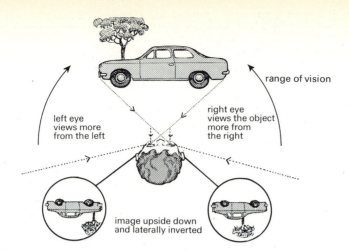

**Fig. 6.17** *Stereoscopic vision*

range of vision

left eye views more from the left

right eye views the object more from the right

image upside down and laterally inverted

Your eyes are approximately 65mm apart and, although this distance seems very small, each eye views the world from a slightly different angle and thus produces a slightly different image (Fig. 6.17). The brain combines these two images to produce a rounded, three-dimensional view. By this means, our eyes help us to judge distance and allocate all objects to their appropriate and relative depths in the "picture" we see of the world.

Stereoscopic pictures are produced by photographing a scene through a camera having two lenses set approximately 65mm apart, so that the two pictures produced are similar to the two different views received by the human eyes. When these pictures are viewed through a stereoscope, each eye is presented with exactly the same view that would have been received when viewing the living scene, and the brain is therefore able to re-create the scene in its full three-dimensional depth.

The same result can be achieved by printing both images on the same piece of paper, one in red and the other in green and slightly offset. When viewed through spectacles composed of one green and one red filter, one eye receives only the red image and the other receives only the green image, and the brain interprets these images as a three-dimensional picture. If the spectacles are reversed, this result cannot be achieved, since each eye will receive the incorrect image.

## 6.6 Persistence of vision

Obtain a cloakroom ticket book. On the back of each ticket, and near the free edge, draw a matchstick man performing a sequence of movements. Figure 6.18 will give you an idea of how to start off. When you have completed the drawings hold the bound edge of the book in one hand and allow the tickets to flicker quickly through the other hand. View the figures on the tickets as they flick by and you will see that they seem to come to life! Your flicker-book is the simplest form of "movie".

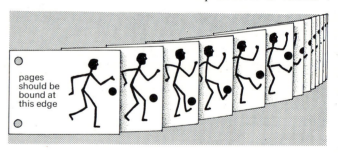

pages should be bound at this edge

**Fig. 6.18** *Exploded diagram of the pages of a flickerbook, showing the sequence of drawings*

This effect depends upon the eye's persistence of vision. The eye retains an image on the retina for about one-tenth of a second.

This means that if two pictures are viewed one after the other within this space of time, then the two images are superimposed. If the same figure is viewed in both pictures with only a slight change in position then that figure appears to move. This drawing out of a sequence, movement by movement, is the basic technique of the animated film.

## 6.7 The movie camera

The construction of a movie camera is very similar to that of the still camera, except for the drive mechanism which moves the film by means of holes (sprockets) through the "gate" past the lens. The number of frames that pass through the gate per second governs the speed at which the shutter must open and close. Some expensive movie cameras provide a range of speeds, expressed as the number of frames per second (fps). Such cameras can run at 12fps, 18fps, 24fps and 36fps. These speeds correspond to $\frac{1}{24}$s, $\frac{1}{30}$s, $\frac{1}{45}$s, $\frac{1}{60}$s and $\frac{1}{84}$s for the still camera.

In making an animated film, each slight movement is photographed one frame at a time. If you can get hold of an 8mm movie camera that can take individual frames (your school may be able to obtain one) try making a short and simple animated film for yourself. The camera must be fixed at the correct distance in front of a board on which you place, in turn, your individual drawings. (You could use pipe-cleaner or plasticine-and-wire figures placed on a flat surface.) Be sure to plan out the movements beforehand. A small section of a movie film is shown in Fig. 6.19.

*QUESTION 2:* In sound movies 24 frames of film pass through the gate of the projector every second. In a silent movie the speed is 16 or 18 frames per second. What would be the effect on the picture viewed on the screen if you passed a silent movie through the projector at a speed of 24 frames per second?

## 6.8 Defects of vision

Some of you are reading this book with the assistance of the simplest form of optical instrument — a single lens. Such lenses correct defects in vision. When they are contained within a frame they are called spectacle lenses; when they rest on the cornea of the eye they are called contact lenses.

What are these defects of the eye that have to be corrected in this way?

a. *Short sight (myopia)*
If a person can see near objects clearly, but not distant objects, then he is said to be short-sighted. This occurs either because the eyeball is too long, or because the ciliary muscles are not able to relax sufficiently to make the lens thinner. The result of both defects is that the rays of light coming from distant objects are always brought to a focus in front of the retina, and therefore form blurred images (Fig. 6.20). A biconcave lens of the required focal length positioned in front of the eye diverges the light rays so that they are focused properly onto the retina (Fig. 6.20).

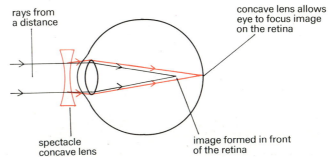

b. *Long sight (hypermetropia)*
A person suffering from long sight is unable to see clearly things that are close to him. In this case either the eyeball is too short or the ciliary muscles are unable to contract

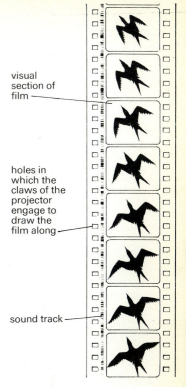

visual section of film

holes in which the claws of the projector engage to draw the film along

sound track

**Fig. 6.19** *A drawing of a section of movie film*

**Fig. 6.20** *Short sight and its correction*

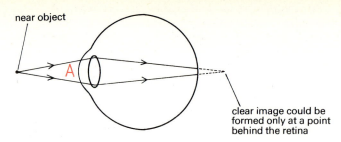

near object

A

clear image could be
formed only at a point
behind the retina

**Fig. 6.21**   *Long sight*

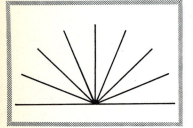

**Fig. 6.22**   *The astigmatic eye is
unable to see every line of this clearly*

enough to make the lens fatter. Thus the rays of light
coming from objects close at hand are not converged
sufficiently to form a clear image on the retina (Fig. 6.21).

*QUESTION 3:* What lens would you put in front of the
eye at point A (Fig. 6.21) in order to correct this defect?
Draw the diagram again, putting in the lens and the
resulting correction of the rays.

c. *Astigmatism and presbyopia*
The arrangement of regular lines shown in Fig. 6.22 can
be used to diagnose the defect of astigmatism. A person
with this optical defect will be unable to see clearly every
line in the drawing. This is because the cornea has
different curvatures in different planes. Cylindrical
lenses are ground to specific requirements in order to
correct the uneven refraction.

   Good eyesight, namely the power of accommodation,
tends to deteriorate with age (presbyopia). Bifocal lenses
are used to correct the older person's inability to see
clearly either at a distance or close to. The top portion of
the lens corrects the short sight, and the bottom portion
corrects the long sight.

*QUESTION 4:* What structure in the eye has been
affected by the onset of old age if the eye no longer
possesses the power of accommodation?

## 6.9   Chromatic aberration in optical instruments

One of the most serious defects in optical instruments is
that of *chromatic aberration* which results in a hazy colouring, or halo, around images viewed through lenses. This
arises because light of short wave-length (blue) is
refracted more than light of long wave-length (red). Thus
the colours in white light are focused to slightly different
points. To overcome this colouring of images, good
optical instruments have lenses that are composed of
different materials, each refracting and dispersing light
by different amounts. Flint glass disperses light less than
crown glass, thus a combination of two such materials
(Fig. 6.23), produces an image with hardly any coloured
halo. A lens made in this manner is termed an *achromatic*
lens.

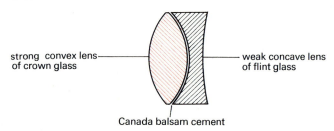

strong convex lens
of crown glass

weak concave lens
of flint glass

Canada balsam cement

**Fig. 6.23**   *An achromatic convex lens*

## 6.10   The microscope

Since the first microscope was invented by Janssen in
1590 it has become an indispensable tool, particularly for
biologists.

   Construct a model microscope in the following way.

   Support a specimen on a microscope slide on a platform of stiff card as in Fig. 6.24(a). A hole should be made
in the card, so that light from a 12V bulb is directed, as
shown, through the area of the specimen. Select a convex
lens of very short focal length, say 5cm, and clamp it
above the slide so that the distance between them is just
about 0.5cm greater than the focal length of the lens.

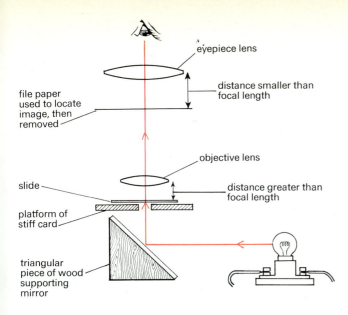

eyepiece lens

distance smaller than focal length

file paper used to locate image, then removed

objective lens

slide

distance greater than focal length

platform of stiff card

triangular piece of wood supporting mirror

**Fig. 6.24** (a) *Construction of a model microscope*

construction diagram showing the passage of light through such a microscope.

Our construction shows us that in order to produce a microscope at least two convex lenses of very short focal lengths (e.g. in the region of 5cm–0.15cm) are required. Such lenses will give magnifications from 2.5 times ($2.5\times$) to 100 ($100\times$). Both the objective lens and the eyepiece lens produce magnification. The final magnification of the specimen is the product of the magnification of each lens. For example, an objective lens of $100\times$ magnification and an eyepiece lens of $20\times$ magnification, when used together in a microscope, give the instrument a magnification of about $2000\times$.

*QUESTION 5:* (a) If you wish to see a specimen the right way up when you look through the microscope, how will you position it under the objective lens? (b) You wish to cover up the left-hand side of the image of the specimen as you view it through the eyepiece lens. You do this by drawing a sheet of paper across the face of the slide, in between the slide and the objective lens. Will you position the paper on the left-hand side or the right-hand side of the slide's face?

### 6.11 The telescope

A simple telescope can, like the microscope, be made from two convex lenses. The difference lies in the focal lengths of the lenses used. The focal length of the objective lens should be greater than 30cm, and the focal length of the eyepiece lens should be in the region of 2.5 to 5cm. As with a microscope, the objective lens should have a wide diameter, for instance 5cm, so that it collects as much light as possible from the object.

To construct the model telescope shown in Fig. 6.25(a), clamp the objective lens in position (or place it in a stand) and direct it towards the distant object that is to

Use a piece of file paper to locate the real, inverted, enlarged image of the specimen that is formed by this *objective lens*. Select another convex lens of short focal length. Clamp this in such a position that the image formed by the objective lens falls within its focal length. Both lenses must have their surfaces parallel to each other, and their optical centres must lie on the same principal axis. In this way, the second lens, the *eyepiece lens*, acts as a magnifying glass and magnifies the initial image. Remove the paper and adjust the positioning of the eyepiece lens until the greatly enlarged, inverted, virtual, final image can be viewed through it. You have constructed a simple microscope. Figure 6.24(b) is a

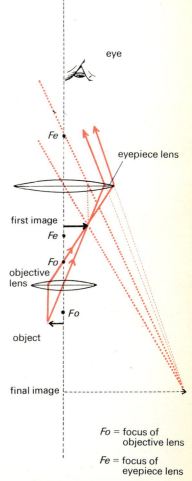

eye

Fe

eyepiece lens

first image

Fe

Fo

objective lens

object

final image

*Fo* = focus of objective lens

*Fe* = focus of eyepiece lens

**Fig. 6.24** (b) *Ray diagram for a compound microscope*

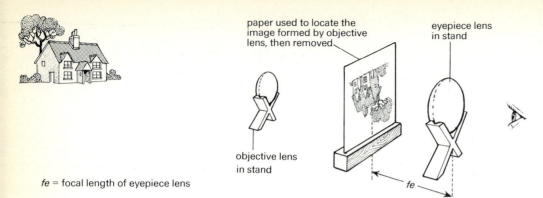

paper used to locate the image formed by objective lens, then removed

eyepiece lens in stand

objective lens in stand

$fe$

$fe$ = focal length of eyepiece lens

**Fig. 6.25** (a) *Construction of a model telescope*

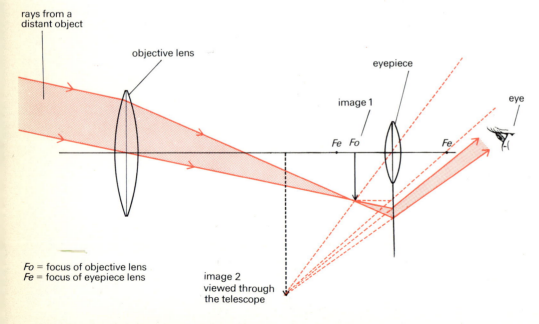

rays from a distant object

objective lens

eyepiece

image 1

eye

$Fe$ $Fo$

$Fe$

$Fo$ = focus of objective lens
$Fe$ = focus of eyepiece lens

image 2 viewed through the telescope

**Fig. 6.25** (b) *Ray diagram for an astronomical telescope*

64

be viewed. Locate the image, and position the magnifying eyepiece lens in the same manner as in the construction of the microscope described in the previous section. Once you know how to set up the telescope try using lenses of different focal lengths and compare the magnification produced by each set of lenses. Do this by directing the telescope towards a brightly lit screen which has horizontal lines drawn on it at equal distances apart. You will find that the objective lens must have a long focal length and the eyepiece lens a short focal length.

Figure 6.25(b) shows the passage of the rays through an astronomical telescope.

It is essential in the study of distant stars that the telescope is able to collect as much light as possible. The refractor with the largest objective aperture (100cm) in the world is to be found at Williams Bay, Wisconsin, USA. Constructional difficulties are so great, however, that it will be a long time before anything larger is made.

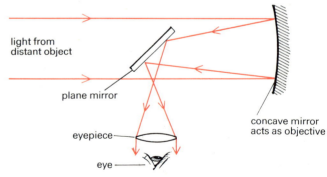

light from distant object

plane mirror

concave mirror acts as objective

eyepiece

eye

**Fig. 6.26** *Reflecting telescope: concave mirror collects and focuses the light, convex lens magnifies the image*

However, a *reflecting telescope* can have a greater light-collecting property than the refracting telescope. Its action is illustrated in Fig. 6.26. Again, because of constructional difficulties, there is a limit to the size

**Fig. 6.27** *Reflecting telescope at Mount Palomar*

of mirror that can be used. Probably the best known of these instruments is the 500cm reflecting telescope at Mount Palomar, USA (Fig. 6.27).

*QUESTION 6:* Is the image formed by such astronomical telescopes erect or inverted? Does this present any difficulties to astronomers?

### 6.12 Terrestrial telescopes and prism binoculars

A third convex lens may be used to correct the inverted image formed by astronomical telescopes. This third lens does not magnify the image. Usually such a telescope is made out of concentric tubes so that it can be collapsed into a shorter and more convenient length when not in use.

Two prisms are used in prism binoculars to correct the inverted image formed by the astronomical lens system. The passage of rays through such binoculars is illustrated in Fig. 6.28. This is a much more compact instrument than a terrestrial telescope.

### 6.13 Projectors

The optical systems of the slide projector and the cine-projector are both based on the construction illustrated in Fig. 6.29(a). The concave mirror and condenser shown in Fig. 6.29(b) are used to collect and concentrate the maximum amount of light through the transparency. In a cine-projector, the shutter closes to shut out all light when the frame is moving into position. The condensers can easily break because they expand in their loose mountings under the considerable heating effect of the lamp. Nowadays, cool quartz iodide lamps can overcome this problem.

*QUESTION 7:* The method of directing all light evenly through the film is shown in Fig. 6.29(b). What is the distance of the concave mirror from the lamp?

The transparency is placed a little further away from the lens than its focal length so that an enlarged image is formed on the screen.

*QUESTION 8:* You are giving a cine film show. (a) Which way would you place the film in the gate of the projector in order to give an image which is erect and the right way round? (b) In what direction would you move the projector in order to form a larger image on the

**Fig. 6.28** *Prism binoculars*

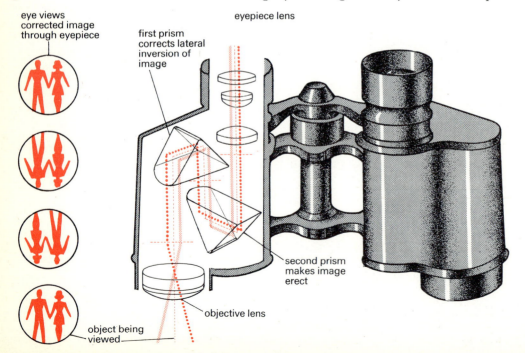

eye views corrected image through eyepiece

eyepiece lens

first prism corrects lateral inversion of image

second prism makes image erect

objective lens

object being viewed

screen? (c) A piece of fluff appears in the top left-hand corner of the image on the screen. Whereabouts on the gate would you expect to find the fluff?

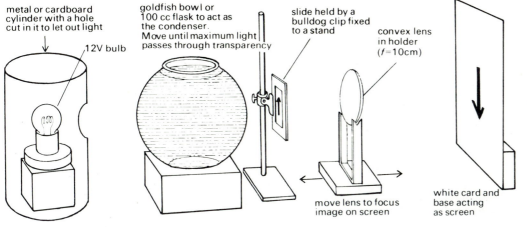

metal or cardboard cylinder with a hole cut in it to let out light

12V bulb

goldfish bowl or 100 cc flask to act as the condenser. Move until maximum light passes through transparency

slide held by a bulldog clip fixed to a stand

convex lens in holder (f=10cm)

move lens to focus image on screen

white card and base acting as screen

**Fig. 6.29** (a) *Construction of a model projector*

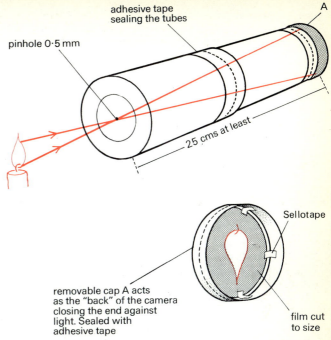

adhesive tape sealing the tubes

A

pinhole 0·5 mm

25 cms at least

removable cap A acts as the "back" of the camera closing the end against light. Sealed with adhesive tape

Sellotape

film cut to size

**Fig. 6.30** *A pinhole camera that can be used to take photographs*

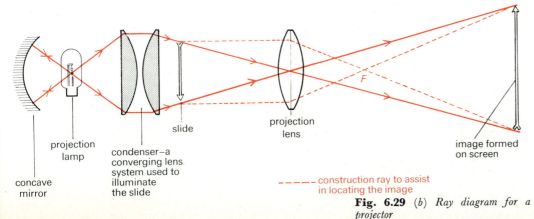

concave mirror

projection lamp

condenser—a converging lens system used to illuminate the slide

slide

projection lens

F

image formed on screen

- - - - - construction ray to assist in locating the image

**Fig. 6.29** (b) *Ray diagram for a projector*

**THINGS TO DO**    A.    Take photographs with a pinhole camera (Fig. 6.30, on page 67).

Remove the screen of the pinhole camera described in "Things to do" Chapter 1, page 10. Seal the two tubes, using black adhesive tape. Cut the second tube so that its open end is at a suitable distance from the pinhole to give a clear image. (Satisfactory results may be obtained with a pinhole of 0.5mm diameter and a screen at a distance of 25cm.) Construct a tightly fitting cardboard cap A to fit over the open end of the tube, thereby forming the back of the camera.

All the following operations must be carried out in a dark room. Cut a piece of film to fit the inside of the cap, hold it in position with strips of sticky tape on the edges if necessary. Fit the cap on the tube. Seal its edges with a strip of black adhesive tape. Place similar tape, of the correct thickness, over the pinhole to prevent light from entering the camera. The camera is then ready to be removed from the dark room.

Direct the front of the camera towards the brightly lit scene which is to be photographed. Remove the tape from over the pinhole and hold the camera perfectly still for an exposure time of 3 to 4 minutes. Replace the tape.

Back in the dark room remove the back of the camera, take out the film and develop it.

A pinhole camera made out of a rectangular box, with the back of the camera constructed in the same manner can also be used.

B.    Convert the pinhole camera to a simple box camera (Fig. 6.31).

Construct the body of the camera in the same manner as described in "Things to do" Chapter 1, page 10. However, use no foil and enlarge the diameter of the hole from 1.5cm to 2 or 3cm. Position a convex lens with a focal length of approximately

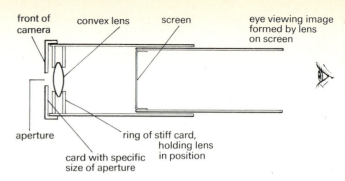

**Fig. 6.31**   *Construction of a simple box camera*

10cm behind the hole inside the tin. Either stick the lens into position or hold it there by a ring of stiff card which has been cut to size. Construct the adjustable screen from greaseproof paper in the same manner as for the pinhole camera construction (page 10).

Construct a range of apertures by cutting circles of card which can be placed, in turn, over the front of the hole in the camera; punch a hole of different size in the centre of each piece of card. The effect of these apertures on the image can be studied by using the adjustable screen. You can take photographs with this camera if you make a back for it similar to that described in the previous section.

C.    Construct a telescope (Fig. 6.32).

The lenses and their positioning should first be determined by the method described in this chapter (page 63). Measure the distance between the objective ($f = 30$–$60$cm) and the eyepiece lens ($f = 5$–$10$cm). Select two postal tubes (or tubes made by yourself from newspaper as described on page 10) of such diameter that they can hold the lenses, and that one tube can slide freely inside the other. The total length of the two tubes laid end to end

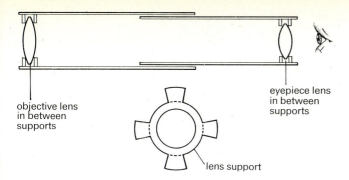

objective lens in between supports

eyepiece lens in between supports

lens support

**Fig. 6.32** *Construction of a telescope*

should be 2 to 3cm greater than the distance between the objective and eyepiece lens. Paint the inside of the tubes matt black. Cut out two rings of cardboard with the same external diameter as the widest tube. Leave four tabs on the outer edge which, when bent over and glued, can anchor the rings in position inside the tube. Position one ring at about 0.5cm inside the open end of the widest tube. Place the objective lens upright inside the tube, hard up against this ring. To keep the lens in position stick in the second cardboard ring. Cut two cardboard rings to the diameter of the second tube and repeat the procedure, as described above, to place the eyepiece lens in position. Check that the lenses are mounted with their surfaces parallel to one another.

Find the best viewing position. Move your eye backwards and forwards until the field of view is as large as possible, i.e. until you see as much of the image as possible. This position is known as the eye ring (the area through which all rays entering the telescope aperture eventually pass). If you have made a good telescope and direct it towards a very brightly lit object, you may be able to see the eye ring on a piece of card placed just outside the focus of the eyepiece.

D. Find your eye's blind spot.

Draw an X on the left-hand side of a piece of paper. At a horizontal distance of 7.5cm to the right of the X draw a black dot ●. Close, or cover, your left eye and look at the X held directly in front of your right eye. Hold the paper at arm's length. Then bring the paper towards you, making sure that your right eye is focused only on the X. Observe that at a certain position the black dot seems to disappear. At this position, the black dot is focused on the blind spot of your right eye.

E. Make a collection of diagrams that illustrate optical illusions. Figure 6.33 gives two examples.

F. Make a stroboscopic top.

Make a top as indicated in Fig. 6.34 and spin it in the light of your TV set with the brightness control turned up and no other lights on in the room. You will see that the lines sometimes go backwards, and if one line replaces another at the same speed as the flashes of light from the TV tube, then the top will appear to stand still. Try drawing figures in the segments.

G. Compile an illustrated project on the history and use of the camera, the telescope or the microscope.

H. Compile an illustrated catalogue showing the different eye structures that are to be found in the animal world. Select one animal from each of the groups of living things (microscopic animals, insects, reptiles, and so on), and indicate how eye structure assists the animal in its way of life.

I. Draw two diagrams, one of a camera, the other a simplified diagram of the human eye. List the similarities and differences of the two instruments.

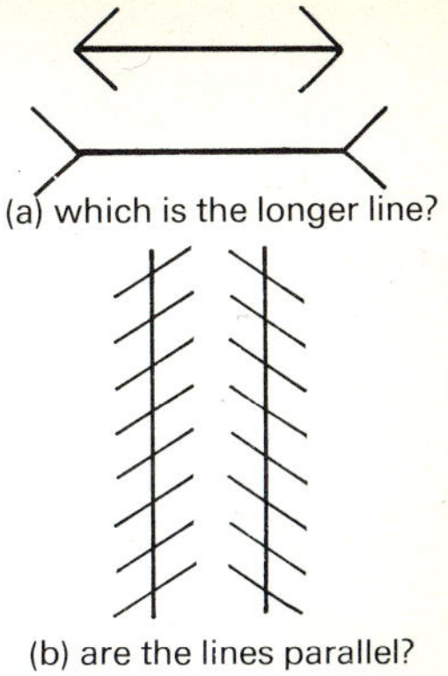

(a) which is the longer line?

(b) are the lines parallel?

**Fig. 6.33** *Examples of optical illusions*

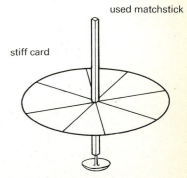

used matchstick

stiff card

drawing pin stuck into base of matchstick

**Fig. 6.34** *A stroboscopic top*

**THINGS TO WORK OUT**

1. Why has each eye a blind spot? Draw a diagram to show roughly the position at which an object would not be visible to your right eye.

2. What would you expect to happen to the pupil of your eye when you move into a darkened room? What else increases the light sensitivity of the eye under such conditions?

3. Explain what you understand by the following: (a) power of accommodation, (b) presbyopia, (c) minimum distance of distinct vision, (d) chromatic aberration.

4. How can you make drawings of pin men "come to life"? Explain why you are able to do this.

5. Explain why some people find it difficult to place a stick through a ring suspended an arm's length away when they have one eye closed.

6. A camera may have a bellows attachment on the lens. What is the purpose of this?

7. Extension tubes are available to photographers. These are screwed in position between the lens and the body of the camera. For what sort of subjects would you find these tubes useful? Explain your answer.

8. Would you use a fast or slow shutter speed to photograph an express train? (a) Give the value of a suitable shutter speed expressed as a fraction of a second. (b) Explain how the amount of light entering a camera is controlled.

9. What is (a) short sight, (b) long sight? How may these defects be corrected? Draw diagrams to illustrate your answers.

10. The convex lens of a camera has a focal length of 15cm. At what distance must the lens be placed from the film in order to photograph an object at (a) infinity, (b) 50cm, (c) 20cm? (Use constructional diagrams where necessary.)

11. Explain the function of the following: (a) ciliary muscles, (b) pupil, (c) retina, (d) choroid coat, (e) visual purple. What is the defect of astigmatism and how may it be corrected?

12. Describe, with a diagram, the action of an astronomical telescope. What extra lens would you require to make the final image erect?

13. How would you use two convex lenses to view a microscopic creature? Draw a diagram to show the passage of light through the instrument. Suggest suitable focal lengths of the lenses.

14. Why is a pinhole camera best used for photographing stationary objects?

15. What is meant by "depth of field"? How is the depth of field of a camera increased?

16. Why is it necessary for the interiors of all optical instruments to be black?

17. A lighted candle is placed 10cm in front of a pinhole in a piece of cardboard. A screen is placed 5cm away from the pinhole and on the opposite side to the candle. If the size of the pinhole is then doubled what difference does this make to the size of the image? Describe what differences you would expect to see in the image if (a) the screen is moved closer to the pinhole, (b) the pinhole is moved closer to the candle, (c) there are two pinholes instead of one. What is the size of the image of the candle flame if the flame is 1cm high?

18. The power of a lens in dioptres (D) is defined by

$$\text{Power} = \frac{1}{\text{focal length expressed in metres}}$$

The focal lengths of three different lenses are (i) 0.1m, (ii) 0.5m, (iii) 5cm. What is the power of each lens? Do you think power is a better way of expressing the "strength" of a lens than focal length? Explain your answer.

# FORCES AND ENERGY

**Part Two**

*The Severn bridge*

# FORCE MASS AND WEIGHT

## 7.1 Force

A stone falls to the ground, a tennis ball is served, a swimmer plunges into a swimming pool. What causes the movement in each of these cases?

In each example the movement was caused by a "push" or a "pull". The stone fell to the ground because it was pulled by the force of gravity (the pull of the earth on the stone); the tennis ball was set in motion by the push of the racket on it, and the swimmer plunged into the water by pushing himself off from the side of the swimming pool.

We use the word *force* to describe such pushes and pulls. In all the above examples a force has changed the speed of an object.

A force may change (a) the speed of an object (don't forget that an object starting from rest is changing its speed), (b) the direction in which the object is travelling, (c) the shape or size of an object.

*QUESTION 1:* Can you think of examples to illustrate statements (b) and (c) above?

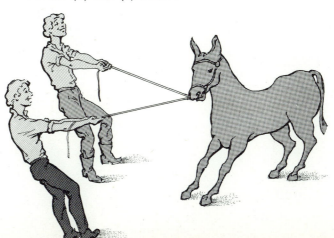

**Fig. 7.1** *The mule does not move: the resultant force on it is zero*

A number of forces may act on one body in such a way that none of the above changes are observed. In such a case the forces balance and the *resultant* force is zero (see Fig. 7.1).

The force that tends to prevent one body sliding over another is called the *force of friction*. The resultant force on a toboggan being pulled along at constant speed is zero, since in this case the forward pull of the rope is exactly balanced by the frictional force between the toboggan and the ground. If the force on the rope is greater than the frictional force the speed of the toboggan increases.

## 7.2 Mass

One characteristic of matter is that it tends to resist any change in motion. This property is called *inertia*.

All bodies, from the size of a speck of dust to the size of the earth, possess inertia. You know, for example, that it is difficult to set a car in motion on level ground by pushing it, but once it is moving it is comparatively easy to keep it moving. The large inertia of the car is again apparent if you try to stop it with your body once you have succeeded in putting it in motion! (But don't try!)

When the brakes are suddenly applied in a car in which you are travelling, you may hit your head on the windscreen if you are not wearing a safety belt (Fig. 7.2). You are not in fact jerked forward as you might at first think. Without a safety belt you are not fixed to the seat, and when the car is braked, no braking force is applied to you and so you continue moving. The inertia of your body keeps you moving forward until the force of the windscreen brings you to rest. Can you explain what is happening in Fig. 7.3?

**Fig. 7.2** *If the car brakes suddenly the safety belt prevents the driver's inertia from carrying her forward into the windscreen*

**Fig. 7.3**

**Fig. 7.4** *Warning! Practise first with plastic cups and saucers—you will find it easiest to do with just one cup and saucer on a baize-topped card table. It is easier if a plastic tablecloth is used.*

The familiar trick of removing a tablecloth and leaving the crockery on the table demonstrates that the crockery possesses inertia (Fig. 7.4). The photograph in Fig. 7.5 was taken an instant after the table itself had been whisked away. What principle is illustrated?

The inertia of a body is measured by its mass (that is, the amount of matter or material in it). A body with a large mass has a large inertia.

**Fig. 7.5** *The table, disappearing on the right, started so rapidly that the dinner service was suspended for an instant in mid-air*

A stone the same size as a football has a greater mass than a football. It is much more painful to set the stone in motion by kicking it than it is to set the football in motion! A bucket with sand in it has a greater mass than an identical empty bucket. If you suspend these two buckets from a beam, the one with sand in it requires a larger force to start it swinging.

**Fig. 7.6** *These masses are used for calibrating force-measuring instruments at the National Bureau of Standards, Washington USA*

*QUESTION 2:* (a) Would that experiment be just as effective if the buckets were standing on the ground? (b) Does the force of gravity have any effect on the experiment? (c) Would it be just as difficult to set the bucket full of sand in motion if the force of gravity were to be removed?

The standard unit of mass is the kilogram (kg) — actually a certain lump of platinum alloy kept at Sèvres, near Paris.

Figure 7.6 shows a woman holding a mass of 1kg, and below it are masses of 4500kg, 9000kg and 13 500kg respectively.

### 7.3  Weight and force

A man has the same mass wherever he is in the universe, because the quantity of matter in his body does not alter. His weight, however, does alter. Why?

Most of you will at some time have seen television pictures of American astronauts demonstrating the weightlessness of objects, including themselves, in outer space. This is the clue to the distinction between mass and weight.

The weight of a body is the force of gravity on it.

The weight of a body on the earth is the force with which the earth attracts the body, and the weight of a body on the moon is the force with which the moon attracts the body. Since the pull of the moon is less than the pull of the earth an object weighs less on the moon than on the earth. The weight of a body depends upon where it is situated in the universe.

The weight of a body also varies over the surface of the earth. One reason for this is that the earth is not a perfect sphere, and its pull on a body depends on the distance of that body from the centre of the earth.

At one time, scientists measured forces by the pull of the earth on a standard mass. The pull of the earth, however, did not prove a satisfactory unit of force because the size of this pull depends on where the body is situated.

We now measure force by measuring the change in speed that a force can produce. For example, the greater the force you apply to a lawn-mower, the faster the lawn-mower will change its speed.

A newton (N) is the force that changes the speed of a kilogram (kg) mass by 1 metre per second every second (m)/²).*

Thus 1kg acted on by a force of 1N will be travelling at 1m/s at the end of 1 second (s). At the end of 2s it will be travelling at 2m/s, and so on. A larger force produces a greater change of speed (i.e. acceleration). The pull of the earth on a mass of 1kg (or any other mass) produces an acceleration of 9.8m/s². The important thing to remember is:

The force exerted by the earth on a mass of 1kg is 9.8N. Therefore a mass of 1kg has a weight of 9.8N.

It will help you if you remember that the weight of an average-sized apple is in the region of 1 newton.

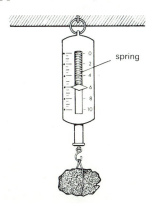

**Fig. 7.7**  *The spring balance measures weight: this one reads 6 newtons*

*A justification of this statement and a fuller discussion of acceleration will be found in Chapter 31.

Weights and forces may be conveniently measured by a spring balance (Fig. 7.7). The extension of the spring depends on the force acting on it. The greater the weight or force, the greater is the extension.

### 7.4  Hooke's law
In the experiment illustrated in Fig. 7.8 known forces are applied to the spring by putting known masses in the scale pan.

| If the mass is | 1kg the force exerted is | | 9.8N |
| --- | --- | --- | --- |
| ,,  ,,  ,, | 1000g | ,,  ,, | 9.8N |
| ,,  ,,  ,, | 100g | ,,  ,, | 0.98N |
| ,,  ,,  ,, | 10g | ,,  ,, | 0.098N |

The extension is measured using the fixed scale at the side. A graph of force against extension (Fig. 7.9) will be found to be a straight line through the origin.

*QUESTION 3:* What can you conclude from the above experiment?

Hooke's law may be stated as follows:
The extension of a spring is proportional to the applied force.

*QUESTION 4:* Spring balances usually have their scales marked in newtons. Would it be possible by examining a spring balance to discover Hooke's law?

### 7.5  A home-made balance
A simple home-made balance is illustrated in Fig. 7.10. A wooden or metal bar is pivoted as shown and a light scale pan hung near the pointed end of the bar. This end is prevented from falling by attaching a rubber band to a hook at the other end, as shown. The balance may be calibrated by putting standard masses in the scale pan or applying known forces by some other means.

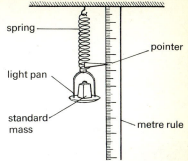

**Fig. 7.8**  *Experiment to investigate the relationship between applied force and extension of a spring*

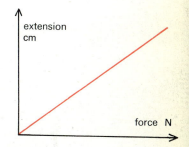

**Fig. 7.9**  *The extension is proportional to the force*

**Fig. 7.10** *A simple home-made balance*

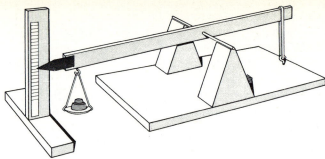

**Fig. 7.11** *A force meter to measure pushes and pulls*

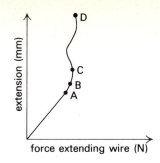

**Fig. 7.12** *Graph of extension of a metal wire against the force extending it*

*QUESTION 5:* Does the balance measure mass or weight?

### 7.6 A force meter

A force meter (sometimes called a dynamometer) is simply an instrument with a spring that extends when a force is applied to it. The greater the force the greater is the extension of the spring. It is usually calibrated in newtons. The one shown in Fig. 7.11 may be used to measure the magnitude, in newtons, of pushes or pulls. You could make one for yourself to measure pulls, by using an elastic rubber strap and calibrating it against known forces.

### 7.7. Extension of a wire

So far you have looked at the extension of springs when a force is applied. Does a similar extension occur in a straight wire? It does, but the loads needed to extend a wire are much greater than those needed to extend a spring. Fig. 7.12 shows a force extension graph for a steel wire. Notice that the wire obeys Hooke's law until the load is very large. Until point A is reached on the graph the extension is proportional to the force extending the wire. The point B, which is usually just beyond A is called the *elastic limit*. Up until this point if the load

extending the wire is removed the wire returns to its original length. Beyond the elastic limit the wire will not return to its original length when the load is removed. C is called the *yield point*. At this point the radius of the wire begins to decrease and we say the wire begins to *flow*. The wire breaks at D.

### 7.8 Prefixes

Prefixes that are commonly used to indicate multiples of basic units are given in the table below:

| Multiple | Prefix | Symbol |
| --- | --- | --- |
| $10^6$ | mega | M |
| $10^3$ | kilo | k |
| *$10^{-2}$ | centi | c |
| $10^{-3}$ | milli | m |
| $10^{-6}$ | micro | $\mu$ |
| $10^{-9}$ | nano | n |

For example, a kilogram (kg) is $10^3$ or 1000 grams (g), a centigram (cg) is $10/^2$ or $\frac{1}{100}$ of a gram, and a milligram (mg) is $\frac{1}{1000}$ of a gram.

*A convenient shorthand notation is used for numbers in scientific work. $10^{-1}$ means $\frac{1}{10}$. $10^{-2}$ means $\frac{1}{100}$. $10^{-6}$ means one millionth 0.000001 which is $\frac{1}{1\,000\,000}$ may be written $10^{-6}$.

A. Make a simple balance like the one shown in Fig. 7.10 and calibrate it.

B. Place a piece of card on a tumbler and place a coin on top of the cardboard. Flick the cardboard off so that the coin drops into the tumbler. Can you explain why this is so easy to do?

C. Get a thin piece of wood about 1m long and tie two pieces of cotton to the ends. Get someone to hold these threads so that the wood is roughly horizontal. Strike the piece of wood in the middle with a strong bar. The wood will break but not the cotton. The experiment may even be done with two long hairs borrowed from a kind lady. Can you explain why the cotton and hairs do not break, but the wood does?

D. Roll ball bearings of different sizes down a triangular shute onto a horizontal table and past a magnet. Does this arrangement enable ball bearings to be sorted according to size? Why does it work?

E. Place a small coin on a tablecloth. Place a tumbler over the coin with its rim resting on two table mats so that there is a gap between the rim of the tumbler and the tablecloth. How can you get the coin out without touching the tumbler? Try scratching the tablecloth. Explain what happens.

F. Find out what is meant by elastic fatigue and why an understanding of it is so vital to aircraft manufacturers.

## THINGS TO WORK OUT

**Fig. 7.13**

1. What is meant by the inertia of a body? Describe two experiments to demonstrate that bodies possess inertia.

2. In Fig. 7.13 a nail is being hammered hard into a piece of wood. The wood is on a piece of stone and the stone rests on a boy's head. Why doesn't the hammering hurt his head?

3. Two boys are riding on bicycles and travelling at the same speed. One boy is twice as heavy as the other boy. They apply their brakes together and the braking force on each bicycle is the same. Will they come to rest at the same spot? Give reasons for your answer.

4. State three different effects a force can have on a body and give examples of each. How could you use a table-tennis ball to illustrate all three effects?

5. You are sitting in a train with your back to the engine. Will it be easiest to stand up when the train is (a) moving with a constant speed, (b) slowing down, or (c) speeding up?

6. a. Is it safer to have luggage racks set across a railway coach or parallel to the length of the coach? Why?
   b. You tend to fall when you are standing in a tube train which accelerates. In which direction do you fall? Why?
   c. Are you more likely to fall over if you step backwards off a moving bus rather than forwards? Why?
   d. Ships require large stopping distances. Why?

7. Explain, using the physical principles discussed in this chapter, why it is advisable to use safety belts in a car.

8. Which would be more difficult to lift, 10kg of feathers or 10kg of concrete?

9. What is the force in newtons on a mass of 10kg, on the earth?

10. On a rocket trip to the moon the engines are not running for most of the journey. Why is this possible?

11. A mass of 1 kg weighs 9.8N. It is put in a deep freeze; what happens to (a) its mass, (b) its weight?

12. Two instruments, one measuring weight and the other mass, are constructed and calibrated on the earth. They are then transported to a space station where the force of gravity is only half the magnitude of the earth's gravity. Would the owner of a grocery store on the space station make a greater profit by selling fruit at 1p per newton weighed on the balance that measures weight, or at 6p per kilogramme on the balance that measures mass? Give reasons for your answer.

13. An empty box is suspended from a rope. In order to set the box swinging a man pushes against it. Why is it easier to do this when the box is empty rather than when it is full?

14. Why is it possible to pull the middle book out of a pile of books without the others falling over? What technique would you use to make sure of the greatest success?

15. Distinguish carefully between mass and weight. In order to manufacture an identical standard kilogramme in a country other than France a scientist uses the following method. He gets the standards laboratory in France to put the standard kilogramme on a spring balance and to mark the extension on the balance. The spring balance is then sent to the scientist who loads it until it is extended by the same amount as it was in France. He then considers the load on the balance to be a kilogramme. Is he correct?

## CHAPTER 8

# DENSITY AND RELATIVE DENSITY

## 8.1 What do we mean by density?

Which is the heavier, wood or brick? Brick is not necessarily the correct answer. A large tree trunk is heavier than a small brick. When people say that brick is heavier than wood, what they really mean is that if we compare equal volumes of brick and wood, the brick is heavier. In order to make such comparisons it is convenient to compare the masses of unit volume of each material. The mass of unit volume is called the *density*.

$$\text{Density} = \frac{\text{mass}}{\text{volume}}$$

If we know the mass and the volume of any substance we can calculate its density from the equation given above.

The density of brick is about 1900 kg/m³ and the density of wood is about 800kg/m³. Any given volume of brick is certainly heavier than an equal volume of wood.

**Fig. 8.1** *Large block of polystyrene —child's play to lift and carry*

A knowledge of density is essential to architects and engineers when they design any large structure. From a knowledge of the size of the parts and their density an engineer can calculate the total weight of the structure, and thus avoid the dreadful consequences of building a structure that would sink into the ground because the foundations were not strong enough.

*QUESTION 1:* If 2m³ (2 cubic metres) of copper have a mass of $1.78 \times 10^4$kg what is the density of copper? What is the mass of 0.5m³ of copper?

The following table gives the density of some common substances. Note the very low density of expanded polystyrene. A large block of this can easily be carried by a child (Fig. 8.1).

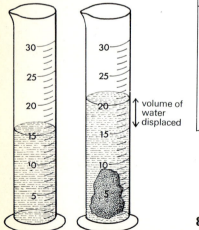

**Fig. 8.2** *When the solid is immersed the water rises 5 cm³; the volume of the solid is 5 cm³*

↑ volume of water ↓ displaced

**DENSITIES OF VARIOUS SUBSTANCES**

| Substance | kg/m³ | g/cm³ |
|---|---|---|
| Platinum | 21 400 | 21.4 |
| Lead | 11 300 | 11.3 |
| Copper | 8 900 | 8.9 |
| Wood | 600–1100 | 0.6–1.1 |
| Ice | 920 | 0.92 |
| Expanded polystyrene | about 20 | 0.02 |
| Mercury | 13 600 | 13.6 |
| Water | 1 000 | 1.0 |
| Paraffin | 800 | 0.8 |
| Air (at 0°C at sea-level) | 1.29 | 0.00129 |

## 8.2 What is relative density?

*Relative density* (or specific gravity as it is sometimes called) is defined by the following equation:

$$\text{relative density} = \frac{\text{density of substance}}{\text{density of water}}$$

It is a ratio and has no units.

The relative density of a substance tells us by how many times that substance is denser than water. For example, the relative density of copper is 8.9 because copper is 8.9 times more dense than water.

Unscrupulous people have sometimes sold milk or beer diluted with water. This practice may be detected by measuring the relative density of the product sold. Diluted milk or beer has a lower relative density than the pure liquid. The instruments used in an experiment of this kind are described on page 107.

*QUESTION 2:* The density of water is 1000kg/m³ and the density of lead is 11 300kg/m³. What is the relative density of lead?

## 8.3 Measurement of density

To determine the density of a substance it is necessary to measure both the mass and the volume of a sample of the material.

For a solid the mass is easily determined by the use of a balance. If it is a cube or rectangular solid, the volume may be determined by measuring the lengths of the sides. For irregular solids the volume may be found by displacement of water (Fig. 8.2). In this method the solid is immersed in a measuring cylinder containing water. The volume of water displaced (i.e. the difference in readings on the measuring cylinder before and after the solid is added) is equal to the volume of the solid.

The density of a liquid may be determined if a beaker is first weighed empty and then weighed again containing a known volume of the liquid (previously measured in a measuring cylinder).

A. Construct a balance capable of weighing a match box full of matches. Make an estimate of the density of the wood from which the matches are made by weighing a box of matches and measuring its volume.

B. Find out how the weight of aircraft or spaceships is kept to a minimum by using materials of low density.

1. What is the difference between density and relative density? Explain why the density of a substance can have a number of different numerical values, but the relative density can have only one numerical value.

2. Use the table on page 82 to answer the following questions, assuming for the purpose of this question that wood has a density of 800kg/m³.
   a. Which solids would float in water?
   b. Which solids would float in mercury?
   c. Which solids would float in water but sink in paraffin?

3. The mass of 10cm³ of a substance is 100gm. What is its density?

4. 2000kg of a substance has a volume of 2m³. What is its density?

5. Use the table of densities on page 82 to calculate the volume of (i) 1.13kg of lead, (ii) 0.89kg of copper, (iii) 1kg of air.

6. Use the table of densities on page 82 to determine the mass of (i) 10m³ of platinum, (ii) 0.2m³ of paraffin, (iii) 1m³ of air.

7. A solid of mass 10g is immersed in a measuring cylinder containing some water. The reading on the measuring cylinder changes from 25cm³ to 29cm³. What is the density of the solid?

8. 20cm³ of copper sulphate (density 1.2g/cm³) are mixed with 80cm³ of water. What is the density of the resulting mixture?

9. The density of air is 1.29kg/m³. What is the mass of air in a room 10m × 8m × 4m? What is the weight of air in the room?

10. Estimate the volume of the room you are in at the moment. Calculate the mass of air in the room if the density of air is 0.00129g/cm³.

11. Suggest a method for measuring the density of air.

12. You are given a box containing 500 tiny steel ball bearings. How would you use kitchen scales (calibrated in kilograms) to determine the mass of one ball bearing? If you have a second box which is said to contain 4000 ball bearings of exactly the same size, how could you check that the number was correct without counting them?

13. A drowning man rises to the surface, shouts for help and sinks. It is often said that this happens three times before he finally drowns. Is there any scientific reason you could give for the truth of the above statement?

14. Complete the following table:

| Object | Mass | Volume | Density |
|--------|------|--------|---------|
| A | 800kg | 1m³ | |
| B | 33 900kg | | 11300kg/m³ |
| C | 5000kg | | 1000kg/m³ |
| D | | 3m³ | 800kg/m³ |
| E | 20g | 1000cm³ | |

Which of the above (a) would float in water, (b) could be of the same material, (c) would be the heaviest to lift, (d) would occupy the greatest space?

**CHAPTER 9**

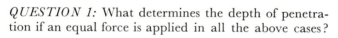

# PRESSURE

## 9.1 Force and pressure

In the experiment illustrated in Fig. 9.1 a 10p piece is being pushed into a piece of plasticine (a) by pressure over the large flat surface, (b) by pressure on the edge. If an equal force is applied in each case, the upright coin penetrates further into the plasticine than the flat coin. Try this experiment with a number of different objects, applying equal force in each case.

*QUESTION 1:* What determines the depth of penetration if an equal force is applied in all the above cases?

The depth of penetration depends not only on the size of the force but also on the area over which it acts. The force acting on unit area (for example, 1m²) is called the *pressure*.

Pressure is the force (or thrust) acting on unit area, i.e.

$$\text{pressure} = \frac{\text{force}}{\text{area}}$$

*QUESTION 2:* Suppose a cube of side 2cm and weight 0.72N is resting on this book. (a) What force does it exert on the book? (b) What pressure in newtons per square centimetre does it exert on the book?

It is not necessary to exert a great force to cut a piece of wood provided a sharp knife is used. The small area of the cutting edge means that a large pressure is exerted on the wood by only a moderate force on the knife.

## 9.2 Pressure in liquids

Figure 9.2 shows a tall can with small holes drilled in its sides at different depths. When the can is filled with water, the water squirts out of the holes. The lower down the hole is, the faster the water squirts out. The pressure in a liquid increases with depth.

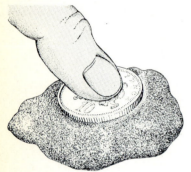

**Fig. 9.1** *Can you explain why the upright coin penetrates further?*

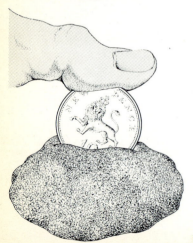

84

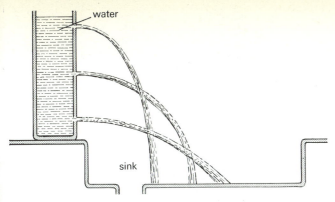

**Fig. 9.2** *Pressure increases with depth*

*QUESTION 3*: If the tank were filled with a liquid of greater density than water would the pressure on the bottom be the same?

The pressure at any point in a liquid acts equally in all directions, and depends on both the depth of the liquid and the density of the liquid.

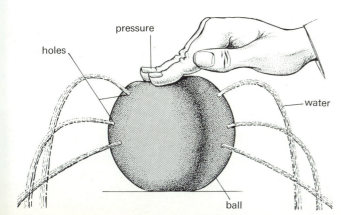

**Fig. 9.3** *Pressure in a liquid acts in all directions*

It may be shown (see Question 17 on page 94) that the pressure due to a liquid column is given by

$$\text{Pressure (N/m}^2) = 9.8 \times \underset{(\text{kg/m}^3)}{\text{density}} \times \underset{(\text{m})}{\text{depth}}.$$

Fig. 9.3 demonstrates the pressure in a liquid acting in all directions. A rubber ball with a number of holes in it is filled with water. When the top of the ball is pressed down the water squirts out through all the holes, thus showing that the water exerts a pressure on every part of the inside surface of the ball.

### 9.3   The public water supply

The starting point of the water supply to your home is probably a reservoir situated on high ground. From there the water flows or is pumped through a system of pipes to storage tanks in the roofs of the houses and factories in the district (Fig. 9.4).

**Fig. 9.4**   *The public water supply*

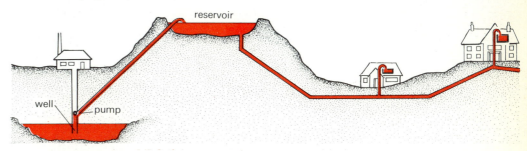

*QUESTION 4:* Is the pressure of water in the ground-floor taps of a house greater than the pressure in those on the higher floors? Why?

### 9.4   **Hydraulic machines**

Try corking a bottle filled with air, and then corking the same bottle filled to the brim with water. It is virtually impossible to cork the latter tightly. This is because air is compressible and water is virtually incompressible. In the

85

experiment illustrated in Fig. 9.3 the pressure applied by the hand is transmitted throughout the water, and the greater the pressure the faster the water squirts out from all the holes.

When any part of a liquid that completely fills a closed vessel is subjected to a pressure, that pressure is transmitted equally to all parts of the containing vessel.

**Fig. 9.5** *The principle of the hydraulic machine*

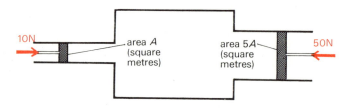

All hydraulic machines work on this principle. In Fig. 9.5 a force of 10N acts on a piston of cross-sectional area $A$ square metres. This means that the pressure exerted by the piston on the liquid in the container is $10/A$ newtons per square metre. This pressure is transmitted throughout the liquid and is the pressure acting on the other piston. Since the larger piston has an area of $5A$ square metres, the force acting on it is $\dfrac{10}{A} \times 5A = 50$N. The machine has multiplied the force by 5. If the ratio of the areas of the pistons had been 10, then the force would have been multiplied by a factor of 10.

Figure 9.6 is a diagram of a hydraulic press. The pressure exerted by the small piston when the handle is depressed is transmitted throughout the liquid, and a much larger force is exerted on the larger piston. This large force may be used to compress bales of paper or other material. If the top part of the press is removed and the cylinder A has a platform fixed to it (Fig. 9.7) we have a hydraulic jack.

**Fig. 9.6** *A hydraulic press*

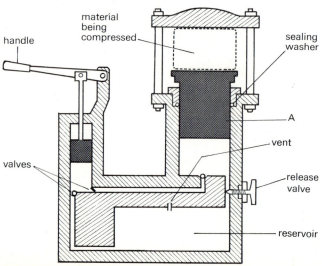

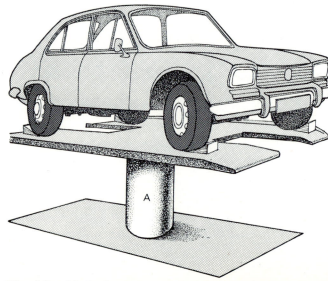

**Fig. 9.7** *A hydraulic jack*

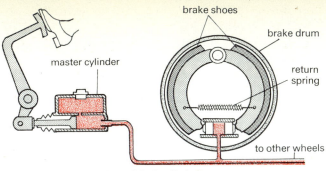

**Fig. 9.8**  *The principle of hydraulic brakes*

Hydraulic brakes are illustrated in Fig. 9.8. The pressure transmitted through the liquid forces the brake shoes against the brake drum and frictional forces bring the car to rest.

## 9.5  Atmospheric pressure

Most people know that we are surrounded by a huge sea of air weighing many newtons, and that the weight of air above us exerts a considerable pressure on us. Why do we not feel this pressure? We are not normally aware of it

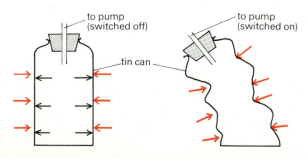

**Fig. 9.9**  *The pressure of the atmosphere crushes the can*

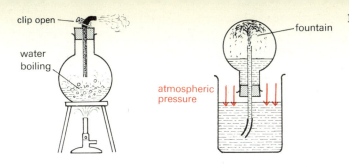

**Fig. 9.10**  *The fountain experiment*

because it acts equally in all directions and the pressure inside our bodies is almost the same as that around us.

The existence of atmospheric pressure is easily demonstrated by the experiments illustrated in Figs. 9.9 and 9.10. As air is pumped out of the can (Fig. 9.9), the pressure of the air on the inside is reduced and the atmospheric pressure acting on the outside crushes the can. If a vacuum pump is not available, the experiment may be performed by boiling some water in the can for several minutes, removing the heat supply and quickly inserting a rubber bung in the neck. The steam has driven the air out of the can, and in a little while, when the steam condenses, the pressure inside the can falls and the can collapses. In the fountain experiment (Fig. 9.10), air is first driven out of the flask by boiling water in it. The screw clip is then closed, the flask inverted as shown, and the clip opened when the neck is under water. When the steam in the flask condenses, the pressure inside is reduced and the atmospheric pressure forces water into the flask, producing a fountain.

## 9.6  A simple barometer

A simple barometer may be constructed by nearly filling a long clean dry tube with mercury, placing your finger over the open end and inverting the tube so that the remaining bubble runs up and down collecting all the small air bubbles trapped in the mercury. The tube is then completely filled with mercury, closed with the finger and inverted under the surface of mercury contained in a bowl. When the finger is taken away the mercury drops to a height of about 76cm above the level of the mercury in the bowl (Fig. 9.11) and remains there no matter how the tube is positioned.

**Fig. 9.11** *The vertical height of the mercury is the same in each position*

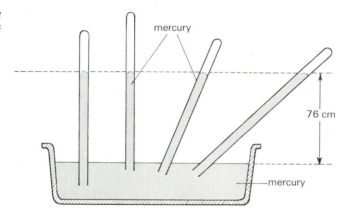

mercury

76 cm

mercury

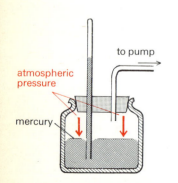

to pump

atmospheric pressure

mercury

**Fig. 9.12** *What does this experiment demonstrate?*

QUESTION 5: (a) What holds the mercury up in the tube? (b) What would happen if you carried the barometer up to the top of a mountain?

The fall in the mercury level when a barometer of this kind is taken up a mountain is evidence that it is atmospheric pressure which is holding the mercury up in the tube. A similar experiment may be done in the laboratory using the apparatus illustrated in Fig. 9.12. The air above the mercury in the bottle is pumped out, thus decreasing the air pressure on the mercury. The level of the mercury in the tube falls rapidly because the lower pressure cannot support such a long column of mercury.

88

If you try this experiment with water and a drinking straw instead of mercury and a barometer tube you will find that it is impossible to drink through the straw when air is being withdrawn by means of the pump.

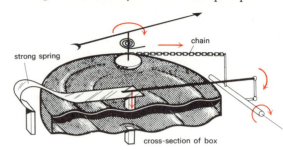

strong spring

chain

cross-section of box

**Fig. 9.13** *An aneroid barometer*

### 9.7 Instruments that measure pressure

a. *The aneroid barometer*

The aneroid (meaning "without liquid") barometer is illustrated in Fig. 9.13. The main feature is a partially evacuated box (i.e. some air has been removed), corrugated to give it extra strength. The box is prevented from collapsing completely by a strong spring. An increase in atmospheric pressure causes the box to cave in slightly, and this movement is magnified by a system of levers. A pointer, indicating air pressures on a scale, is moved by a chain attached to the last lever.

QUESTION 6: (a) Why is it possible for an aneroid barometer with a suitable scale to be used as an altimeter? (b) Would such an altimeter in an aeroplane read height above sea-level or height above the ground?

b. *The Bourdon pressure gauge*

This gauge works in a similar manner to a blow-out toy which is often seen at Christmas parties (Fig. 9.14). When the gas or liquid whose pressure is to be measured

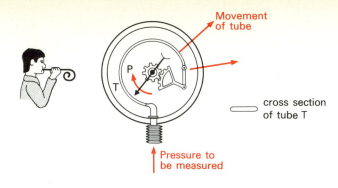

cross section
of tube T

is connected to the gauge, the tube tends to straighten out and the pointer P rotates. The greater the pressure the further the tube unwinds.

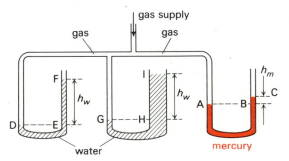

**Fig. 9.15**  *Measurement of gas pressure. The height $h_w$ does not depend on the cross-sectional area of the tube*

c.  *The U-tube manometer*

This is illustrated in Fig. 9.15. The diagram shows the gas pressure being measured. The two U-tubes on the left hand side contain water. Notice that *the height $h_w$ which measures the pressure does not depend on the cross-sectional area of the tubes*. The pressure in newtons per square metre may be calculated from the formula given on page 85. Also using information on page 85 you

should be able to explain why the height of mercury held up by the gas pressure ($h_m$) is less than the height of water held up ($h_w$).

## 9.8  Breathing

Our lungs are situated in the chest cavity at the bottom of which is a flexible membrane called the diaphragm. They are connected to the throat by the trachea (windpipe). Figure 9.16 shows a model to illustrate breathing. When the rubber membrane at the bottom of the bell jar is pulled down, the pressure inside the bell jar is decreased. Hence the pressure on the outside of the balloons is decreased, and atmospheric pressure forces air into them. As the rubber membrane is moved up and down the balloons inflate and deflate, in much the same way as our lungs inflate and deflate following upon the movement of the diaphragm.

The cabins of high altitude aircraft are pressurized, that is, the pressure is increased so that it is above that of the air outside. If this were not done the low pressure at high altitudes would result in considerable discomfort for the passengers because of the decreased supply of oxygen.

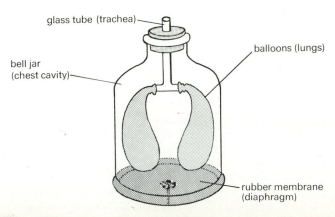

**Fig. 9.16**  *Model to illustrate breathing*

**Fig. 9.17** *Apollo 11 astronauts, the men who accomplished the first moon landing, on their way to the launch pad, 16 July 1969*

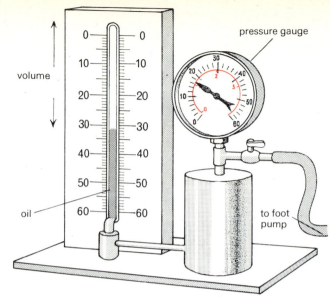

**Fig. 9.18** *Apparatus for testing Boyle's law*

The "popping" of the ears during aircraft take-off is caused by the pressure difference between the air in the middle ear and that in the outer ear and results in an uncomfortable distortion of the ear-drum (see also Fig. 29.21 and page 351). The pressures can often be equalized by swallowing, and this is why air hostesses sometimes offer passengers sweets to suck during take-off.

Astronauts must always wear spacesuits (Fig. 9.17), unless they are in a pressurized cabin. These suits supply oxygen and also maintain a suitable pressure. An unprotected human body could not survive out in space where there is no atmospheric pressure: water and blood would boil and probably explode into the surrounding vacuum.

## 9.9 Boyle's law

The apparatus illustrated in Fig. 9.18 enables us to investigate how the volume of a gas changes with the pressure. The pressure of the trapped air in the apparatus is read directly on the pressure gauge. The volume of the trapped air is measured by the length of the air column. The foot pump enables the pressure on the air to be altered. When the pressure is doubled it will be found that the volume of the trapped air has halved.

Boyle's law states that the volume of a fixed mass of gas is inversely proportional* to the pressure, provided that the temperature is kept constant.

* "Inversely proportional" **is** a way of saying that if one of the quantities is doubled the other is halved.

Stated mathematically this is:

$$p \propto \frac{1}{v} \text{ or } pv = \text{constant}$$

If $p_1$ and $v_1$ represent the initial pressure and volume, and $p_2$ and $v_2$ the final pressure and volume, then

$$p_1v_1 = p_2v_2$$

## 9.10 Pumps

a. Syringe. Some of you will have discovered that doctors' syringes make very good water pistols. The syringe (Fig. 9.19) may be filled with water by putting the narrow nozzle underneath the water and drawing back the piston. The piston is a tight fit in the barrel. As it is drawn back the air pressure below the piston is reduced, and the atmospheric pressure forces water up into the barrel. When the piston is pushed down again water is forced out of the nozzle.

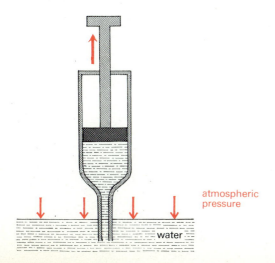

**Fig. 9.19** *A syringe*

b. Force pump. The force pump (Fig. 9.20) is designed to maintain a constant jet of water. It is essentially a syringe with two valves added. When the piston is raised the valve B is closed and the atmospheric pressure forces water through A. When the piston is pushed down, the water forces valve A to shut, and forces valve B to open. Simple force pumps are used in many modern caravans to pump water.

*QUESTION 7:* What do you think is the purpose of the chamber marked X in Fig. 9.20 below?

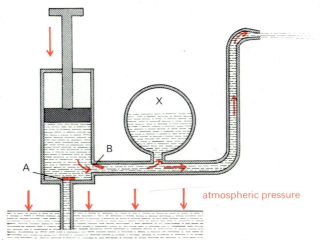

**Fig. 9.20** *A force pump*

c. Bicycle pump. The piston within the bicycle pump has a leather washer on its end. When the handle is pulled out the reduced air pressure in B (Fig. 9.21) results in the leather washer curling inwards as air rushes past it from A into B. When the handle is pushed in, the air in B is compressed and the washer is forced against the sides of the pump, so that air cannot pass it. The air in B is forced out of the end of the pump through the valve on the inner tube of the bicycle tyre.

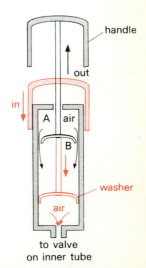

**Fig. 9.21** *A bicycle pump. The leather washer presses against the barrel when the handle is pushed in*

**THINGS TO DO**

A. Reverse the rubber washer on a bicycle pump and measure the force needed to pull the handle out when the hole in the end is sealed. This may be done by attaching a spring balance to the handle of the pump.

B. Push two suction caps together and measure (using a spring balance or force meter) the force required to pull them apart. Try to get suction caps of different sizes and carry out experiments to investigate how the force required depends on the area of the suction caps. Estimate the atmospheric pressure from your readings.

C. Make a collection of newspaper and magazine advertisements showing machines and devices that depend on air pressure for their operation.

D. Fill a washing-up-liquid bottle with water, and replace the top. Make a few small holes in the bottom. Hand it to a friend and tell him to take the top off and smell the liquid in it.

E. Find out, where possible by direct examination, how the following work: (a) a bicycle pump, (b) a vacuum pump, (c) a vacuum cleaner, (d) a fountain pen, (e) a fire extinguisher, (f) a rubber suction cap (the type found on the end of toy darts or used for fixing objects to glass or smooth walls).

Draw a diagram of each and describe how it works, paying particular attention to the part played by atmospheric pressure.

F. Find out if your dentist's chair is hydraulic. Visit your local garage and find out how many hydraulic machines are in use.

G. Find out why car manufacturers recommend certain pressures for their tyres, and why the recommended pressure for front tyres is usually lower than that for back tyres. Is this always the case? Does the size of the car determine the recommended pressure? To answer these questions visit your local garage and write to motor-car manufacturers.

1. Explain why (a) a good knife must have a sharp blade, (b) drawing pins have sharp points, (c) liquid pours better from a tin with two holes rather than one pierced in it, (d) putting on skis prevents you from sinking into snow, (e) a rubber suction cap works better when wet than when dry.

2. Which of the following are units of pressure: (a) grammes per square centimetre, (b) newtons per square metre, (c) kilogrammes per square metre?

3. A cube of side 1cm is resting on a horizontal surface. If the cube weighs 0.6N, what is the pressure on the surface (a) in newtons per square centimetre, (b) in newtons per square metre?

4. A pressure of 5N/m² acts over an area of 2m². What is the force acting on the surface?

5. A rectangular uniform solid has dimensions 0.1 × 0.05 × 0.02m.

    If its density is 9000kg/m³ and it rests on a horizontal surface, what are the maximum and minimum pressures it can exert on the surface?

6. (a) A woman weighing 600N is wearing shoes with stiletto heels. The area of each heel is 0.5cm² and the area of each sole is 60cm². What pressure does she exert on the floor when she is standing (i) on her heels only, (ii) on one foot, (iii) with her weight distributed evenly over both feet?

    (b) An elephant weighs 50 000N and each of its feet has an area of 0.06m² in contact with the ground. What pressure does the elephant exert on the ground if (i) he stands on one foot, (ii) he stands on all four feet?

    Which would do more damage to floors, an elephant or a woman wearing stiletto heels?

7. If taps upstairs and downstairs are turned fully on, will more water per second come out of the upstairs taps or out of the downstairs taps?

8. A hovercraft has a mass of 500kg. If the area of the base is 7500mm², what is the average pressure exerted by the craft on the ground? Express your answer in newtons per square metre.

9. Figure 9.22 is a diagram of a lift pump. Study the diagram carefully and then describe how the pump works.

10. In a small hydraulic press a force of 30N is applied to a piston of area 4cm². The area of the other piston is 60cm². (a) What is the pressure that is transmitted through the liquid? (b) What is the force on the other piston?

11. Does a child usually find it less painful than an adult to walk barefoot on stones? Explain.

12. Figure 9.23 is a diagram of a vacuum brake on a train. The evacuated pipe at the bottom runs the length of the train and is connected to each brake mechanism. The brake mechanisms are connected to the rod X and the brakes are on when the rod X is up. Describe what has to happen for the brakes to be applied. Why are the brakes automatically applied if a carriage breaks loose from the rest of the train?

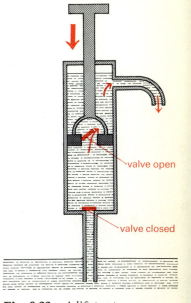

**Fig. 9.22** *A lift pump*

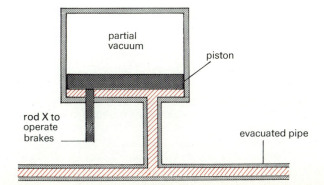

**Fig. 9.23** *Vacuum brake on a train*

93

13. If you press your hands tightly together when you are washing them, then a little force is needed to separate them. Why is this? Explain the noise that is heard when the hands separate.

14. Why are skis made long and narrow? Would any other shape be as good for skiing?

15. What will happen to a stewardess wearing an inflatable bra when the cabin of her jet plane is depressurized?

16. Jane goes skating on a day when the ice is not sufficiently thick, and falls through the ice.
Jack lies flat on his stomach and slowly works his way towards Jane. John says, "Don't be such a fool! She'll be drowned by the time you get there!" and rushes onto the ice in an attempt to get to Jane quickly. Sarah takes a ladder lying nearby and puts this over the ice. Who has the best chance of rescuing Jane first, and why?

17. In this question we are going to calculate the pressure $h$ metres below the surface of a liquid of density $d$ kilogrammes per cubic metre.

    Consider a cylinder with a base of area 1m² (Fig. 9.24) and of length $h$ metres.
    a. What is the volume of liquid in the cylinder?
    b. What is the mass of liquid in the cylinder?
    c. What is the force in newtons acting on the liquid in the cylinder? (A mass of 1kg has a weight of 9.8N: see page 77.)
    d. What is the force acting on the base of the cylinder?
    e. What is the pressure on the base of the cylinder?
    f. What is the pressure at a depth $h$ metres below the liquid surface?
    g. Sea water has a density of 1150kg/m³. What is the pressure 100m below the surface?

18. Suppose you were to fill a bicycle pump with water and put your finger over the end before pushing on the handle. What difference would there be

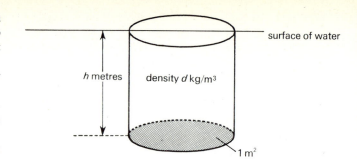

**Fig. 9.24**

between doing this and doing the same thing with the pump full of air? What does this tell you about the way water behaves and the way air behaves? What can you say about (a) the density of water near the surface of the sea and the density of water at greater depths, (b) the density of air near the earth and the density of air higher up in the atmosphere?

19. A fixed mass of gas has a volume of 600cm³ when the pressure is $10^5$ N/m². If the temperature is kept constant what will the pressure be if the volume is decreased to 150cm³.

20. How high would an "atmosphere" of mercury have to be if it were to exert the same pressure as the actual atmosphere? (a mercury barometer normally stands at a height of about 76cm). How high would an "atmosphere" of water have to be? (The density of mercury is 13.6g/cm³).

# KINETIC THEORY OF MATTER

## 10.1 Molecules

In this chapter we shall describe a number of experiments that illustrate the present-day belief that matter consists of tiny particles, or *molecules*, in a continual state of motion. These molecules are themselves made up of even smaller particles called *atoms*.

It will help our understanding if we have some idea of the very small size of molecules. 1cm³ of the air you are breathing at the moment contains some $10^{19}$ molecules. Molecules are so small that if each of the molecules of a speck of dust were magnified to the size of a table-tennis ball and placed side by side, they would encircle the earth many times. If the whole population of the world started counting the number of molecules in 1 gramme of air, and continued counting until they died, they would all be dead before the task was accomplished! You can now understand why molecules are invisible even under a powerful microscope.

Since we cannot see molecules we must do a little detective work and look for the clues that tell us of their existence.

*QUESTION 1:* Suppose air does consist of these tiny particles that are too small to be seen, all of them moving about in different directions. What would you be likely to observe if you were to watch the behaviour of a minute speck of dust in the air that was just visible under a microscope? (Hint: what would happen to a tennis ball if a crowd of people formed a large circle with the ball at the centre and all hurled marbles at the ball as fast as they could?)

## 10.2 Brownian movement

The experiment suggested in Question 1 may be carried out with the aid of a smoke cell (see Fig. 10.1). The transparent cell C is filled with smoke before the cover

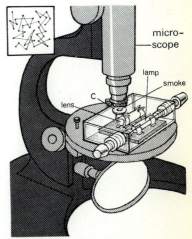

**Fig. 10.1** *Observation of Brownian motion; the smoke particles move in a random manner as air molecules collide with them*

glass is placed on top. The smoke is illuminated by light focused on it by means of the lens. On looking through the microscope the smoke particles are seen to be jostled about in a haphazard and random manner as the invisible particles, or molecules, of air collide with them. This is called Brownian movement or motion, and gets its name from the botanist Robert Brown, who first observed it.

Brownian movement may also be observed in a liquid by using a very dilute solution of Indian ink or aquadag (a graphite paste). When viewed under a microscope the carbon particles can be seen to move about in the same random manner.

*QUESTION 2:* Why is it not possible to observe Brownian movement when a table-tennis ball is hung up in the air?

*QUESTION 3:* If the bottom gas-jar (Fig. 10.2) contained a coloured liquid giving off a coloured vapour, and the top gas-jar contained air, what would you observe when the glass plate was removed?

The experiment referred to in Question 3 can be carried out using bromine liquid, which gives off a brown vapour. But bromine is too dangerous a substance for you to handle by yourselves: it can cause severe burns and, if inhaled, can damage the lungs.

It takes some minutes in this experiment for the coloured vapour to fill the top of the gas-jar and we must now ask ourselves why the diffusion is so slow. One possible reason is that the vapour molecules move very slowly. Can you think of any other reason? Could it be that the air molecules act as a sort of buffer, and that the molecules of the coloured vapour bounce back as they collide with the air molecules?

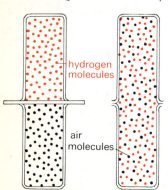

**Fig. 10.2** *Diffusion: the molecules are moving at about 1600 km/h*

hydrogen molecules

air molecules

## 10.3 Diffusion

The smell of a fragrant flower or the smell of a rotten egg quickly spreads throughout a room. The smell is caused by a rapidly moving vapour. This tendency of a substance to spread out of its own accord is called *diffusion*, and may be demonstrated by the experiment illustrated in Fig. 10.2. A gas-jar filled with hydrogen is placed on top of a similar jar filled with air, the two jars being separated by a glass plate. When the plate is removed the *lighter* hydrogen diffuses *downwards* into the air, and the *heavier* air diffuses *upwards* into the hydrogen. If after a few minutes a lighted splint is plunged into either jar the resulting explosion shows that the jar contains a mixture of air and hydrogen.

The results of the above experiment are just what we would expect if air and hydrogen consist of molecules in motion.

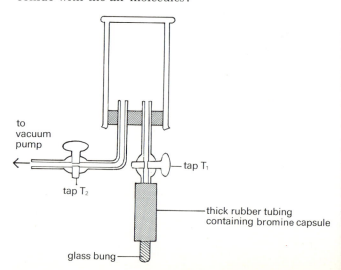

to vacuum pump

tap T₁

tap T₂

thick rubber tubing containing bromine capsule

glass bung

**Fig. 10.3** *Apparatus used to discover the effect of air molecules on rate of diffusion*

It is possible to check this by removing the air molecules and then seeing if the coloured vapour still diffuses at the same rate. The apparatus used is illustrated in Fig. 10.3. A capsule containing bromine is inserted into the rubber tube and the tube is sealed with the glass bung. The taps $T_1$ and $T_2$ are opened and the apparatus evacuated. Then tap $T_2$ is closed and the bromine capsule is broken using pliers. The brown colour immediately fills the whole tube. If the experiment is done without first evacuating the jar, the bromine diffuses slowly.

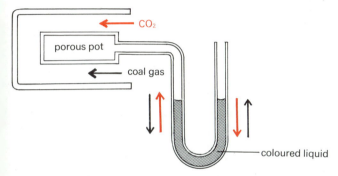

**Fig. 10.4** *Coal gas diffuses into the pot faster than air diffuses out and the pressure inside the pot increases; the reverse happens with carbon dioxide*

Another experiment to illustrate diffusion of gases is shown in Fig. 10.4. A porous pot connected to a U-tube containing coloured liquid is surrounded first by coal gas and then by carbon dioxide. When the coal gas surrounds the pot the liquid in the U-tube moves in the direction shown by the black arrows, because the fast-moving hydrogen molecules (in the coal gas) diffuse into the pot faster than the air diffuses out. The reverse happens when the slower-moving carbon dioxide molecules surround the pot, the liquid in the U-tube moving in the direction shown by the green arrows.

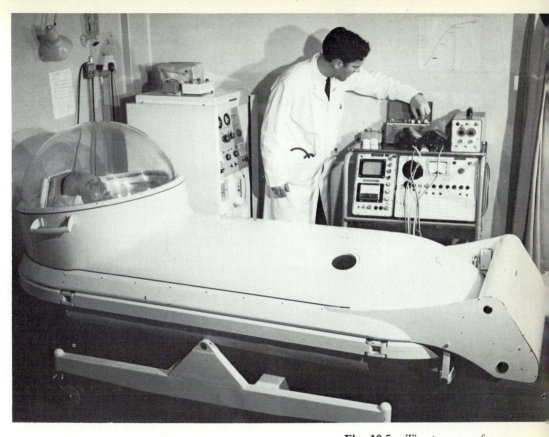

**Fig. 10.5** *The pressure of oxygen round the body is increased in order to diffuse it more quickly into the body*

Figure 10.5 is a photograph of a hyperbaric oxygen bed being used in a hospital. The bed diffuses high-pressure oxygen into the blood and other body tissues. Doctors believe that deaths from acute heart attacks can be considerably reduced by such methods.

Diffusion in liquids can be seen by putting distilled water, copper sulphate, and sugar solution into a gas-jar

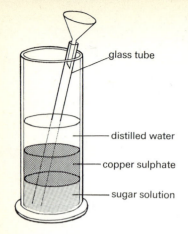

glass tube

distilled water

copper sulphate

sugar solution

**Fig. 10.6** *The blue copper sulphate diffuses upwards and downwards*

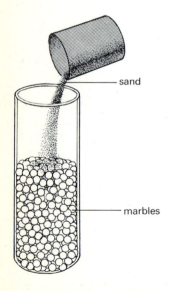

sand

marbles

**Fig. 10.7**

98

as shown in Fig. 10.6. If the different liquids are introduced carefully there will be a sharp dividing line between them. After some hours it can be seen that the blue copper sulphate has diffused upwards into the water and downwards into the sugar solution.

A very simple way of observing diffusion is to let a small drop of ink fall into a glass of water. What would happen?

### 10.4 Dissolving salt in water

Fill a tumbler with water and slowly add salt, stirring carefully all the time. Does the volume occupied by the water increase? If not, where are the particles of salt going? If matter were continuous, and did not consist of tiny molecules, we would expect the volume occupied by the salt and water together to be equal to the volume of water plus the volume of salt added. In fact the volume occupied by the water changes very little when the salt is added. This is because the salt molecules are filling the spaces between the water molecules. Figure 10.7 shows a simple analogy with marbles and sand. The sand occupies the space between the marbles.

### 10.5 Crystals

If a small crystal of copper sulphate is suspended in a saturated solution of copper sulphate and left for a week a large crystal results. If you study a number of crystals of copper sulphate grown in this way you will notice that they all have the same shape, that is, the angles between the faces are always the same. Crystals of one substance all have exactly the same shape. It would be difficult to explain this without assuming that all the crystals were built up from the same smaller units.

**Fig. 10.8** *Atoms of tungsten on a fine metal point*

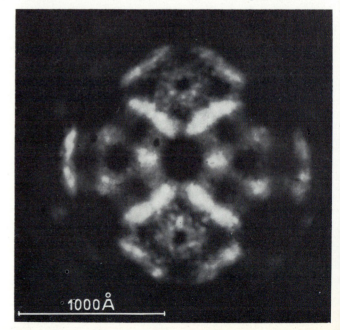

1000 Å

**Fig. 10.9** (a) *Barium atoms on a tungsten point*

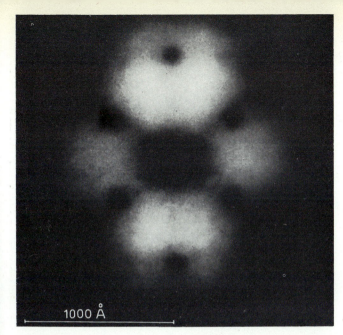

**Fig. 10.9** (b) *Image of a pure tungsten monocrystal point: the bright and dark regions are due to the different binding energies of the electrons*

### 10.6   The evidence of photographs

Figures 10.8, 10.9(a) and 10.9(b) are photographs of atoms taken with a powerful microscope called a field emission microscope. Such a microscope forms an image on a screen and we "see" the atoms in the same sense that we see a person on a television screen. Notice the beautiful orderly arrangement of the atoms.

### 10.7   Models of molecule arrangements in a solid, liquid and gas

Figure 10.10 shows a simple model of the arrangement

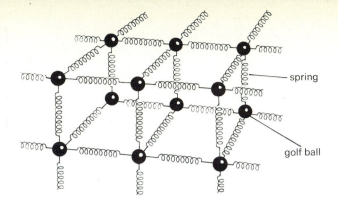

**Fig. 10.10**   *Molecular model of a solid*

of the molecules in a solid. Golf balls, representing the molecules, are connected together by weak springs and vibrate to and fro about a mean position but do not move far from this mean position. The whole model wobbles rather like a jelly. When a solid is heated the molecules vibrate with greater amplitude (i.e. the distance they move from their mean position gets greater), but the position about which they are vibrating does not alter.

When the molecules in a solid are vibrating with such large amplitude that they move away from their near neighbours, the solid has begun to melt. The attractive forces between the molecules are no longer great enough to give the material a definite shape.

Marbles joggled about in a sloping tray (Fig. 10.11) provide a simple model of molecules in a liquid. Some of the faster molecules (like the marble marked A) will break through the surface of the liquid. Most of the molecules that break through the surface will be attracted back to the liquid by the attractive force of the other molecules. A few will be moving so fast that they escape from the liquid and never return. We say then that the liquid is evaporating.

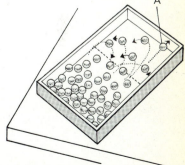

**Fig. 10.11**   *Model of molecular motion in a liquid*

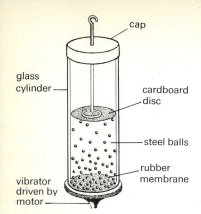

**Fig. 10.12** *Model of gas molecules in motion*

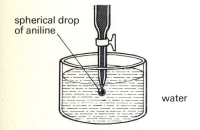

**Fig. 10.13** *A drop of aniline in water is spherical*

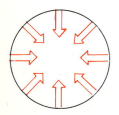

**Fig. 10.14** *The forces of cohesion in a drop of liquid result in a spherical surface*

*QUESTION 4:* The greater the average speed of the molecules the higher is the temperature of the liquid. What will happen to the temperature of a liquid when it is evaporating? (Hint: the most energetic molecules escape.)

In a gas the molecules are very much farther apart than they are in a liquid, and are moving fast. This is why a gas introduced into a container quickly fills it. When the temperature of a gas is increased the average speed of the molecules increases.

The apparatus shown in Fig. 10.12 is a model of molecules in a gas. The steel balls represent the molecules which are moving about at random in all directions. In the model this is achieved by the vibrator pushing the rubber membrane up and down at the bottom of the glass tube. When the speed of the vibrator is increased the balls move faster, and the cardboard disc is pushed higher up the tube, i.e. the volume of the gas has increased.

This apparatus may be used to demonstrate Boyle's law (page 90). Remove the cap from the top of the tube and drop a second cardboard disc on top of the first and maintain a constant vibrator speed. The weight of this second disc is such that the total weight the moving balls are supporting is doubled.

*QUESTION 5:* What would you expect to happen to the volume occupied by the balls?

## 10.8 Forces of attraction between molecules of a liquid

Have you ever watched drops of water dripping from a tap? What shape are they? To slow down the rate of fall and to reduce the effect of gravity, aniline may be dripped into water as illustrated in Fig. 10.13 and the shape of the drops observed.

*QUESTION 6:* The aniline drops are spherical. What does this tell you about the forces between the molecules? Remember that for a given volume of liquid the shape with the smallest surface area is a sphere.

The forces of attraction between molecules of the same substance are called forces of *cohesion*. There are also short-range repulsive forces when the molecules are very close together, and these forces make it impossible for a liquid to be compressed.

Figure 10.14 shows how the forces of cohesion draw a drop into spherical shape.

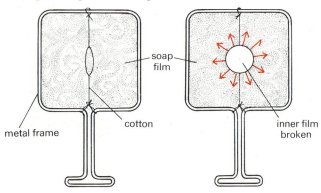

**Fig. 10.15** *The loop is pulled to a circular shape as a result of surface tension forces*

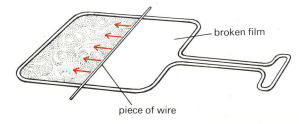

**Fig. 10.16** *The force resulting from surface tension pulls the wire along*

Two experiments to illustrate the forces of cohesion between soap molecules are illustrated in Figs 10.15 and 10.16. A loop of cotton is tied by its two ends to the top and bottom of the frame as shown. The whole frame is immersed in soap solution. It is then removed from the solution and the soap film in the middle of the loop is broken. The loop is immediately drawn into a circular shape by the cohesive forces of the surrounding soap film. The cotton is now removed and the frame again immersed and withdrawn from soap solution. A small piece of wire is gently placed across the middle of the soap film. When the right-hand side of the film is broken the forces of cohesion pull the wire across the frame to the left.

### 10.9 The skin effect of surface tension

The forces of cohesion between molecules give rise to a skin effect on the surface of a liquid. Because of this it is possible to fill a tumbler with water above its brim, and also to float a razor blade or needle on water. Try these for yourself, but when you try to float the needle or razor blade on water make sure that the water is clean. If you find it difficult to place the needle on the water, float it on a piece of blotting paper and when the blotting paper sinks the needle will remain on the surface.

### 10.10 Soap and detergents

The attraction between molecules of different substances is known as the force of *adhesion* (this is how adhesives get their name). When water spills out on a wax surface it forms drops because the forces of cohesion (the attraction of water molecules for each other) are greater than the forces of adhesion (the attraction of the wax for the water molecules). In contrast, water spilt on glass forms a pool on the surface (Fig. 10.17) because the adhesive forces are greater than the cohesive forces. A drop of mercury on glass, however, forms a spherical shape because of the strong cohesive attraction of the mercury molecules for each other. In the same way, water on certain fabrics (Fig. 10.18) forms droplets because the cohesive forces within the drop are greater than the adhesive forces between the water and the fabric. If a drop of detergent is added to the water the cohesive forces are reduced and

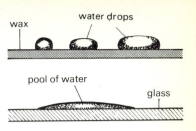

**Fig. 10.17** *Water "wets" the surface of glass but not the surface of wax*

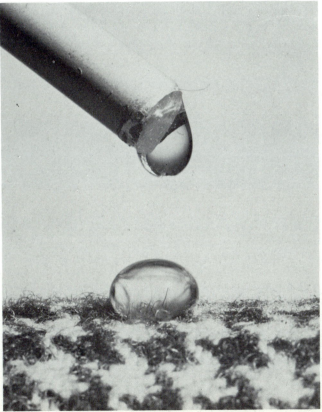

**Fig. 10.18** *A globule of water on a piece of fabric*

the water enters the surface. This is why detergents are effective in the washing of greasy garments. They enable the water to penetrate the fibres and release the dirt caught there.

*QUESTION 7:* Can you explain why a razor blade or needle floating on the surface of water will sink when a drop of detergent is added carefully to the water?

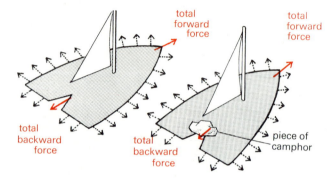

*Fig. 10.19   The camphor reduces the surface tension of water*

When a small model boat is floated on water, it does not move forwards or backwards because the total forward force and the total backward force due to surface tension are the same (Fig. 10.19). But if a small piece of camphor is attached to the back of the boat, at the water surface, the boat will move forwards and appear to be driven along by the camphor. In fact the boat is pulled along by the surface tension forces at the front, because the camphor dissolving at the back reduces the surface tension and hence the backward pull on the boat.

### 10.11   Capillary rise

If one end of a glass tube of narrow bore (a capillary tube) is dipped in water the water-level rises up inside the tube (Fig. 10.20). This is because the adhesion

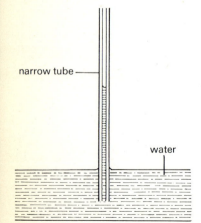

*Fig. 10.20   The rise of water in a capillary tube*

between the water and the glass is greater then the cohesion between the water molecules.

Blotting paper consists of many fine fibres with narrow spaces between them. When dipped into a liquid the liquid rises up between the fibres as a result of capillary attraction.

It is because of capillary attraction that tea rises up into sugar lumps held at its surface. Capillary attraction also causes the rise of water up the brick walls of a house. To prevent this rise of water from the ground every house is now built with a damp course, i.e. a layer of non-porous material incorporated in the wall just above the ground-level.

Capillary rise is an important factor in the movement of water through the soil and of fuel up a wick. Perhaps you can think of many more examples.

A. Put three drops of water on a piece of card so that they form a triangle of side about 2cm. Put a drop of detergent in the middle of the triangle. Explain what happens.

B. Build two imitation brick walls on a large plate with lumps of sugar. On one of the walls put a thin piece of plastic between the first and second row of lumps. Pour some coloured water onto the plate. Watch the colour rise up the sugar lumps, and observe the effect of the plastic layer. Find out all you can about damp courses and see if you can locate the damp course around your house.

C. Find out why (a) touching the inside of a tent when it is raining will cause the tent to leak, (b) reproofing will often stop a tent from leaking, (c) oil is sprayed on water where mosquitoes are known to lay their eggs, which develop into larvae. What disease is man fighting in this instance?

1. Why does observation of (a) diffusion and (b) Brownian motion lead you to believe that matter consists of tiny particles in motion?

2. A balloon blown up with air goes down after some days. A balloon filled with hydrogen goes down overnight. Explain (a) why the balloons go down and (b) why the hydrogen-filled balloon goes down faster than the air-filled balloon.

   Why doesn't the air outside an air-filled balloon diffuse into the balloon at the same rate as the air inside diffuses out?

3. Do the shapes of crystals provide any clue as to the structure of matter? Explain your answer.

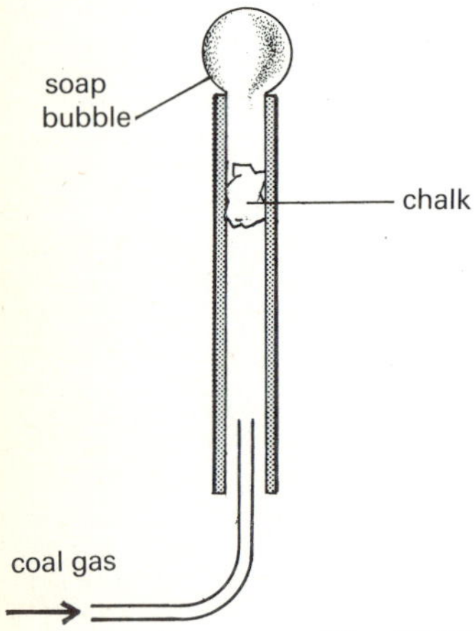

soap bubble

chalk

coal gas

**Fig. 10.21**

4. What evidence is there that hydrogen molecules travel faster than air molecules?

5. Give an example of a substance with strong cohesive forces and a substance with weak cohesive forces.

6. Look at Fig. 10.12 again. Explain why the cardboard disc stays where it is. What is supporting its weight? Explain how the apparatus could be used to produce a model of the atmosphere.

7. A piece of chalk is jammed into a polythene tube (Fig. 10.21) and one end is dipped in a soap solution. When the tube is removed from the solution and a slow stream of coal gas is directed onto the other end, a soap bubble starts to form. Explain this.

8. Uncle George is intelligent but knows very little about physics. How would you convince him that matter consists of molecules in motion?

9. A test-tube full of water to which a crystal of potassium permanganate has been added goes a deep purple colour when the crystal dissolves. Does the fact that this solution may be diluted many hundreds of times before the purple colouring becomes invisible provide any evidence about the size of a molecule?

10. Explain what is meant by adhesion and cohesion. Explain what makes a good adhesive on a molecular level (i.e. by referring to forces acting on the molecules).

11. Why is it difficult to write with a fountain pen if the paper is a bit greasy?

12. Lead shot is made by pouring molten lead through a sieve at the top of a high tower. At the bottom the shot is collected in cold water. Explain why the pellets are spherical.

13. Describe how you would picture (a) a solid, (b) a liquid and (c) a gas in terms of its molecules.

14. Explain why (a) the whole of a wick remains moist when one end is in paraffin, (b) the horizontal layer of slate or bitumen-coated felt running right round a house between two layers of bricks is set so near the ground, (c) a good toothpaste must have a low surface tension, (d) the rubber hose connections on a car radiator system sometimes leak when antifreeze is added, (e) if you examine an umbrella you can see lots of little holes in it, yet it does not let the rain through, (f) some fountain pens have no lever or plunger; all you do to fill them is to dip them in the ink.

15. Air contained within a cylinder with a piston attached exerts a pressure on the piston because the moving molecules are striking the piston and rebounding from it. How does this explain (a) the fact that when the volume of air in the cylinder is halved the pressure is doubled, if the temperature remains constant; (b) that when the air in the cylinder is heated the pressure or volume, or both, will increase?

16. Uncle George has a problem. In a model of the atmosphere constructed with little balls, these balls have to be continually agitated or they stop moving, yet the molecules of the atmosphere, on the other hand, are said to be in constant motion but are not apparently agitated by anything. Write a letter to Uncle George explaining what "agitates" the air molecules.

17. In the experiment illustrated in Fig. 10.1 on page 95, (a) What are the little 'specks' seen through the microscope? (b) What makes them move? (c) Describe the motion that is seen (d) Could Brownian motion be observed using a table tennis ball suspended in the air? Give a reason for your answer.

# BUOYANCY AND FLOTATION

## 11.1 Archimedes' principle

"Iron floats in water," insisted the girl. "Everyone knows that ships made of iron float on water."

Why is it that a lump of iron sinks if it is thrown into water, whereas a ship floats?

Why does a swimmer trying to climb into a boat find it easier to get the top half of his body out of the water than to get his legs out?

The following experiment (shown in Fig. 11.1) will help to answer these questions.

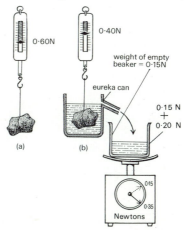

**Fig. 11.1**  *The stone loses 0.20N when under the liquid, and the weight of liquid displaced is 0.20N*

A stone is hung from a spring balance and its weight is recorded (a). The stone, still attached to the spring balance is then weighed while it is immersed in a liquid (b). The stone seems to weigh less when immersed in the liquid. In fact the liquid appears to be pushing up on the stone with a force, or upthrust, of 0.20N. This apparent loss of weight is known as *buoyancy*.

QUESTION 1: Imagine a cube immersed in a liquid with its top face horizontal. Can you explain in terms of the pressure on the top and bottom faces why there is an upthrust on it?

Look again at Fig. 11.1. The liquid in which the stone is weighed is contained in a special eureka can. This type of can has a spout on its side, and the liquid is put into the can up to this level. The stone is then immersed in the liquid, and the liquid displaced by it flows out of the spout. This displaced liquid is collected and weighed. The table (right) shows a series of readings for different liquids used:

| Liquid used | Water | Paraffin | Copper sulphate solution |
|---|---|---|---|
| Wt of stone in air (N) | 0.60 | 0.60 | 0.60 |
| Wt of stone in liquid (N) | 0.40 | 0.44 | 0.36 |
| Upthrust (N) | 0.20 | 0.16 | 0.24 |
| Wt of liquid displaced (N) | 0.20 | 0.16 | 0.24 |

**Fig. 11.2** *This balloon carries research instruments into the upper atmosphere*

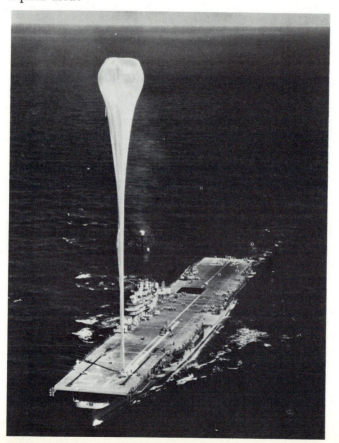

*QUESTION 2:* Study the table carefully. What may we conclude from the readings?

Archimedes' principle states the conclusion you must have arrived at:

When a body is completely or partially submerged in a fluid the upthrust is equal to the *weight* of fluid displaced.

The word "fluid" is used because the principle applies to gases as well as to liquids. When a balloon filled with hydrogen is released in air, it rises because the weight of air displaced by the balloon (i.e. the upthrust) is greater than the weight of the balloon (hydrogen plus fabric). Figure 11.2 is a photograph of the release of a balloon.

## 11.2 Principle of flotation

If a beach ball is held under the surface of the sea and released, it rises to the surface and shoots up into the air.

*QUESTION 3:* Why does the ball rise when it is under the water? Why does it fall back into the water when it is in the air?

When the ball is floating on the water the upthrust (equal to the weight of fluid displaced) is equal to the weight of the ball, i.e. the total or resultant force acting on the ball is zero. We may state the principle of flotation as follows:

**When a body floats it displaces a weight of fluid equal to its own weight.**

A boat which is gradually lowered into water will sink until it displaces a weight of water equal to its own weight. The upthrust is then equal to the weight of the boat and the boat floats.

The principle of flotation may easily be verified with the apparatus illustrated in Fig. 11.3. A tin can of known

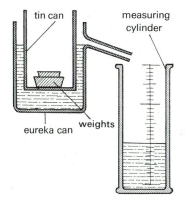

**Fig. 11.3** *What principle does this experiment demonstrate?*

weight is lowered into a eureka can full of water. The water displaced is collected in a measuring cylinder. Weights are added to the can and the water displaced by the addition of each weight is recorded. The weight of water displaced is easily calculated from the volume of water displaced. The experiment shows that the total weight of water displaced is always equal to the total weight of the can.

## 11.3  Hydrometers

A hydrometer (Fig. 11.4) is an instrument designed to float upright in a liquid in order to measure its density.

The depth to which it sinks depends on the density of the liquid.

*QUESTION 4:* Will it sink further in a liquid of low density than in one of high density? Give reasons for your answer.

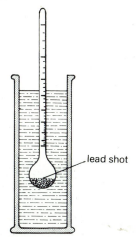

**Fig. 11.4**  *A hydrometer*

The stem of a hydrometer has a scale on it from which the density of the liquid in which it is floating may be read.

Fig. 11.5 shows a hydrometer designed to measure the density of the acid in a car battery. On squeezing the bulb and then releasing it acid enters the glass container which contains the hydrometer. The density of the acid may be read on the hydrometer floating in the acid. Hydrometers are used extensively in the brewing industry as a means of determining the density of beer. Fig. 11.6 shows a hydrometer being used to measure the density of beer. Hydrometers for this purpose are used by people who brew their own beer at home, and are on sale in many chemists; you may even have one in your home.

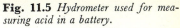

**Fig. 11.5** *Hydrometer used for measuring acid in a battery.*

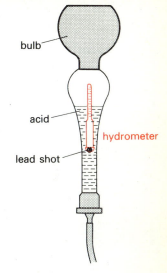

**Fig. 11.6**  *Measuring the relative density of beer*

You can make a hydrometer and test your answer to Question 4 by sealing one end of a drinking straw with sealing-wax, and dropping in lead shot until the straw floats upright when placed in water. The length of the stem immersed when the hydrometer is floated in different liquids may easily be compared.

*QUESTION 5:* Fig. 11.7 shows girls floating in the Dead Sea. The water in the Dead Sea contains large quantities of dissolved salt and is much denser than normal sea water. Explain why the girls can float so easily.

**Fig. 11.7** *Why is it so easy to float in the Dead Sea?*

## 11.4 Submarines and the Cartesian diver*

Submarines have large buoyancy tanks built into their hulls. When these tanks are empty, the submarine floats on the surface. When a submarine submerges, its tanks are opened, and for every kilogramme of water that

**Fig. 11.8** *A submarine of the US Polaris fleet surfacing*

enters the tanks, the submarine sinks until it displaces another kilogramme of water. Once the submarine is submerged, the rate at which it sinks with intake of water increases, since the upthrust remains constant. Compressed air is used to blow the water out of the tanks when the submarine is brought to the surface (Fig. 11.8).

The toy diver illustrated in Fig. 11.9 sinks to the bottom of the bottle when pressure is applied to the cork, and rises to the surface again when the pressure is released. The diver has holes in his body, and when the

* Named after its inventor, the French philosopher, mathematician and physicist René Descartes (1596–1650).

pressure inside the bottle is increased water is forced into these holes; the weight of the diver is therefore increased and he sinks. When the pressure inside the bottle is reduced, the water comes out of the holes, thus reducing the diver's weight and causing him to rise.

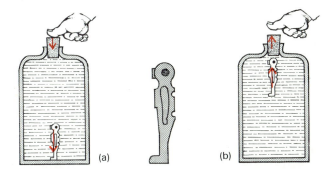

**Fig. 11.9** *Can you explain how the Cartesian diver works?*

## 11.5 Plimsoll mark

During the eighteenth and nineteenth centuries many lives were lost at sea because greedy traders overloaded their ships and many ships sank in storms. The owners claimed the insurance which was sometimes more than the value of the ship. Samuel Plimsoll persuaded Parliament in 1876 to pass a law which made it compulsory for all ships to have safety lines marked on them. The lines are officially known as the "load lines" and it is against the law to load a ship so that the water comes above the load line. Some of the most important markings are shown in Fig. 11.10. FW stands for Fresh Water, S for Summer and W for Winter. Why do you think the FW line is higher than the S line and that the S line is higher than the W line?

## 11.6 Calculations

We shall now use the readings in the table on page 106 to calculate the density (see Chapter 8) of the stone and the density of paraffin. To calculate the density of the stone:

Weight of stone in air $= 0.60N$
Mass of stone (since 9.8N is the force on 1kg, see page 77) $= \dfrac{0.60}{9.8} = 0.0612kg = 61.2g$
Weight of water displaced by stone $= 0.20N$
Mass of water displaced by stone $= \dfrac{0.20}{9.8} = 0.0204kg = 20.4g$
Volume of water displaced (since 1g of water occupies 1cm³) $= 20.4cm^3$

But the volume of water displaced by the stone is equal to the volume of the stone.

$$\text{Therefore volume of stone} = 20.4cm^3$$

$$\text{Density of stone} = \frac{\text{mass}}{\text{volume}} = \frac{61.2}{20.4} = 3.0g/cm^3$$

To calculate the density of paraffin:

Weight of paraffin displaced by stone $= 0.16N$
Mass of paraffin displaced by stone $= \dfrac{0.16}{9.8} = 0.0163kg$
$= 16.3g$

But the volume of this displaced paraffin is equal to the volume of the stone, i.e. 20.4cm³, since it is the same stone of volume 20.4cm³ which displaces the paraffin.

$$\text{Density of paraffin} = \frac{\text{mass}}{\text{volume}} = \frac{16.3}{20.4} = 0.8g/cm^3$$

(Note: the arithmetic can be simplified by leaving the masses and volumes as fractions. In the last line of the calculation 9.8 will cancel.)

*QUESTION 6:* Use the readings in the table on page 106 to calculate the density of the copper sulphate solution.

**Fig. 11.10** *Load lines on a ship, called the Plimsoll mark*

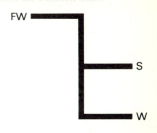

**THINGS TO DO**

A. Make a hollow submarine out of plasticine. Attach a rubber or plastic tube to it in such a way that air can be blown into it. Leave a small hole in the bottom. Float the hollow plasticine in water and suck out some air. Water will then enter the submarine and it will sink. To raise it blow air into the submarine.

B. Make a hydrometer by adding sealing wax and lead shot to the bottom of a drinking straw.

C. Make a toy diver like the one shown in Fig. 11.9. A medicine dropper will make a good diver. Adjust the amount of water in it until it only just floats.

D. Find out all you can about Samuel Plimsoll and the Plimsoll line. What other markings are used other than those shown in Fig. 11.10? (Fig. 11.10 does not show all the possible lines.)

(Where needed take the earth's gravitational field as 10N/kg.)

1. Explain the following statements:
   a. A ship rises in the water as it passes from a river into the sea.
   b. A rowing boat sinks deeper in the water as you step into it.
   c. A submarine is able to travel either on the surface or under the surface of water.
   d. An egg floats in salt water but sinks in fresh water.
   e. It is easy to lift a boulder from the bottom of a swimming pool to the surface, but it is much harder to lift it out and place it on the side of the pool.
   f. When you run over a shingle beach into the sea, the stones hurt much less as you get deeper into the water.
   g. A ship made of iron can float, but a solid lump of iron sinks.
   h. Icebergs are very dangerous to shipping.

2. A picnic party in a boat in the middle of a lake drop their food hamper overboard (it may be assumed to be non-porous). Does the boat rise or sink in the water? What happens to the level of water in the lake?

3. Sunken ships are sometimes raised by filling them with expanded polystyrene balls. Explain why this is possible.

4. A balloon filled with air falls to the ground, but a balloon filled with hydrogen rises. Explain.

5. An object weighs 0.20N in air and 0.15N when immersed in water.
   a. What is the upthrust on the body? In what direction does it act?
   b. What is the weight of water displaced by the body?
   c. Would the upthrust be greater if the object were immersed in salt solution? Give a reason for your answer.
   d. What is the volume of the object?
   e. What is the density of the object?

6. If the object in the above question weighs 0.10N when immersed in a liquid, what is the density of the liquid?

7. A piece of wood displaces 10cm³ of water when it is floating. What is the weight of the piece of wood? If the volume of the wood is 18cm³, what is the density of wood?

8. If the density of a certain metal is 9g/cm³ and its volume is 10cm³, (a) what is its mass, (b) how much will it weigh when completely immersed in water, and (c) how much will it weigh when completely immersed in a liquid of density 0.8g/cm³?

9. A tin box with vertical sides is floating in water. The cross-sectional area of the base is 500cm². What change will occur in the level of the water if a 100g mass is placed in the box?

10. If a solid of volume 100cm³ is immersed in the following liquids, by how much will its weight appear to change in each case: (a) paraffin of density 0.8g/cm³, (b) water, (c) salt solution of density 1.2g/cm³?

11. A cylinder of mass 100g and volume 11cm³ is immersed in water.

   a. What is the volume of water displaced?

   b. What is the weight of water displaced?

   c. What is the upthrust on the cylinder?

   d. What is the apparent weight of the cylinder when it is in the water?

   e. What would the answers to questions a to d be if instead of water a liquid of density 0.8g/cm³ was used?

12. A piece of metal weighs 1N in air and 0.89N when immersed in water.

   a. What is the upthrust on the metal when it is immersed in water?

   b. What is the weight of water displaced by the metal?

   c. What is the volume of the piece of metal?

   d. What is the density of the metal?

13. A block of wood has a volume of 200cm³ and floats in water with 150cm³ underneath the surface of the water. What is the weight of water displaced by the wood? What is the density of the wood? If it were floating in salt solution of density 1.2g/cm³ what volume of the wood would be under the water?

14. Two identical inflated rubber toy balloons are hung on the ends of a beam and the beam is balanced so that it is horizontal. One of the balloons is then punctured. Is the balance of the beam upset? Does the experiment demonstrate that air has weight? Explain your reasoning carefully.

15. In a novel by H. G. Wells a character called Pyecroft loses all his weight but retains his original volume. He then floats up through the air fully clothed to the ceiling. Is this a possible consequence? Give reasons for your answer.

16. Three solid spheres X, Y and Z are all the same size and are painted black so that they all look the same. One is made of plastic, another of copper and the third of lead. The following observations are made: (a) X and Y together weigh less than Z, (b) X floats in water with half of its volume below the surface, (c) Y weighs 0.90N in air, 0.80N when immersed in water and 0.82N when immersed in paraffin.

   Now answer the following questions: (i) What is Z made of? (ii) What is X made of? (iii) What is the density of X? (iv) What is the volume of each sphere? (v) What is the density of paraffin?

17. A balloon has a volume of 1m³. What is the weight of air displaced by the balloon? What is the upthrust? If the balloon is filled with hydrogen what is its lifting power? Neglect the weight of the fabric and take the density of air as 1.3kg/m³ and the density of hydrogen as 0.09kg/m³.

18. Using the values for the density of hydrogen and air given in question 17, calculate the lifting power of a balloon which, filled with hydrogen, has a fabric of mass 2kg. The volume of the balloon is 2m³.

19. A toy boat is floating in a bath: (a) A bar of soap is put in the boat, (b) the same bar of soap is dropped in the water. In which case did the water level rise more? Give reasons for your answer.

20. Explain why a swimmer tends to sink in the water when he breathes out.

21. If a spacecraft of the future could suddenly shield itself from the earth's gravity, would it suddenly lift off? Explain your answer.

# WORK POWER AND ENERGY

## 12.1 The meaning of "work"

"I put energy at your command." So runs the advertisement for a well-known breakfast cereal.

In order to do work we need energy. If we were unable to get this "food energy", or *chemical energy* as the scientist calls it, we could not work and would gradually get weaker and weaker.

Energy is the ability to do work.

What does the scientist mean by "work"? In general usage the word covers a great variety of occupations and jobs; but in scientific usage a word must be limited to one definite meaning. As you try to understand what you are reading at this moment you would say that you are doing work; but are you working in the same sense as if you were dragging a huge lump of concrete along the road? Would you need the same amount of energy to perform each of these tasks?

*QUESTION 1:* Does a professional football player need to get more energy from food than a typist?

**Fig. 12.1** *Egbert and Chinstrap are not doing* work *even if they go on exerting themselves for hours; Flatfoot is doing all the work*

Scientists say that work is done whenever a force causes movement. In Fig. 12.1 the man moving the block of concrete is doing work, but the other two are not, because the force they are exerting is not moving the object on which it is acting. Neither the wall nor the weights are moving. The weight-lifter could easily slip the weights onto the shelf behind him and they could remain there for ever without anyone using up energy to keep them there.

**Fig. 12.2** *The boat exerts a constant force on the skier. How would you calculate the work done in pulling the skier along?*

The boat (Fig. 12.2) is exerting a constant force on the skier. Work is being done by this force, because the skier is being pulled along (i.e. the force is moving the object on which it is acting).

*QUESTION 2:* Work is done when a force moves its point of application. Is work done (a) when you pinch your brother's ear and go on pinching while he goes on squealing, (b) when you lift a brick up from the ground?

## 12.2   How can we measure work?
The man in Fig. 12.3 is lifting a brick, which is exerting a force of 1N on his hands, onto shelves 1m and 2m above the ground.

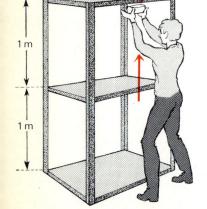

**Fig. 12.3**

114

*QUESTION 3:* (a) If he lifts the mass onto the shelf 2m above the ground will he do more work than if he lifts it onto the shelf 1m above the ground? How much more work? (b) If he now lifts 10 bricks onto the shelf 1m above the ground will he do more work than if he lifts one brick onto that shelf?

The quantity of work done depends on both the size of the force and how far the force moves. We define work as follows:

Work = force × distance moved in the direction of the force                                    (equation 1)

A force of 1N pulling an object through 1m does 1 newton-metre of work. A newton-metre is given the name *joule*.

*QUESTION 4:* How much work is done in moving a force of (a) 4N through 2m, (b) 2N through 2m?

1 Joule (J) of work is done when a force of 1N moves through 1m.

In general,
Work (J) = force (N) × distance (m)        (equation 2)

## 12.3   Energy
We have already stated that energy is the ability to do work. Suppose we want to lift a heavy load; are there any ways of doing it other than using up the body's chemical energy by lifting it with our hands? We could, for example, use a crane driven by an electric motor or a steam engine. In the one case we would be using electrical energy, and in the other case heat energy.

Some other forms of energy are listed below:
*Nuclear energy* is produced by changes in the centre (or nucleus) of the atom. Nuclear power stations produce energy in this way (see page 378).
*Chemical energy*, as we have seen, may be obtained from food. It results from chemical changes. It is also released

**Fig. 12.4** *The Kariba dam. What sort of energy has the water which is behind the dam?*

when a fuel is burnt; and in a battery chemical energy is turned into electrical energy.

*Kinetic energy* (KE) results from the movement of a body. For example, the kinetic energy of moving water (Fig. 12.4) can be used to drive a dynamo in order to produce electrical energy (hydro-electric power). When the flywheel in a friction-driven toy car is set rotating, its kinetic energy drives the car along.

*Potential energy* (PE) is energy resulting from the position or state of a body. When the weights of a pendulum clock are wound up (i.e. lifted) they possess potential energy which, as the weights slowly fall, is released and converted into other forms of energy which keep the clock going for days at a time.

Wound-up clock springs and stretched elastic bands possess potential energy resulting from change of state (i.e. shape or size). In each case the stored potential energy may later be released.

The water behind the dam in Fig. 12.4 has tremendous potential energy. This on release is converted into kinetic energy which, in turn, produces electrical energy. Fig. 12.5 shows an aircraft landing on a flight deck. The kinetic energy of the aircraft becomes potential energy in the stretched wires, thus bringing the aircraft to rest.

*Sound* energy and *light* energy are produced in addition to *heat* energy when, for example, a bomb explodes and its chemical energy is converted into heat, light and sound.

**Fig. 12.5** *The stretched arrestor wires possess potential energy*

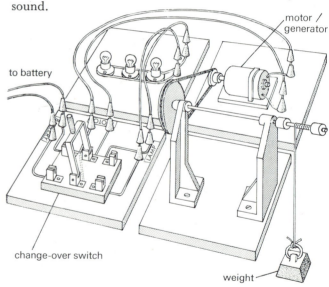

**Fig. 12.6** *The battery is used to drive the motor which lifts the weight giving it potential energy. As the weight falls it drives the generator which lights the bulbs. While the weight is being lifted the energy changes are: chemical (of battery) → electrical → kinetic (of motor) → potential (of weight). What are the energy changes as the weight falls?*

*QUESTION 5:* What sort of energy has (a) a stretched elastic luggage strap, (b) a tennis ball which has just been served?

All the available evidence points to the fact that energy cannot be created and it cannot be destroyed. This principle, known as the conservation of energy, may be stated as follows:

Energy cannot be created or destroyed, although it may change from one form to another.

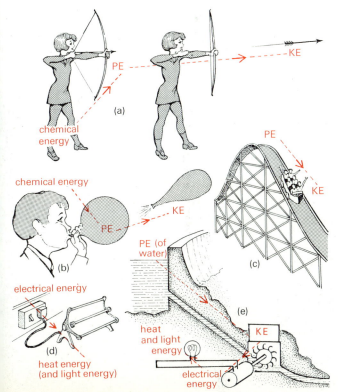

**Fig. 12.7** *Some energy transformations*

**Fig. 12.8** *One of Britain's atomic explosions*

Some examples of such changes are illustrated in Figs 12.6 and 12.7. Tremendous energy is released when an atomic bomb explodes (Fig. 12.8). What changes of energy occur in such an explosion?

**Fig. 12.9** *Which car is the more powerful? What do we mean by power?*

## 12.4 Power

Which of the two cars in Fig. 12.9 is the more powerful?

The word "power" in physics means much the same as it means in ordinary life. A Jaguar car is more powerful than a BMC Mini. Because it is more powerful it can climb hills faster, that is, it can do work faster.

Power is the rate of doing work; or power $= \dfrac{\text{work done}}{\text{time taken}}$
(equation 3)

Suppose there are two cranes available for loading a ship. No matter which is chosen to do the job of loading, the work to be done remains the same. But if one crane could load the ship twice as fast as the other one, that crane would have twice the power.

*QUESTION 6:* Consider the definition of power given above and then suggest a unit in which it could be measured.

A rate of working of 1 joule per second is called a watt(W)
1 kilowatt = 1000 watts

These units of power and energy do not only apply to machines, but are used for the measurement of all forms of energy and power. For example, electrical energy is measured in joules and electrical power in watts (though, as the watt is a small unit, kilowatts or megawatts are met with more frequently).

## 12.5 Early types of engine

When James Watt invented his steam engines, people naturally wanted to know how these compared with the horses previously used to do the same sort of work. How many horses would one of his engines replace? Watt therefore decided to express the power of his engines in terms of the rate at which a horse could do work. To do this he got a horse and measured the rate at which it could pull a known weight up a mine shaft (Fig. 12.10).

*QUESTION 7:* If a horse could walk at 1m/s when raising a load of 500N, how much work would it be doing every second?

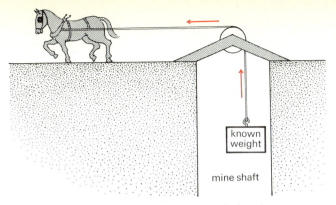

known weight

mine shaft

**Fig. 12.10** *James Watt's experiment to determine the rate at which a horse can do work*

By means of a very similar calculation to the one you have just done, James Watt estimated the horsepower of his engines. 1 horsepower is now defined as 746 watts.

People in the days of James Watt thought that his engines were powerful. What would they think of the engines in use today? For example, the offshore power-boat in Fig. 12.11 develops about 110kW (kilowatts);

**Fig. 12.11** *Offshore power boat developing about 110 kW*

and the rocket in Fig. 12.12 over 2000MW at take-off and a thrust of over 1 million newtons. More powerful engines still will need to be developed before inter-planetary travel becomes a possibility.

**Fig. 12.12** *Eldo-Europa launch vehicle with* Blue Streak *as its first stage being launched at Woomera. It is developing about 2000 MW. The take-off thrust is over $10^6$ N.*

## 12.6 Friction

The force that opposes the motion of one surface over another when they are in contact is called the *force of friction*.

Rub your hands together. Can you feel the frictional force? Does the force increase as you push your hands harder together? Can you feel the heat energy produced?

Tie a piece of string to a block of wood and pull the block across a table (Fig. 12.13 — the spring balance is not essential). Does piling books on top of the wood make any difference to the force needed to pull the wood along? Is the force needed to start the wood moving greater than the force needed to keep it moving once it has been set in motion?

**Fig. 12.13**    *Measuring    frictional    force*

*QUESTION 8:* (a) Does the friction depend on the force pressing on the surface? (b) Is the force of starting friction greater or less than the force of sliding friction (the force needed to keep the wood moving)? (c) Do the forces (sliding and starting friction) change if the block is turned on its side?

The frictional force between an object and the surface over which it is sliding depends on the weight of the object and the nature of the surfaces. It does not depend on the area of contact.

## 12.7    Reducing friction

There are two main ways of reducing friction:
1. By lubrication. If you don't oil the moving parts of your bicycle it will not run so well, because of increased friction between these parts. If a motor car, for example, is run without oil it will seize up. That is, the moving metal parts get so hot because of friction that they begin to melt.

120

2. By using rollers or ball bearings. Figure 12.14 illustrates the use of these. A book on rollers slides along the table much more easily because the friction between the book and the table has been reduced.

## 12.8    The usefulness of friction

In machinery friction is usually a nuisance and results in energy being lost as heat; but in the absence of friction, starting and stopping would present considerable difficulties. Without friction it would be impossible to hold things with your fingers — we all know how difficult it is to pick up a slippery or greasy object. And how difficult it is to walk on ice! Where there is very little friction, as on ice, our foot slips backwards as we try to move forwards. Normal walking is only possible because of friction between the foot and the ground. If there were no friction, the world would be a very different kind of place: you could push your friend with your finger and he would go sliding right across the room, nails would not stay in place and car and bicycle brakes would not work!

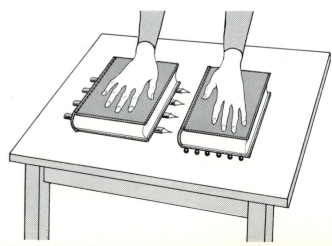

**Fig. 12.14**    *Reducing friction by means of rollers and ball bearings*

A. **How powerful are you?**

Find a flight of steps or stairs, the higher the better. Borrow a watch with a second-hand and get a friend to see how long it takes you to run up to the top of the steps. Time your friend too. Weigh yourselves. Calculate the total height of the stairs by measuring the height of each stair. Calculate the work done in lifting your weight up to the top of the stairs (your weight × the height of the steps). Use equation 3 on page 118 to calculate your power. Show your calculation to your teacher.

B. Measure the power of a toy electric motor by seeing how long it takes to lift up a weight.

C. Get two identical toy cars and a sloping piece of wood for them to run down. Stuff paper into the front wheels of one and the back wheels of the other, to lock them. Before you release the cars try and decide which one will turn round as it runs down the incline. Think carefully, most people get it wrong!

D. Make a miniature hovercraft, like the one shown Fig. 12.15. A cork is stuck in the middle of the non-shiny side of a piece of hardboard, and a hole is drilled through the cork and the hardboard. An inflated balloon is fixed to the cork. The air issues from the bottom of the hardboard. The hovercraft works best on a shiny surface.

E. Collect a variety of roller and ball bearings. Make a classroom display of them.

F. Find out all you can about James Watt and his engines.

G. Find out all you can about the reduction of friction in machines. Visit a garage and find out how a ball race reduces friction. Reduce the friction on a stiff drawer by rubbing the sliding surfaces with pencil lead.

H. One disadvantage of friction is that it makes the soles of shoes wear out. On the other hand, without friction cars would skid on roads. Make a list of the advantages and disadvantages of friction in everyday life.

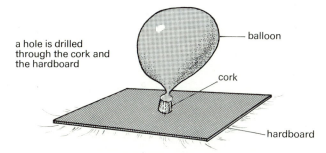

a hole is drilled through the cork and the hardboard

balloon

cork

hardboard

**Fig. 12.15** *A toy hovercraft*

## THINGS TO WORK OUT

1. What does a scientist mean by work? State in which of the following cases work is being done: (a) a girl holding on to a tent pole to prevent the tent being blown away by a strong wind, (b) the piston of a bicycle pump being pushed in, (c) a car moving along a road, (d) a boy pushing a car which has its brakes on, (e) a hall porter standing with the guest's case in his hand, (f) the same porter carrying the case upstairs, (g) a cricketer throwing a ball, (h) the ball being caught by the wicket keeper.

2. What is the difference between work and energy? Give examples.

3. Describe the energy changes that take place when (a) an arrow is fired horizontally, (b) an arrow is fired vertically upwards, (c) water flowing down a mountainside is used to drive a dynamo, (d) an electric torch is switched on, (e) a bullet is fired from a rifle, (f) electrons hit the screen of your television set, (g) a pendulum is swinging to and fro, (h) a toboggan runs down a slope and hits a tree at the bottom.

4. You have been asked to set up some demonstrations to illustrate energy changes. How would you demonstrate the following: (a) kinetic energy to potential energy, (b) potential energy to kinetic energy to heat energy, (c) electrical energy to heat energy, (d) chemical energy to light and sound energy, (f) chemical energy to kinetic energy and heat energy, (g) light energy to electrical energy (hint: photographers do this)?

5. John and Jean, on their way back from the fair, were discussing the big dipper (switchback). John said that each summit must be lower than the previous one. Jean said he was talking nonsense and that as long as the first one was the highest it didn't matter what height the others were. Discuss their statements.

6. An acrobat is standing on one end of a see-saw. Another acrobat jumps from a height onto the other end. The first acrobat flies up into the air and lands on the shoulders of a third acrobat. Discuss the energy changes that have taken place.

7. Comment on the following statements: (a) all other forms of energy can be completely converted into heat energy, but heat energy cannot be completely converted into other forms of energy; (b) all other forms of energy tend ultimately to be converted into heat energy.

8. Give some examples of machines that use friction to convert unwanted energy to heat energy.

9. Why is rubber a good substance to use on bicycle brakes but not on car brakes?

10. How much work is done when (a) 10N is moved through 5m, (b) 100N is moved through 20m, (c) 20N is moved through 40m, (d) 10N is moved through 60m? (e) a dog is taken for a walk; the tension in the lead is 15N and the distance walked is 1km?

11. How much work is done per second when (a) 10N is moved through 75m in 2s, (b) 150N is moved through 200m in 20s, (c) 300N is moved through 10m in 6s, (d) 100N is moved through 750m in 5s? In each case calculate the power.

12. A cable car is pulled up a mountain in 6 minutes, and during the journey it travels at a constant speed of 4m/s. If the tension in the cable is $5 \times 10^3$N, calculate (a) the work done in getting the car up the mountain, (b) the work done in getting the car to the top if it travels at 2m/s, (c) the power developed by the engines in each case.

13. Calculate the power developed by an athlete weighing 650N when he runs up (a) a staircase 8m high in 6s, (b) a hill to a vertical height of 40m above his starting point in 36s.

14. A woman lifts a 4kg bag of sugar (remember that the force of the earth on a mass of 1kg is 9.8N) from a table 1m high onto a shelf 2m high. How much work does she do? What sort of energy has the bag of sugar got when it is on the shelf? How much energy was given to the bag of sugar by the woman? The sugar falls off the shelf onto the table. What sort of energy has it got just before it hits the table? What happens to this energy when the sugar bag hits the table?

15. The Niagara falls are about 50m high. It is estimated that 100 million kilogrammes of water pour over the falls every second. If 50 per cent of this energy could be harnessed, how much power would be available?

16. A boy using his arms can produce a power output of 20 watts. How much work can he do in 1 hour? If his work consists of lifting boxes from the floor onto a conveyor belt 1m from the ground, what is the total mass he can lift in 1 hour?

17. "All machines would work better if there were no such thing as friction." Is this always true? Discuss.

18. On wide staircases where people walk up and down always keeping to the right, one side is usually worn much more than the other. Which side gets most wear and why?

19. "Friction occurs between two metal surfaces because clusters of molecules near the surface of the metal project above the surface, and this means that a surface is not perfectly smooth." Use this explanation to work out (a) why "welding" occurs at points when metal surfaces rub together, and (b) why frictional forces are reduced by lubricating the surfaces.

20. Explain (a) why car tyres have treads but aeroplane tyres do not, (b) why the melting of the ice below ice-skates makes skating easy.

21. Does friction make the experiment illustrated in Fig. 7.4 easier or more difficult?

22. Why can a hovercraft travel faster than a boat?

23. Discuss various possible methods of getting about easily on ice and snow.

24. Two car manufacturers claim that their car has the best brakes. How would you test their claims?

25. Rubber can be stretched or compressed. (a) Does it store energy when it is stretched? If so, what sort of energy? (b) Explain in terms of energy why a rubber ball dropped from a certain height never rebounds to the height from which it was released. What happens if it is thrown down? Why?

26. A swimmer weighing 700N dives from a diving board which is 3m high. Describe the energy changes which take place from the time when he is standing on the board ready to dive until the time he returns there ready to dive a second time. Is energy conserved? Explain your answer. What is the diver's kinetic energy the instant before he enters the water?

27. The earth is in constant motion. Would you consider this to be a good example of perpetual motion? Explain your answer.

28. A man pushes a car at constant velocity for 200 metres along a level road. Describe the energy changes that are taking place. What else would you need to know in order to calculate how much work he did?

**CHAPTER 13**

## 13.1 Levers in everyday life

When the lid of a paint pot is very difficult to get off what method can be used to release it? I expect you have all used a screwdriver or spoon handle to do this. Would you need a screwdriver to open a coffee tin if the lid were tight, or would a coin be just as good? Have you ever seen poor old Dad struggling to loosen the wheel nut on his car when the garage mechanic has tightened it using a wheel brace with a very long arm (Fig. 13.1)? What do you think would happen if someone tightened the nut using the spanner shown in Fig. 13.2? Have you thought why door handles are placed on the opposite edge from the hinge?

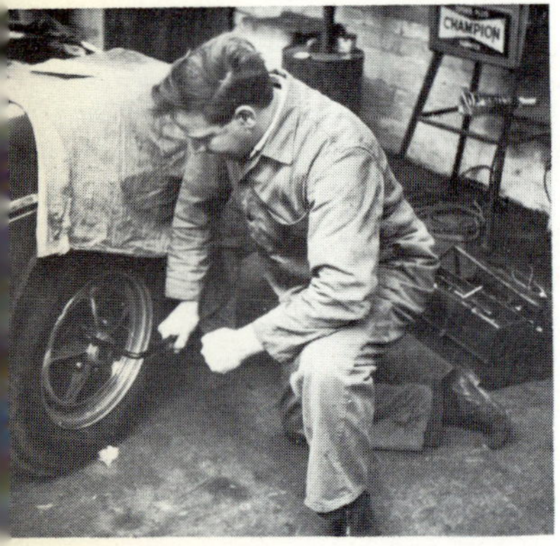

*QUESTION 1:* Which spanner (Fig. 13.3) would you use to get the nut as tight as possible?

**Fig. 13.1**   *A wheel brace being used to tighten a nut*

**Fig. 13.2**   *What is the danger in using too long a spanner?*

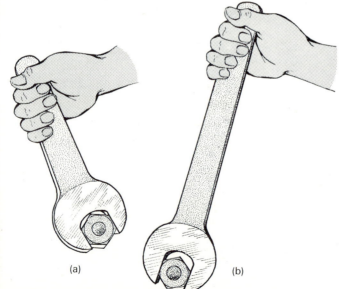

(a)

(b)

**Fig. 13.3**   *Which spanner will get the nut tighter?*

# LEVERS AND CENTRES OF GRAVITY

## 13.2 A simple experiment

Balance a half-metre rule (or any other beam) as shown in Fig. 13.4. Do this first without any masses on the ruler. When the beam is balanced it is said to be in *equilibrium*. Now put a 100g mass at a fixed distance (say 20cm) from the *fulcrum* (point of pivot). Move another known mass *m* along the other side of the ruler until the ruler is as nearly balanced as you can get it. Note the distance *y* between *m* and the fulcrum. Keeping the 100g mass fixed obtain a number of values for the distance marked *y*, using different known masses *m*.

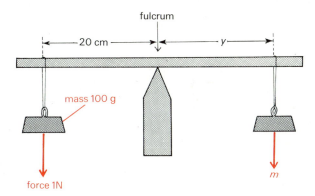

fulcrum

—20 cm—

—y—

mass 100 g

force 1N

*m*

**Fig. 13.4**

You can do this experiment at home with a piece of wood and some 10p pieces for masses; take the weight of one coin as one unit of weight.

When you have obtained your results look at the following table. The figures in the second column are obtained by multiplying the mass by 0.01. This is because the pull of the earth on 1000g is approximately 10N (see page 77). Therefore the weight of a mass of 100g is 1N, and so on.

| Mass (m) g | Force (F) N | Distance (y) cm |
|---|---|---|
| 100 | 1 | 20 |
| 200 | 2 | 10 |
| 80 | 0.8 | 25 |
| 400 | 4 | 5 |

*QUESTION 2:* Can you see any connection between the force and the distance in the above table?

The placing of the 100g mass 20cm from the fulcrum tends to turn the beam anticlockwise.

The product of force × perpendicular distance which measures the turning effect is called the *moment of the force*. Notice that the moment of the force turning the ruler clockwise ($F \times y$) is equal to the moment of the force turning the ruler anticlockwise ($1 \times 20$).

*QUESTION 3:* Fig. 13.5 shows a beam in equilibrium. Write down (a) the clockwise moment of the 1N force about the fulcrum, (b) the anticlockwise moment of the 0.2N force, (c) the anticlockwise moment of the 0.5N force. Add up the answers to (b) and (c). What is the connection between this sum and the answer to (a)?

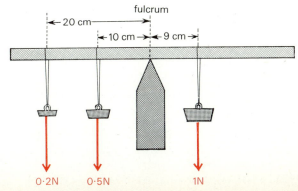

fulcrum

—20 cm—

—10 cm—

—9 cm—

0·2N

0·5N

1N

**Fig. 13.5**

125

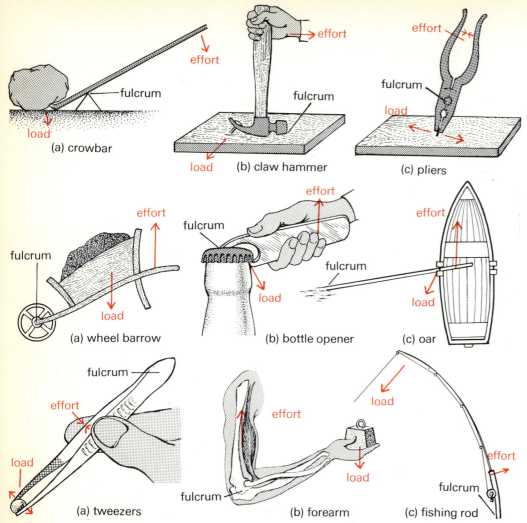

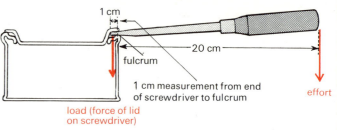

## 13.3 Law of moments

The result of the above experiment is quite general and may be summarized as follows:

For a body in equilibrium, the sum of the clockwise moments about any axis equals the sum of the anticlockwise moments about the same axis.

1 cm
20 cm
fulcrum
1 cm measurement from end of screwdriver to fulcrum
load (force of lid on screwdriver)
effort

**Fig. 13.6** *A screwdriver being used to open a tin lid*

## 13.4 Useful applications of levers

When levers are used a force called the *effort* overcomes a force called the *load* acting at some other point on the lever. If a screwdriver is used to prise open a lid (Fig. 13.6) the effort is made by the hand and the load is the force of the lid on the tip of the screwdriver. Suppose a force of 1000N is needed to prise off the lid, then the load is 1000N. Taking moments about the fulcrum:

clockwise moments = anticlockwise moments
effort $\times$ 20 = 1000 $\times$ 1
effort = 50N

In other words an effort of 50N exerted on the handle of the screwdriver will result in a force of 1000N being exerted on the lid.

Others examples of levers are illustrated in Fig. 13.7. Notice that the fulcrum is not always between the load and the effort. Look carefully at the illustration of a rowing boat. Note the position of the fulcrum. Ask your

effort
effort
effort
fulcrum
load
(a) crowbar
fulcrum
load
(b) claw hammer
fulcrum
load
(c) pliers

effort
fulcrum
load
(a) wheel barrow
effort
fulcrum
load
(b) bottle opener
effort
load
(c) oar

fulcrum
effort
load
(a) tweezers
effort
fulcrum
load
(b) forearm
load
effort
fulcrum
(c) fishing rod

**Fig. 13.7** *Various types of lever. Those in the first row are known as* first order *levers; those in the second row as* second order *levers; and those in the third row as* third order *levers. Notice the position of the fulcrum in each case.*

teacher to explain if you cannot understand why the fulcrum is in the water.

*QUESTION 4:* Suppose that in the wheelbarrow (Fig. 13.7) the distance from the fulcrum to the effort is 1m and the distance from the fulcrum to the load is 0.5m. If the load is 400N what is the effort needed?

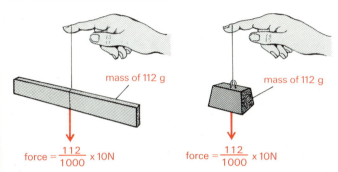

**Fig. 13.8** *The ruler behaves as if its weight were concentrated at its centre of gravity*

## 13.5   Centre of gravity

Get a friend to shut his eyes, then tell him that you are going to hang on his finger either a metre ruler or a mass having the same weight as the ruler (Fig. 13.8). He will not be able to tell without looking which one you have hung on his finger. The ruler behaves as if all its weight were acting at its mid-point. What is true for a ruler is true for all bodies, namely that there is a point in the body (not usually at the centre) through which its whole weight appears to act. The point is called the *centre of gravity of the body.*

The centre of gravity (c.g.) of a body is the point through which its whole weight may be considered to act.

The centre of gravity of a piece of cardboard may be determined by the method illustrated in Fig. 13.9. The weight or bob of the plumb-line will always come to rest

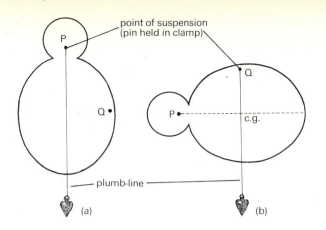

**Fig. 13.9** *Determination of the centre of gravity*

immediately below the point of suspension. Since the body behaves as if all its weight acted at its centre of gravity, the centre of gravity of the body will also be vertically below the point of suspension. The cardboard is suspended from P and a line drawn showing the position of the plumb-line. It is then suspended from Q and the new position of the plumb-line marked.

Where the lines cross is the position of the centre of gravity. The position may be checked by using a third point of suspension. When you have found the position see if you can balance the cardboard on a sharp point.

## 13.6   Stability

A rowing boat will upset much more easily if the occupants are standing up (Fig. 13.10) than if they are sitting down. Have you noticed how much more *unstable* a rowing boat becomes if two occupants stand up and change places? If you try it, make sure you are in shallow water and have not got your best clothes on! While you are sitting down the boat is fairly stable; that is, it is not likely to upset.

**Fig. 13.10** *Why do parents tell children not to stand up in a boat?*

When a body is in stable equilibrium it returns to its original position if it is displaced slightly and then released.

A simple experiment on stability may be done with a matchbox and a piece of wood (Fig. 13.11). The wood must not be too smooth; a covering of emery paper makes a good surface. Stand the matchbox upright, as shown, and gradually tilt the wood. Observe the angle of tilt when the matchbox topples over. Now open the matchbox, stand it on the wood, and again tilt the surface. Is the angle of tilt about the same when the matchbox falls over? Put the matchbox on its side and try again.

**Fig. 13.11** *Which matchbox is the more likely to tip over?*

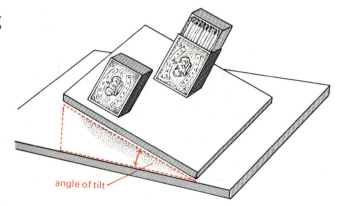

QUESTION 5: For maximum stability does the centre of gravity have to be as high as possible or as low as possible? What can you say about the area of the base in relation to maximum stability?

In the experiment illustrated in Fig. 13.12 a similar board is used for tilting a wooden block. A plumb-line is fixed at the point where the diagonals of the block cross. The angle of tilt of the board is increased until the block topples over. At the instant when the block topples the plumb-line hangs outside the base.

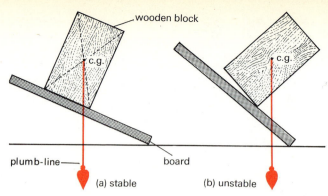

**Fig. 13.12** *The block will not tip over whilst the plumb-line falls within the base*

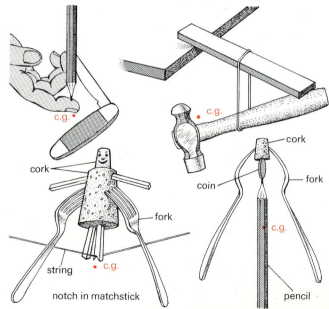

**Fig. 13.13** *In each case the centre of gravity falls directly below the point of support*

Look again at Fig. 13.9(b). If the suspended card is displaced slightly to one side and then released, it returns to its original position. A suspended body is always in stable equilibrium if the centre of gravity is directly below the point of suspension. Figure 13.13 shows a number of balancing tricks for you to try. In each case the centre of gravity is vertically below the point of support.

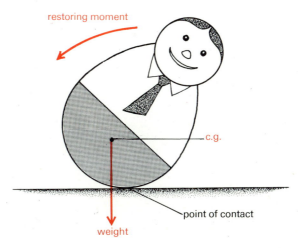

**Fig. 13.14** *A self-righting toy*

You can make a self-righting toy (Fig. 13.14) by cutting a rubber ball in half, or sawing a wooden sphere in half, and gluing a paper figure on the top. When the toy is displaced, the weight acting through the centre of gravity results in a restoring moment about the point of contact. The toy is therefore in stable equilibrium.

The stability of a body may be increased (a) by increasing the base area, (b) by lowering the centre of gravity.

**THINGS TO DO**

A.   Stand with your back to a wall, and try to touch your toes while your heels and the backs of your legs remain close to the wall. Why is this difficult?

B.   Design balancing tricks of your own similar to those illustrated in Fig. 13.13.

C.   Cut the shape of a parrot out of cardboard (Fig. 13.15). Be sure to give it a long tail. Attach a coin or a small weight to the bottom of its tail. If you design your parrot carefully it will stand upright on a pencil or stick held horizontally.

D.   Make a self-righting toy like the one shown in Fig. 13.14 (the instructions are in the text on the same page).

E.   Cut out two paper cones and stick them together as illustrated in Fig. 13.16, or use two large filter funnels fixed together with sticky tape. Set two rods as shown. The cones appear to defy the law of gravity and move up the slope. Can you explain why?

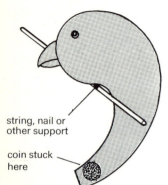

string, nail or other support

coin stuck here

**Fig. 13.15**   *Why does the parrot balance?*

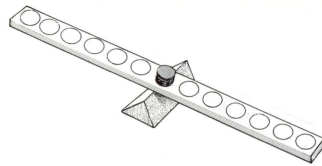

**Fig. 13.17**   *The game of sym*

F.   The game of sym (for two or more players).

Get a beam and mark a series of equally spaced circles (each the diameter of a 10p piece) as shown in Fig. 13.17. Balance the beam at its centre. Put four coins in the circles on the beam so that the beam still balances, but make the arrangement as unsymmetrical as possible (you can put more than one coin in a circle). Draw a sketch of the arrangement and remove the coins and pile them on the centre over the fulcrum. Each of the players in this game then moves two coins at a time, placing them in circles so that the beam balances after each move, until he arrives at the original pattern. The winner is the one who gets there in the smallest number of moves. As you become more skilful, increase the number of coins you use.

G.   Construct a balance like the one illustrated in Fig. 13.18 and use it to test the strength of various fibres.

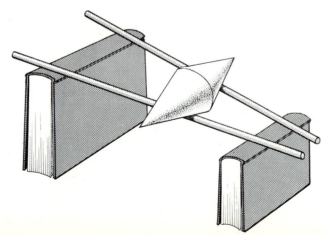

**Fig. 13.16**   *The cone appears to run up the hill. Can you explain why?*

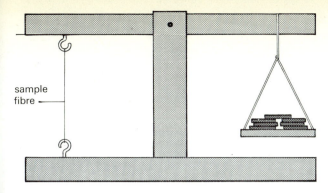

sample fibre

**Fig. 13.18**

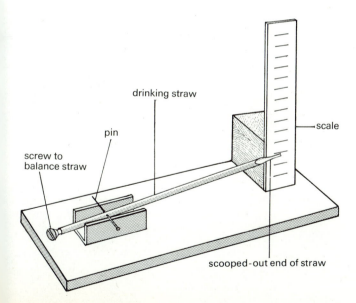

drinking straw

pin

scale

screw to balance straw

scooped-out end of straw

**Fig. 13.19**  *A microbalance*

H.   Construct a very sensitive balance using a drinking straw and a pin (Fig. 13.19). A very light piece of paper placed in the scoop will cause a deflection.

I.   How could you calibrate the scale?
     Stick a fairly large piece of plasticine on the inside of a cylindrical tin. Place the tin on an incline. If placed so that the plasticine is in a certain position, the tin will run up the incline. Why? You can baffle your friends if you have a lid on the tin.

J.   Find out (a) how double-decker buses are tested for stability, (b) how racing cars are designed to give them a low centre of gravity.

K.   Find out all you can about how bones are made to act as levers in your body by the pulling action of muscles. Draw diagrams to illustrate this action, and in particular to show how the biceps and triceps work in the movement of the arm.

1. In what ways can manufacturers of glass tumblers make their products more stable?

2. A film stuntsman has to jump from a moving vehicle and make a getaway without falling over. Discuss in terms of the movement of his centre of gravity how he can do this.

3. Explain (a) why most chairs have legs that slope outwards, (b) why standing with legs apart makes it easier to keep one's balance in a moving vehicle, (c) why a wardrobe is more stable if the bottom drawer is full.

4. How can levers be used to increase movement? A rod expands by 0.5mm when heated. Design a system of levers sensitive enough to show this expansion. The pointer of the lever system should move at least 5cm when the bar expands by 0.5mm.

5. At one of the sideshows at a fair John and Jean are throwing balls at coconuts. John says that you will be more likely to knock the coconuts over if you hit them near the top, but Jean says that it is better to hit them near the middle. Who is right and why?

6. What is meant by stable equilibrium? If a body when displaced slightly and then released remains in its new position it is said to be in neutral equilibrium. One example of this is a ball resting in a new position on a horizontal surface. Give two other examples of neutral equilibrium. What do you think is meant by a body being in unstable equilibrium?

7. The model ballerina in Fig. 13.20 may be pushed in any direction but will not fall over. Explain.

8. (a) Why are the men in Fig. 13.21 leaning over the side of the boat? What determines how far they lean over? What force provides the anticlockwise moment and what force provides the clockwise moment? (b) Which glass (Fig. 13.22) do you think is most stable? Which is least stable? Give reasons for your choice.

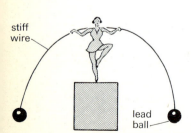

stiff wire

lead ball

**Fig. 13.20**

**Fig. 13.21**  *Why are the men leaning out of the boat?*

**Fig. 13.22** *Which glass is the most stable?*

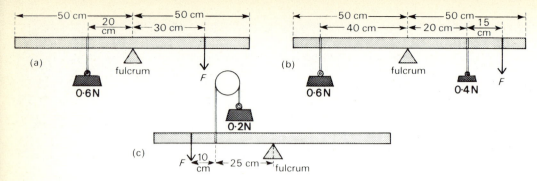

Fig. 13.23

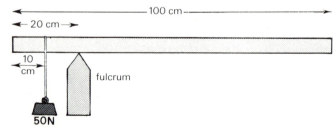

Fig. 13.24

9. In Fig. 13.23 the beams are 1m long and balanced at their midpoint. Calculate the magnitude of the force $F$ needed in (a), (b) and (c) for the beams to be in equilibrium.

10. A metre rule is balanced on a fulcrum at its centre point. (a) A weight of 1N is hung from a point 20cm from the fulcrum. Where must a 5N weight be hung if the ruler is to remain horizontal? (b) Another 1N weight is now hung 30cm from the fulcrum on the same side as the other 1N weight (which is left in position), where must the 5N weight now be hung in order for the ruler to remain balanced horizontally? (c) The 1N hung 20cm from the fulcrum is now changed to a weight of 2N and the weight 30cm from the fulcrum left in position. Where must the 5N weight be hung to keep the ruler balanced horizontally?

11. A see-saw is balanced when no one is sitting on it. John who weighs 480N sits 3m from the point of balance. Where must Jean who weighs 360N sit so that the see-saw will balance? If Jean moves 2m nearer to the centre, where must Jill, who weighs 240N, sit in order to balance the see-saw?

12. A plank is balanced with its fulcrum at the centre. A man weighing 800N sits 0.5m from the centre. Where must a 200N weight be placed to balance the plank horizontally? If this weight is left in position but the man moves to 1m from the centre, where must a 100N weight be placed to keep the plank horizontal?

13. A uniform beam 1m long is balanced when the fulcrum is 0.2m from one end and a weight of 50N is hung 0.1m from the same end (Fig. 13.24). What is the weight of the beam? (Hint: remember the beam may be treated as though all its weight acts at its centre of gravity.)

14. A uniform plank 5m long and weighing 5N rests on a fulcrum which is 2m from one end. A cat weighing 20N walks onto the end of the plank that is resting on the ground and walks up the plank. How far can the cat walk before the plank starts to tip?

15. A decorator painting a room sets up a plank weighing 300N on two trestles. The trestles are placed 1m from each end, and the total length of the plank is 5m. Can a man weighing 800N stand on the end of the plank without tipping it?

16. A plank is balanced horizontally with its centre resting on a stone. A very heavy man sitting near the centre on one side is just balanced by a small boy sitting at the end on the other side. What adjustment must be made to keep the plank balanced horizontally when the man moves away from the fulcrum?

134

# MACHINES MAKE WORK EASIER

## 14.1 Fundamental definitions

We live in an age of machines. This book was produced by the use of machines and many of the things you see around you at this moment were made by machines.

Complicated machinery is made up of many simpler units. The lever, pulleys and gears are examples of these simpler units. But before finding out more about machines we must first learn the meaning of one or two frequently used terms.

A machine is basically a device by means of which a force at one point may be overcome by applying a force at another point. The lever illustrated in Fig. 14.1 is a simple example of a machine. Study the diagram carefully. How far does the effort move when the load is raised 1cm?

We call the ratio of these two distances the velocity ratio.

$$\text{Velocity ratio } (VR) = \frac{\text{distance moved by effort}}{\text{corresponding distance moved by load}}$$

(equation 1)

*QUESTION 1:* What is the velocity ratio of the lever illustrated in Fig. 14.1?

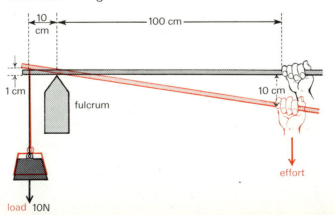

**Fig. 14.1**

load 10N

Calculate the magnitude of the effort needed to raise the load by taking moments about the fulcrum. You will find that an effort of 1N will raise the load of 10N. The ratio $\dfrac{\text{load}}{\text{effort}}$ is called the *mechanical advantage*.

$$\text{Mechanical advantage } (MA) = \frac{\text{load}}{\text{effort}} \qquad \text{(equation 2)}$$

*QUESTION 2:* (a) What is the *MA* of the lever illustrated in Fig. 14.1? (b) How much work is done on the load when it is raised through 1cm? (c) How much work is done by the effort in raising the load? (Hint: remember *work is force × distance moved.*)

By using chemical energy (from our food) we can raise a load. What sort of energy has the load now got (see pages 114–16)? We have changed some chemical energy into potential energy. But has all the work done by the effort been transferred to energy stored by the load? From the answers to the questions above it would seem so. The work done by the effort was 0.1J and the work done on the load was 0.1J. Can you think of anything we have forgotten?

One thing we have forgotten is the friction at the fulcrum: we can never eliminate friction in machines. This means that the effort needed in the above example will be slightly greater than 1N, and hence the work done by the effort is more than the work done on the load. This is true for all machines. We can never get more energy out of a machine than we put into it (if you invent such a machine you will make your fortune).

$$\text{Efficiency of a machine} = \frac{\text{work got out}}{\text{work put in}} = \frac{\text{work done on load}}{\text{work done by effort}}$$
$$\text{(equation 3)}$$

*QUESTION 3:* Suppose that in the above example the effort needed, allowing for friction, was 2N. What is the efficiency of the lever? Can the efficiency of a machine ever be greater than 1?

effort slightly greater than W

W

**Fig. 14.2**  *Why is the effort greater than W?*

## 14.2  Simple pulleys

Pulleys are used:

1. *To change the direction of a force* (Fig. 14.2). It is very much easier to pull downwards than to lift the bucket up without using the pulley. The force needed to raise the bucket using the pulley is slightly greater than the weight of the bucket. This is a result of the friction on the pulley bearings and also because a force greater than the weight of the bucket is needed to set the bucket in motion owing to its inertia (see page 74).

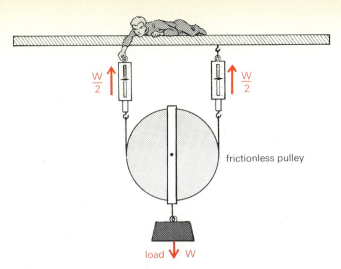

**Fig. 14.3**  *A pulley used in this way reduces the effort needed to raise the load*

2. *To reduce the force needed to raise a given load* (Fig. 14.3). Neglecting friction, the force needed to raise a load of W is W/2. The reason for this is that each of the strings supports half the load.

### 14.3  Pulley systems

A pulley system consisting of two pulleys is illustrated in Fig. 14.4. Which of the two pulleys simply changes the direction of the force?

*QUESTION 4:* If the pulleys (Fig. 14.4) were frictionless and weightless what effort would be needed to raise a load of 1N?

The effort needed to raise the load may be determined by hanging weights on the end of the string. Using different loads, determine in each case the effort needed just to raise the load. The following table shows a series of readings obtained in this way.

**READINGS FOR PULLEY SYSTEM SHOWN IN FIG. 14.4**

| Load in newtons | Effort in newtons | $MA\left(\dfrac{\text{load}}{\text{effort}}\right)$ |
|:---:|:---:|:---:|
| 0.5 | 0.40 | 1.25 |
| 1.0 | 0.77 | 1.30 |
| 1.5 | 0.97 | 1.55 |
| 2.0 | 1.18 | 1.70 |
| 2.5 | 1.43 | 1.75 |
| 3.0 | 1.66 | 1.80 |

Discuss, with your friends and your teacher why the *MA* increases as the load increases.

*QUESTION 5:* Can you suggest reasons why the *MA* never reaches the value of 2 obtained by calculation in question 4?

The velocity ratio may be determined by using a ruler to measure how far the effort moves when the load rises 1m.

*QUESTION 6:* Calculate the *VR* by reference to Fig. 14.5.

We can work out the efficiency as follows:
If the load moves up 1m then the effort moves 2m. For the first reading in the table:

Work done on load $= 0.5 \times 1 = 0.5$J
Work done by effort $= 0.4 \times 2 = 0.8$J

$$\text{Efficiency} = \frac{0.5}{0.8} = 0.625 = 62.5 \text{ per cent}$$

A similar calculation for the other readings will show that the efficiency is always half the *MA* for this system of

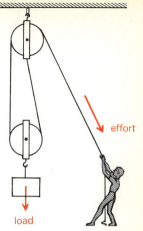

**Fig. 14.4**  *A simple block and tackle*

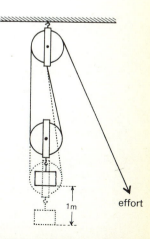

**Fig. 14.5**  *What is the velocity ratio?*

**Fig. 14.6** *The block and tackle may be used to raise large loads*

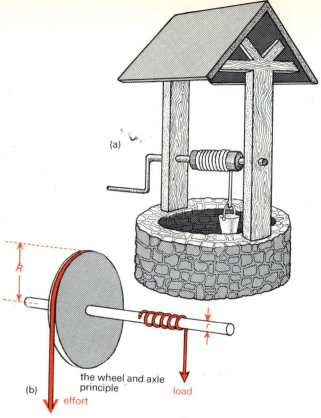

**Fig. 14.7** *A windlass*

pulleys. It may be shown for all pulley systems (see question 17, page 144) that

$$\text{Efficiency} = \frac{MA}{VR}$$

The pulley system (known as a "block and tackle") shown in Fig. 14.6 is capable of lifting loads of up to 2 million newtons.

## 14.4 The wheel and axle

The windlass used to raise a bucket from a well (Fig. 14.7) or to draw stage curtains (Fig. 14.8) is an example of a wheel and axle. A simple wheel and axle may be made at home using a cotton reel and a piece of dowelling. You can obtain a series of readings similar to those in the table on page 137.

138

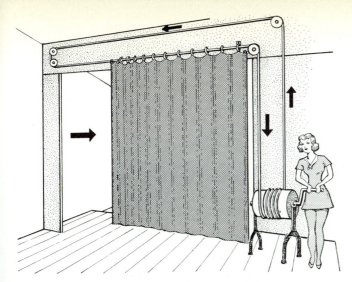

**Fig. 14.8**  *Wheel and axle system*

**Fig. 14.9**  *The inclined plane makes the work easier*

If the wheel makes one complete revolution the distance travelled by the effort is $2\pi R$ (Fig. 14.7). The distance gone by the load is $2\pi r$. Hence

$$VR = \frac{2\pi R}{2\pi r} = \frac{R}{r}$$

QUESTION 7: (a) If $R = 10$cm and $r = 2$cm, what is the VR? (b) What sort of value would you expect for the MA?

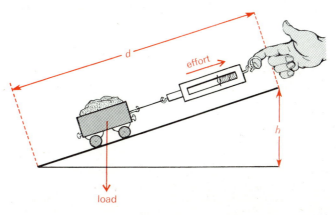

**Fig. 14.10**  *Determining the MA of an inclined plane*

## 14.5   The inclined plane

You have probably all seen beer barrels being pushed up an incline, and so loaded onto a lorry (Fig. 14.9). The mechanical advantage of an inclined plane may easily be measured by the experiment illustrated in Fig. 14.10. A toy truck loaded with sand is being pulled up a slope using a spring balance. The load is first determined by weighing the truck. When the effort moves the truck a distance $d$ along the plane, the weight of the load, *acting vertically downwards*, moves a distance $h$ upwards.

QUESTION 8: What is the velocity ratio of the inclined plane?

**Fig. 14.11** *Mountain pass, Trollstegveien, Norway*

A mountain pass (Fig. 14.11) is an example of an inclined plane. If the road did not zig-zag and slope up gradually, cars would not be able to climb up the

mountain. Other examples of the use of the inclined plane are to be found in a chisel, an axe, a penknife and a screw.

Can you see why?

If one of your friends told you he could lift up ten men with his little finger you would probably not believe him. But the boy in Fig. 14.12 is lifting up a car. He can raise it using only his little finger.

**Fig. 14.12** *A small effort can raise a large load*

Study Fig. 14.13 and see how many different kinds of machine are being used.

## 14.6 Gears

A common way of altering the velocity ratio (and hence the mechanical advantage) of a machine is to use gears. In Fig. 14.14 one gear has 10 teeth and the other 20 teeth. For each complete revolution of the big wheel the small wheel makes two complete revolutions.

*QUESTION 9:* What is the *VR* of the gear system in Fig. 14.14?

*QUESTION 10:* A spanner tightening a nut (Fig. 13.3) is a combination of what two machines?

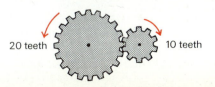

20 teeth    10 teeth

**Fig. 14.14** *What is the velocity ratio?*

141

A.   Study the gear system on Mum's rotary egg whisk. Find out what its velocity ratio is.

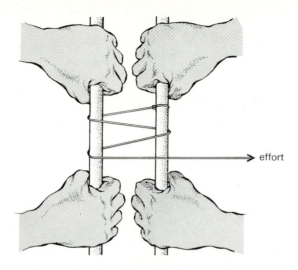

effort

**Fig. 14.15**

B.   Get two friends to hold on to two sticks as shown in Fig. 14.15. Tell them however hard they pull you will be able to pull the two sticks together. Can you see why this is similar to a block and tackle? Is it easier for you to do it if you wind the cord round a few extra turns?

C.   Make a collection of photographs showing levers, wheel and axle, inclined plane, pulleys and gears in use. Draw diagrams showing how they work.

D.   Devise an experiment to measure the *MA* and *VR* of your bicycle.

E.   Find out how many examples of the wheel and axle there are in your home. Make a list of them, stating carefully which part is the wheel and which part is the axle.

F.   Find out how the pyramids of Egypt were constructed. What sort of machines were used?

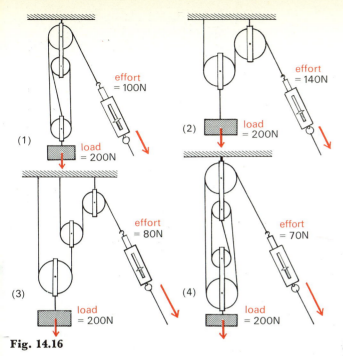

(1) effort = 100N    load = 200N

(2) effort = 140N    load = 200N

(3) effort = 80N    load = 200N

(4) effort = 70N    load = 200N

**Fig. 14.16**

5. The driver gear has 20 teeth and the follower gear has 50 teeth. If the driver gear rotates twice every second, how many times does the follower gear rotate?

6. Illustrate with diagrams any examples of (a) the wheel and axle, (b) the lever, that can be found on your bicycle.

7. A man uses a machine of velocity ratio 5 to raise a load of 1000N through a distance of 5m. If the efficiency of the machine is 50 per cent what is (a) the work done on the load, (b) the work done by the man, (c) the effort he makes?

8. Draw a diagram of a wheel and axle having a velocity ratio of 4. The machine is used to raise a load of 1000N through 2m. The effort required is 300N. What is (a) the *MA*, (b) the work done in raising the load, (c) the work done by the effort, (d) the efficiency?

9. Study Fig. 14.17. Is the screw an example of a type of machine discussed in this chapter? If so, what machine?

1. Study carefully the diagrams of the pulley systems shown in Fig. 14.16. For each system what is (a) the *MA*, (b) the *VR* and (c) the efficiency?

2. Draw a diagram of pulley systems having velocity ratios 4, 5 and 6 respectively. Explain why the systems you have drawn have these velocity ratios.

3. Why do racing cyclists usually have winged nuts fitted to their wheel hubs instead of ordinary nuts?

4. Would you expect (a) the velocity ratio and (b) the mechanical advantage of a wheel and axle to vary with the load? Do they vary in the case of a block and tackle? Give reasons for your answer.

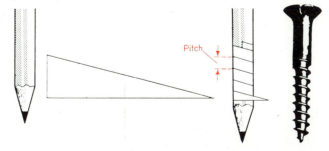

Pitch

**Fig. 14.17**

10. A screw of pitch 1cm is worked by a lever of length 2m. A load of 16 000N may be raised by a force of 20N applied to the end of the lever. What is (a) the mechanical advantage, (b) the velocity ratio, (c) the efficiency of the system?

11. A trolley which has a weight of 50N is pulled up an inclined plane with a constant velocity by a force of 20N. Use the measurements shown in Fig. 14.18 to answer the following questions:

**Fig. 14.18**

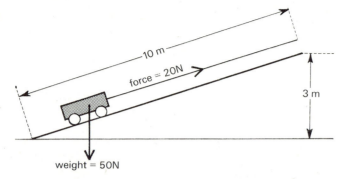

10 m

force = 20N

3 m

weight = 50N

a. What is the work done on the trolley in moving it 10m up the plane?

b. What is the work done by the person pulling it 10m up the plane?

c. What is the efficiency of the system?

d. What is the mechanical advantage and velocity ratio of the inclined plane?

12. A system of pulleys is used to raise a load of 1200N through 10m. In order to do this an effort of 300N is required. If the effort moves through 50m, (a) how much potential energy is transferred to the load, (b) how much energy is transferred by the effort in raising the load, (c) what is the efficiency of the machine?

13. A pulley block and tackle which has three pulleys on each block is used to raise a load of 100N through 1 metre. The effort required is 25N. (a) How much potential energy is transferred to the load? (b) How far does the effort move when the load is raised 1 metre? (c) How much energy is transferred by the effort in raising the load? (d) What is the efficiency of the system?

14. An electric motor draws energy from the mains at a rate of 500W. If the motor is 50 per cent efficient, how much power is available to raise a load? If the load is 50N, over what distance can the motor raise it in 2 seconds?

15. A small dynamo is driven by a load of 10N falling through a height of 2 metres. Assuming the load falls at constant velocity and takes 4 seconds to fall, what is (a) the energy transferred by the load to the dynamo in 2 seconds, (b) the power transferred by the load to the dynamo, (c) the electrical power available if the dynamo is 80 per cent efficient?

16. A block and tackle consists of a fixed block and a movable block, each with three pulleys. The weight of the lower block is 200N. A load of 1000N can be raised by an effort of 220N. What is (a) the energy transferred to the load when it is raised 1m, (b) the energy transferred by the effort in raising the load, (c) the $MA$ of the system, (d) the $VR$ of the system, (e) the efficiency of the system?

What would be the efficiency of the system if the frictional forces were negligible?

17. Write down equation 3 on page 136 and substitute "load × distance load moves" for "work done on load." Make a similar substitution for "work done by effort". Hence show that efficiency $= \dfrac{MA}{VR}$.

# COMBINING VELOCITIES AND FORCES

## 15.1 Adding velocities

If you were the pilot of an aircraft like the one shown in Fig. 15.1, would a knowledge of the aircraft's speed or a knowledge of the speed and direction of the wind be more important to you? If you only knew your speed and nothing about the direction of the wind the chances of reaching your destination would be slight. In this chapter we shall be learning about the law that enables us to deal with problems of this kind.

## CHAPTER 15

**Fig. 15.1**

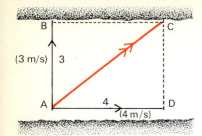

**Fig. 15.2** *Velocities must be added by the parallelogram law*

QUESTION 1: Suppose you are swimming in a stream at a speed of 3m/s on a day when the current is flowing at 4m/s. What will your speed be if you swim (a) downstream, (b) upstream, (c) across the stream?

Figure 15.2 shows us how we can solve (c) in the above question. The line AB, 3 units long, drawn across the stream, represents the distance you would travel in 1 second if there were no current. The line AD, 4 units long, represents the distance you would drift downstream in 1 second if you stopped swimming. Thus, swimming in still water you would be at B at the end of 1 second, and drifting in the stream without swimming you would be at D in 1 second. If both happen at once, i.e. if you swim across and are carried down, you will finish at C by following the path AC.

Hence to solve a problem of adding two velocities we must draw a diagram similar to the one shown in Fig. 15.2 and measure the length and direction of the line AC. If you do this you will find the answer is 5m/s at an angle of 37° with the direction of the stream. Do you think we can use the same method for adding forces?

## 15.2   Adding forces

QUESTION 2: Do forces of 3 units and 4 units always produce a combined force of 7 units?

Consider the experiment illustrated in Fig. 15.3; you can all do this together in the classroom. A ring is attached to one end of a spring S, and the other end of the spring is firmly fixed to the side of the blackboard. Two spring balances are attached to the ring, and the spring S is extended by pulling on the spring balances. When the ring is steady mark the position of its centre A on the blackboard, and draw lines to show the directions

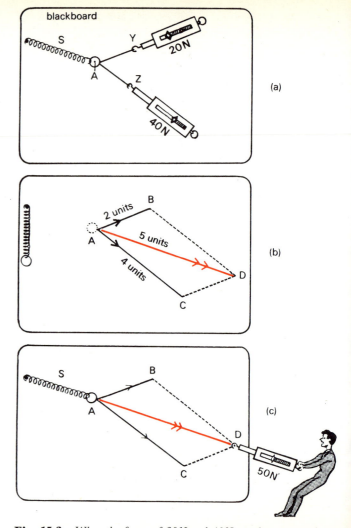

**Fig. 15.3** *When the forces of 20N and 40N are drawn to scale in the correct direction, the diagonal of the parallelogram represents the resultant force.*

146

in which the spring balances are pulling — AY and AZ in Fig. 15.3(a). Record the readings on the spring balances and then untie them. Suppose the readings were 20N and 40N. Draw a line AB 2 units long, as in Fig. 15.3(b), to represent the pull of 20N. This line must be drawn in the direction AY. Draw the line AC 4 units long in direction AZ to represent the pull of 40N. (Note: the lengths of these lines have no connection with the lengths of the strings AY and AZ.) Complete the parallelogram ABCD. The diagonal of the parallelogram is 5 units long. You may have guessed that this is the resultant of the two forces, but how can we verify this?

Figure 15.3(c) shows how we can do this. Tie one spring balance to the ring and pull until the ring is in exactly the same position as before. You will now find that the spring balance reads 50N and that the direction of pull is the direction AD.

We may summarize our conclusions as follows:

If two forces (or velocities) are represented in magnitude and direction by the adjacent sides AB and AD of a parallelogram ABCD, then the resultant force (or velocity) is represented in magnitude and direction by the diagonal AC.

Quantities that have direction as well as magnitude are called *vectors*. Quantities that have only magnitude are called *scalars*. Force and velocity are examples of vectors. Temperature and mass are examples of scalars.

### 15.3 Examples

*a. Tugs pulling ships*    Figure 15.4 shows two tugs pulling a ship. The ship moves forwards because the resultant of the two forces is in the forward direction.

**Fig. 15.4**  *The ship moves in the direction of the resultant of the two forces*

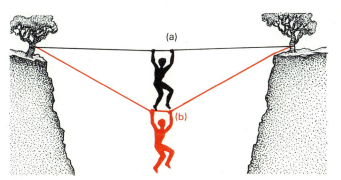

*b. Crossing a ravine*    The man in Fig. 15.5 is crossing a ravine by a rope tied to trees on either side.

*QUESTION 3:* Which rope, (a) or (b), is the more likely to break?

*c. Campers carrying a heavy water can*    The weight of the heavy water carrier is balanced by the resultant of the two forces $F_1$ and $F_2$, which is equal to $F$ (Fig. 15.6).

**Fig. 15.5**  *In which case is the rope more likely to break?*

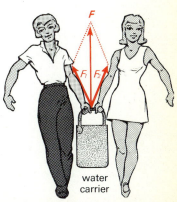

water carrier

**Fig. 15.6**  *The force* **F** *is the resultant of forces* F$_1$ *and* F$_2$

147

**THINGS TO DO**  A.  Get a roller skate and electric fan and a piece of old sheet to make a sail. Attach a sail to the roller skate and use the fan to make the roller skate move (a) in the direction of the wind, (b) at right angles to the wind, and (c) slightly into the wind. Draw a diagram showing what force acts on the roller skate in (c) to make it go forwards. The results of this experiment should help you to answer the questions in B below.

B.  Find out how the wind propels a yacht. How can a yacht sail into the wind? Would it be possible to sail into the wind without a keel? What does it mean when a yacht is said to be sailing "close hauled to the wind", and what forces are acting on it when it does?

1. Determine the resultant of the following pairs of forces: (a) 200N and 300N acting at right angles to each other, (b) 200N and 300N acting at 50° to each other, (c) 50N and 20N acting at 60° to each other.

2. If the angle between the directions of rope on either side of the man in Fig. 15.5 is 120°, what is the tension in the rope if the man weighs 700N?

3. Three dogs, Pluto, Fido and Scrap, are pulling at a bone. Pluto pulls with a force of 100N and Fido with a force of 150N. The angle between Pluto and Fido is 80°. If the bone is stationary how hard is Scrap pulling?

4. An aeroplane is flying due north at 200km/h and the wind is blowing at 80km/h from the south-east. Determine the resultant velocity of the aeroplane.

5. A train is travelling slowly over a section of straight track at 30km/h. A passenger walks up and down the corridor at a speed of 5km/h. What are his velocities relative to the track? He now turns and walks into a compartment. What is his velocity now?

6. How could you determine the resultant of three forces acting at a point?

7. Explain why the wire supporting a tennis net must be capable of standing a much greater tension than the weight of the tennis net.

8. Design an experiment you could conduct at home to demonstrate to Mum that there is less chance of her clothes-line breaking if she keeps it slack than if she keeps it taut.

150N

700N

**Fig. 15.7**

9. A water skier (Fig. 15.7) is being pulled along with a constant force of 150N. His weight is 700N. What is the resultant of these two forces and in which direction does it act? What force balances this resultant and in which direction does it act? How much work is done in pulling him along a distance of 100 metres?

10. John, Jane and Jean are in a speedboat which is towing a water skier. John says the skier is kept up because of Archimedes' principle. Jane says it is because the water is at atmospheric pressure and this pressure is keeping the skier up. Jean says there is an upward force because the skis are kept at an angle to the water. Discuss each of these answers, pointing out how much truth there is in each one.

# MAGNETISM AND ELEMENTS OF ELECTRICITY

## Part Three

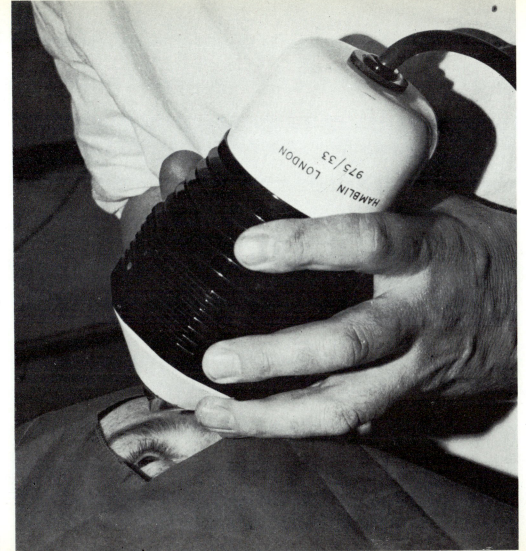

*Removing a splinter from the eye with the aid of an electromagnet*

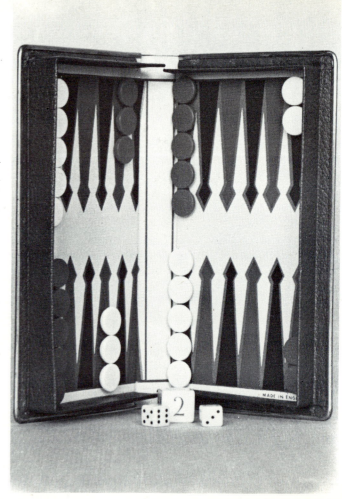

**Fig. 16.1** (a) *This magnet is used to get a tin out of boiling water*

**Fig. 16.1** (b) *A magnetic backgammon board*

# MAGNETISM

## 16.1 The useful magnet

Can you think of any situation in the home when a magnet would be useful? If you have ever upset a large box of pins, and had the tedious task of picking them up, your answer to the question will certainly be "Yes". You can pick up pins very quickly with a magnet because they cling to the regions known as *poles* at either end of it. Other uses of magnets are illustrated in Fig. 16.1.

Could a magnet be used to pick up raisins that had fallen out of the packet? Use a magnet to see what sort of materials it does attract. Substances attracted by a magnet are said to be *magnetic materials*.

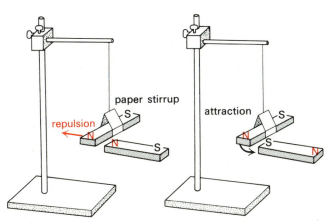

**Fig. 16.2** *Like poles repel; unlike poles attract*

## 16.2 Some simple experiments, using a magnet

a. Suspend a bar magnet so that it can rotate about a vertical axis. Leave it until it comes to rest. Rotate the magnet again and observe what happens when it is released. Repeat this procedure.

b. Get another bar magnet and bring it up towards the end of the suspended magnet as shown in Fig. 16.2. Turn the magnet round and bring up the other end. Try the experiment illustrated in Fig. 16.3.

**Fig. 16.3** *The north pole of the right-hand magnet is kept in mid-air by the repulsion of like poles. If it is depressed slightly and released it will bob up and down for some time.*

c. Bring a nail up to the suspended magnet. Get some other materials such as copper and wood, and bring them up to the suspended magnet.

The results of the above experiments enable us to state three properties of magnets:

1. A magnet that is free to rotate about a vertical axis will always come to rest pointing in the same direction. Because of this property magnets can be used as direction-finding compasses. The end that points north is the north-seeking pole (usually abbreviated to north pole).
2. Like poles repel and unlike poles attract each other.
3. A magnet attracts magnetic substances such as iron and steel, but has no effect on most other materials.

### 16.3 Making a magnet by means of magnetic induction

Hold an unmagnetized nail near some iron filings and bring up a bar magnet as shown in Fig. 16.4(a). When the magnet is near the nail, the nail attracts iron filings,

thus showing that it has become magnetized. When the bar magnet is removed most of the iron filings fall away from the nail, showing that it has lost its magnetism. The nail is said to have become a magnet by *induction*. Figure 16.4(b) shows two nails that have been magnetized by induction. The bottom ends may be seen to repel each other (i). The fact that they are both repelled by the north pole of another magnet (ii) shows that they are north poles.

Figure 16.4(c) shows another example of magnetic induction. An unmagnetized steel knitting needle becomes magnetized as it is stroked with a bar magnet. The polarity of the induced poles is shown in the diagram. You should now be able to explain why unmagnetized, iron and steel are attracted by a magnet.

*QUESTION 1:* Suppose you have a piece of metal and you want to discover whether or not it is magnetized, how would you find out?

### 16.4 Breaking a magnet

Obtain a piece of clockspring (or a thin steel rod that can easily be cut with a pair of pliers) and magnetize it by stroking it with a magnet.

*QUESTION 2:* How can you check which end of the clockspring is the north pole?

Mark the north pole. Break the clockspring in half and test each half for magnetism and polarity by bringing it up to a suspended magnet. Mark the poles. Continue breaking each piece of clockspring in half and test each piece of spring produced for magnetism and polarity. It will be found that each piece of spring is a magnet with the poles as shown in Fig. 16.5. Notice that poles always occur in pairs. It is impossible to get an isolated pole.

**Fig. 16.4**  *Magnetizing by induction and stroking*

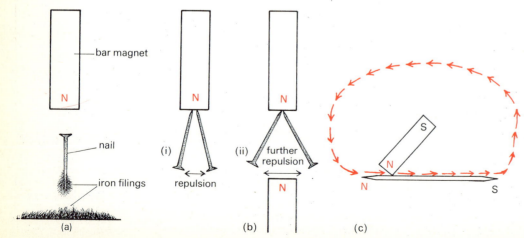

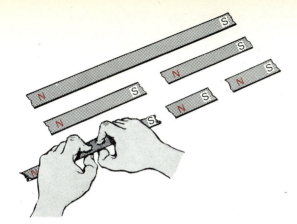

Fig. 16.5 *When a magnet is broken each resulting piece is a magnet*

## 16.5 The theory of elementary magnets

We saw in the previous experiment that when we cut up a magnet into smaller and smaller pieces each of the little pieces was a small magnet. Think of the magnet as being made up of lots of very tiny "elementary" magnets, all aligned. Would this explain our last experiment?

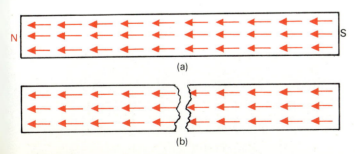

Fig. 16.6 *In a magnetized bar all the elementary magnets are aligned*

Look at Fig. 16.6 in which the elementary magnets are shown as arrows, the arrowheads representing the north poles. The left-hand end is all north poles so this end is the north pole of the magnet. When the magnet is broken, the left-hand end of each piece is all north poles and the right-hand end of each piece is all south poles. However many times the spring is broken, the left-hand end will always be the arrow-head end, i.e. a north pole. Figure 16.7(a) shows an unmagnetized bar. The elementary magnets here are so arranged that the north pole of one is neutralized by the south pole of another. In (b) the bar is partially magnetized and in (c) it is fully magnetized.

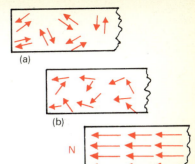

Fig. 16.7 *An unmagnetized bar, a partially magnetized bar and a saturated bar*

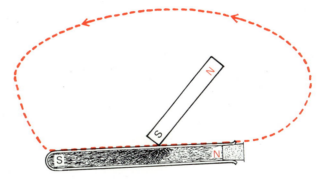

iron filings

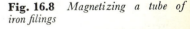

Fig. 16.8 *Magnetizing a tube of iron filings*

The experiment shown in Fig. 16.8 is an illustration of this theory. A test-tube full of iron filings is stroked with a bar magnet. The tube becomes magnetized with the polarity shown in the diagram (test this by bringing it up to the poles of a suspended magnet). If the test-tube is then shaken up and again tested for magnetism, it will be found to be unmagnetized.

Another illustration of the theory is shown in Fig. 16.9. If a group of small compasses (plotting compasses) is arranged as shown in (a), the compass needles arrange themselves in a closed chain. If a magnet is brought near, the compass needles become aligned (b). When the magnet is removed the compass needles again become a closed chain.

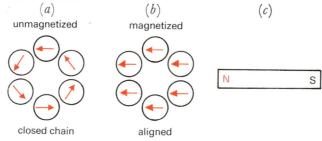

Fig. 16.9  *In an unmagnetized bar the elementary magnets form closed chains; in a magnetized bar the elementary magnets are aligned. The polarity of the magnetized bar is shown in (c).*

But what are these elementary magnets? There is considerable evidence (see for example Fig. 16.10) that the elementary magnets in iron are groups, or *domains*, of millions of atoms, each group about a millionth of a cubic centimetre in volume. The alignment of the atoms within the domains is thought to cause the effect known as magnetism. In an unmagnetized piece of iron each

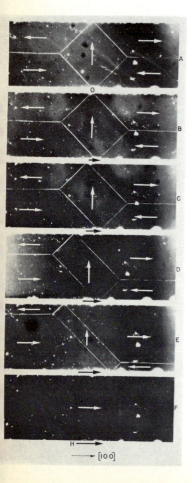

Fig. 16.10  *Direct evidence for the existence of domains. This series of photographs shows the surface of a single crystal of iron viewed through a microscope. The crystal width is 0.28mm. The lines are boundaries between domains, which have been made visible by a special technique. The direction of magnetization is also marked. When a magnetic field going from left to right is increased, the domain pattern changes until the sample is eventually saturated. In normal samples of iron which are not single crystals, the domain pattern and the changes in magnetization are more complex than those illustrated here.*

156

domain has a different axis or direction of magnetization, but when the iron is placed in a strong magnetic field the axes of magnetization in the domains rotate and become aligned with the direction of the field.

Let us see if we can use this theory to answer the following questions about magnetism.

*QUESTION 3:* (a) Will a magnet made by stroking become stronger and stronger as it is stroked more and more, or is there a limit to its strength? (b) When a substance is heated its atoms vibrate with greater energy. What would you expect to happen to the magnetism of a magnet when it is heated? (c) Why does the nail in Fig. 16.4(a) become magnetized by induction? (d) Why is it necessary to make a big sweeping circle when magnetizing a bar by stroking it with a magnet (Fig. 16.8)? (e) Why does a magnet lose power if it is dropped or banged about a lot?

## 16.6  Keepers

Magnets tend to lose their magnetism when stored for long periods. The atomic motion and the force of attraction between the north and south poles at each end tend to destroy the alignment of the elementary magnets. To prevent this, *keepers* (pieces of soft iron) are placed over the ends. The soft iron becomes magnetized by induction

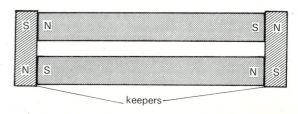

Fig. 16.11  *Keepers are used to help retain magnetism during storage*

and the elementary magnets form closed chains. There are no longer any free poles and the magnetism is preserved. Magnets, therefore, should always be stored in pairs with keepers affixed as in Fig. 16.11.

## 16.7  Magnetic materials

Materials that make good permanent magnets are obviously materials that are reluctant to change their magnetic direction: the elementary magnets, once aligned, do not easily lose their alignment. Such materials are said to be "hard". Materials that are easily magnetized but readily lose their magnetism are said to be "soft". Steel is a much harder magnetic material than soft iron. Some modern alloys are so hard that they can only be successfully magnetized when in the molten state. When such alloys are used in the manufacture of modern magnets they are solidified while in a very strong magnetic field. Mumetal is an example of a very soft magnetic alloy. It is in fact so soft that the earth's magnetic field turns it into a magnet.

## 16.8  Magnetic fields

We have already seen that a magnet affects magnetic substances placed near it.

The region around a magnet where its influence may be detected is called a *magnetic field*, and contains something we call *magnetic flux*.

The direction of the flux may be investigated by placing a large number of plotting compasses round the magnet (Fig. 16.12). The compasses set along curved paths running from the north pole of the magnet to the south pole. An alternative way to trace the direction of the *lines of flux*, or *magnetic field lines*, is to place a magnet on a piece of paper and sprinkle iron filings onto the

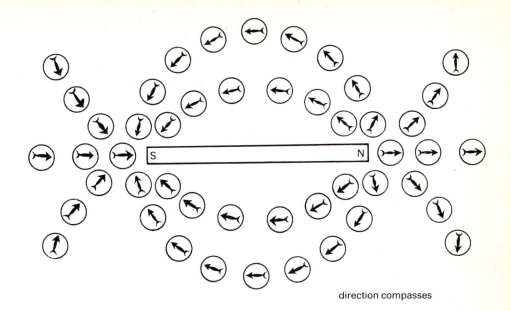

direction compasses

paper, tapping the paper all the time. The iron filings become induced magnets and set along the lines of flux just like the compass needles.

Magnetic lines of flux are lines indicating the direction in which compass needles would set if placed on these lines. The direction of the flux (indicated by arrows on the diagram) shows the direction in which the north pole of the magnet points.

Figure 16.13 illustrates an experiment where a north pole moves along a line of flux. A magnetized knitting needle, north pole upwards, is floated vertically on water. Because the needle is long, its south pole is a long way away from the bar magnet and the force on it is small and may be neglected.

**Fig. 16.12**  *Compass needles set along the lines of flux*

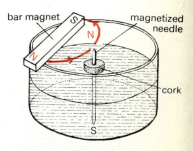

**Fig. 16.13**  *The north pole of the needle moves along the line of flux*

The magnetic flux patterns of a horseshoe magnet in the earth's magnetic field (see section 16.9), and a bar magnet in the earth's field with its north pole pointing north, are shown in Fig. 16.14.

**Fig. 16.14** *Magnetic flux patterns from (a) a horseshoe magnet, (b) a bar magnet in the earth's field*

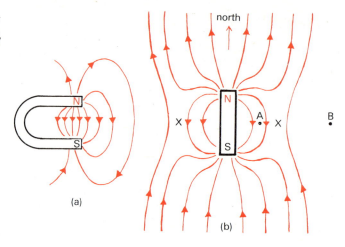

(a)

(b)

*QUESTION 4:* In what direction will a compass needle set if it is placed at A and then at B in Fig. 16.14(b)?

The points marked X in Fig. 16.14(b) are where the earth's field and the magnet's field are equal and opposite. There is no field at these points; they are known as *neutral points*.

### 16.9   The earth's magnetic field

At any point on the earth's surface a compass will always set in a particular direction (see page 154). This is why a compass helps us to find direction when we are not sure of it, and why it is such a useful instrument in navigation at sea. When the compass is used in this way

158

slight adjustments have to be made because the north-seeking pole points to the *magnetic north pole*, and this is not at the same place as the geographical north pole.

The vertical plane containing the direction in which a compass needle sets at any point on the earth's surface is called the *magnetic meridian* at that point.

The angle between the magnetic and geographical meridians is called the *angle of declination* or *variation* (Fig. 16.15).

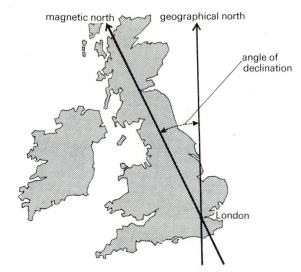

**Fig. 16.15**   *The angle between the geographical north and magnetic north is the angle of variation or declination*

So far we have only considered a magnet free to rotate in a horizontal plane. The magnet in Fig. 16.16 is free to rotate in a vertical plane. If the needle is placed in the magnetic meridian in England it comes to rest with its north pole pointing downwards at an angle of about 65° to the horizontal. A compass needle mounted in this way is called a *dip circle*.

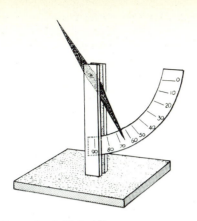

**Fig. 16.16** *The angle of dip is 65°*

The angle which a dip needle makes with the horizontal in the magnetic meridian is called the *angle of dip* (or *inclination*).

Various theories have been put forward to explain the earth's magnetic field. The most likely explanation is that it is caused by electric currents circulating in the liquid core at the centre of the earth.

A simple picture of the earth's field may be obtained if we consider that the electric currents in the liquid core give rise to a magnetic field (see Chapter 19) similar to that of a bar magnet with its south pole pointing to magnetic north.

The north magnetic pole of the earth is a south-seeking pole because it attracts north-seeking poles.

Figure 16.17 shows how the angle of dip and the angle of declination vary over the surface of the earth. There are many local variations. Further the position of the north magnetic pole is continually changing.

*QUESTION 5:* Which pole of a dip needle will point downwards when the dip circle is in the southern hemisphere?

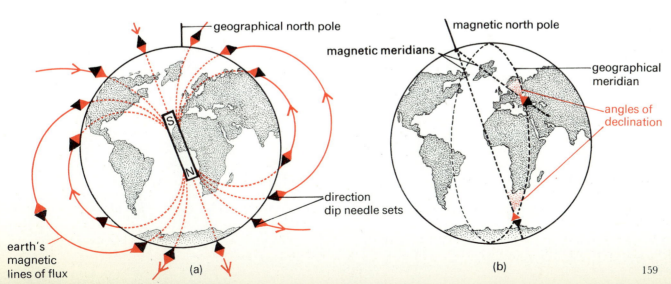

**Fig. 16.17** (*a*) *The angle of dip on the magnetic equator is zero; at the magnetic poles it is 90° (b) The angle between the direction of the compass needle and geographical meridian is the angle of declination.*

**THINGS TO DO**

A. Make some permanent patterns of lines of flux using waxed paper (the bags in many cereal packets are suitable). Place a magnet under the paper and at its centre, then sprinkle some iron filings onto the paper. Tap the paper and when the lines are formed gently warm the paper (you can do this over the hot plate of an electric cooker). As soon as the wax begins to melt remove the paper from the heat. When the wax solidifies you will have a permanent pattern of lines of flux.

B. Make a compass by sticking two bar magnets on the bottom of a circular piece of cardboard, the magnets being parallel to each other and both north poles pointing in the same direction. Pivot the cardboard at its centre (the magnets should be so placed that the cardboard will balance when pivoted at its centre). Mark the points of the compass on the top of the cardboard circle.

C. Make an estimate of the angle of declination near where you live. How can you find the direction of the geographical north?

D. Magnetize a number of needles with north poles at the eye and others with south poles at the eye. Fix paper ships on corks with the needles as shown in Fig. 16.18. Put the fleets in a plastic bowl and control them from underneath with magnets.

E. Make a *Who's Who* file on William Gilbert, known as the father of magnetism.

F. (a) Find out about lodestone and how was it used, many years before Christ, by the Chinese. (b) Find out about ferrites and their properties. Where are they used in transistor radios?

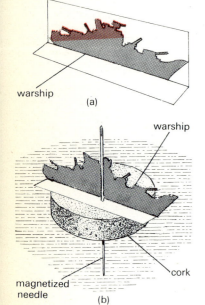

warship

(a)

warship

magnetized needle

cork

(b)

**Fig. 16.18**

1. Why does hammering tend to destroy magnetism? Can you suggest how hammering an unmagnetized nail can turn it into a magnet? (Hint: the nail is in the earth's magnetic field.)

2. Figure 16.12 shows how a large number of plotting compasses may be used to show the patterns of lines of flux around a magnet. How could you do this if you only had one plotting compass?

3. You are given a sheet of metal which you hold vertically in the earth's magnetic field. You bring up a plotting compass to the bottom of the metal sheet and find that the north pole of the compass is repelled. You test the top end of the sheet and find that it repels the south pole. You turn the sheet upside down and the bottom end still repels the north pole. Explain. (Hint: see page 157.)

4. You are given a thin sheet of metal in the form of a ring, which you are told is magnetized, yet you can detect no poles and iron filings do not adhere to it. Explain how it is magnetized. How could you test your explanation?

5. Amundsen, the Arctic explorer, found that his compass needle did not settle along any specific axis or point in any particular direction when he was at Felix Boothea (70°N 97°W). What can you deduce from this?

6. When a steel ship is constructed it becomes magnetized. Explain.

7. When a bar magnet is placed on a piece of cork and floated on water why is it not attracted to the north end of the vessel containing the water?

8. How could you demonstrate the chief differences in the magnetic properties of iron and steel?

9. You are given a metal rod. What experiments would you carry out to determine whether the rod is made of magnetic material? If it is, how can you tell whether or not it is magnetized?

10. George says, "Like poles attract because the north pole of the earth attracts the north pole of a magnet." Explain to George why he is wrong.

11. What evidence is there for believing in the theory of "elementary magnets"?

12. Sketch the lines of flux which result when a bar magnet is placed in the earth's field (a) with its north pole pointing north, (b) with its south pole pointing north.

13. Briefly describe the main features of the earth's magnetic field. How can these be explained?

14. Describe how you would magnetize a nail so that it has a north pole at the pointed end.

15. How could you magnetize a long knitting needle so that it had a north pole at both ends? Sketch the resulting lines of flux.

16. What difference would you notice in the setting of a dip needle if you lived south of the equator?

17. What is the domain theory of magnetism? Use the theory to explain (a) why a magnet broken in half produces two magnets, (b) why heating destroys magnetism, (c) why an iron ring can be magnetized so that it has no poles, (d) why hammering destroys magnetism, (e) the phenomenon of saturation, (f) why magnetism is strongly felt only at the poles and not at the centre of a magnet.

18. Flour is passed through a magnetic separator (Fig. 16.19) before being packed. Why do you think this is necessary? Study Fig. 16.19 and describe how the magnetic separator works.

19. The captain of a ship wants to steer (a) due north, (b) north-west. What compass course must he set in each case if the angle of declination is N 10°W.

20. The following table shows the angle of declination in London in various years. Draw a graph to illustrate the changes. When was the declination 8°E? When will it next be 8°E? When will it next be zero?

| Year: | 1590 | 1622 | 1652 | 1700 | 1760 | 1816 | 1900 | 1936 | 1962 |
|---|---|---|---|---|---|---|---|---|---|
| Declination: | 11°E | 6°E | 0° | 11°W | 19°W | 24°W | 17°W | 12°W | 8°W |

21. How would you distinguish between three rods, one of iron, one of brass and one a steel magnet, painted so that they looked exactly alike. Nothing else is provided.

22. A mumetal rod AB is placed in an east-west direction with its end A to the side of the N-pole of a plotting compass which is pointing north. With the end A in the same position the rod is turned so that it points north-south, and it is observed that the compass needle moves towards the east. With the end A still in the same position the rod is moved so that it is vertical, and it is observed that the compass needle moves towards the west. Explain these observations. (Hint: mumetal is magnetically very soft.)

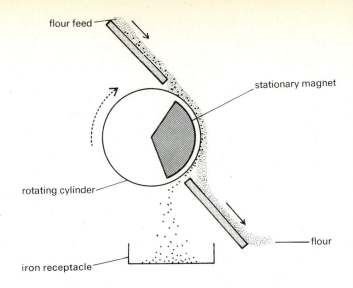

**Fig. 16.19**   *A magnetic separator*

162

## CHAPTER 17

# ELECTROSTATICS THE STUDY OF ELECTRICITY AT REST

## 17.1 Electrons and electricity

We are all familiar with the many advantages electricity brings to the home. At the movement of a switch we can immediately get heat energy and light energy. The basic reasons for the usefulness of electricity are (a) that electric currents provide a clean and quick means of transferring energy from one place to another, and (b) that electrical energy is fairly easily converted into mechanical energy (by an electric motor), heat energy (by an electric fire) and light energy (by an electric light bulb).

You may know that an electric current is a flow of electrons. The electrons come from the tiny atoms of which all matter is composed. We can picture an atom as a central *nucleus*, containing *positively charged protons*, surrounded by *negatively charged electrons*. In a *neutral (unchanged)* atom the number of protons in the nucleus is equal to the number of electrons surrounding the nucleus. For example, the simplest atom, a hydrogen atom, consists of a nucleus of one proton surrounded by one electron (Fig. 17.1); an oxygen atom has eight protons in the nucleus and eight electrons surrounding the nucleus. All nuclei, except the nuclei of hydrogen, also contain uncharged particles called *neutrons*. The protons and neutrons which together make up the nucleus are referred to as *nucleons*.

To set out experimental evidence for the above statements is beyond the scope of this book, but some of the evidence is discussed in later chapters (see in particular Chapters 28 and 30).

In this chapter we shall begin by considering the evidence for believing that there are two different kinds of charge.

## 17.2 Two kinds of charge only: positive and negative

Rub a transparent cellulose acetate strip with a duster and place the rubbed strip on an upturned watch glass or

**Fig. 17.1** *A hydrogen atom consists of a nucleus of one proton with one electron moving rapidly round it; the electron forms a shield that is difficult to penetrate*

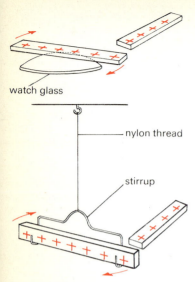

watch glass

nylon thread

stirrup

**Fig. 17.2** *Like charges repel*

suspend it in a stirrup (Fig. 17.2). Bring up another similarly rubbed strip near to it. The two rubbed strips repel each other, the one on the watch glass or stirrup rotating as indicated in the diagram. If the experiment is repeated with two rubbed polythene strips, it is again found that the strips repel each other. But if a rubbed acetate strip is brought up to a rubbed polythene strip the two strips attract each other.

*QUESTION 1:* Look again at the description given in the previous section of the structure of the atom. Can you explain how rubbing an acetate strip with a duster produces a positive charge on the strip? (Hint: what must happen to some electrons surrounding the nucleus?)

In the case of the rubbed polythene strip, some negatively charged electrons pass from the duster to the strip, and therefore the polythene strip becomes negatively charged. The duster, on the other hand, has lost some electrons and therefore some protons are no longer neutralized, and there is a resulting positive charge on the duster.

The mass of a proton is over 1800 times that of an electron. It is the outer electrons (those furthest away from the nucleus) that move from one substance to another in the above experiments. We can draw three conclusions from our experiments:

1. The polythene's electrons are more tightly held than the duster's electrons.
2. The duster's electrons are more tightly held than those of the cellulose acetate strip.
3. Two negatively charged strips repel each other and two positively charged strips repel each other, but two strips with different charges attract each other (Fig. 17.3).

Like charges repel. Unlike charges attract.

*QUESTION 2:* Suppose we have two strips, which we will call X and Y. X, when rubbed, attracts a charged

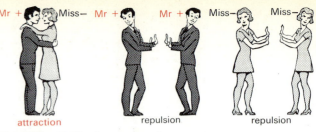

attraction    repulsion    repulsion

**Fig. 17.3** *Unlike charges attract!*

**Fig. 17.4** *The attraction of unlike charges makes this boy's hair stand on end*

164

polythene strip and repels a charged acetate strip. (a) What sort of charge is on X? (b) Would you be surprised if after rubbing the strip Y, you found that it attracted both a charged polythene strip and a charged acetate strip, but repelled another charged strip of Y?

In Fig. 17.4 the boy has just taken off his sweater which he had been wearing over a nylon shirt. It has become charged by friction and because of the attraction of unlike charges the boy's hair is standing on end.

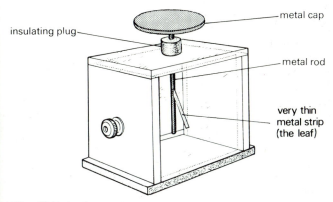

insulating plug

metal cap

metal rod

very thin metal strip (the leaf)

**Fig. 17.5** *An electroscope.*

### 17.3 An instrument for detecting charge
Figure 17.5 is a diagram of an electroscope. It consists essentially of a metal cap and a metal rod, to the bottom of which is connected a leaf of very light metal foil (usually Dutch metal, a substitute for gold leaf). The rod is supported by insulating material (i.e. something that electrons cannot flow through) and suspended inside a box with glass sides.

If a charged acetate strip is brought towards the cap, the leaf diverges (i.e. moves away) from the rod. When the strip is taken away the leaf collapses again. We can understand the reason for this if we think in terms of our description of atomic structure given in section 17.1.

In metals the electrons furthest from the nucleus are not very tightly held by the attractive force of the positive nucleus, and are "free" to wander through the metal. (It is because of all these "free" electrons that electricity can easily flow through metals.) When the positively charged acetate strip is held near the cap of the electroscope, some of the "free" electrons are attracted from the rod up to the cap (Fig. 17.6). The leaf and bottom of the rod now contain atoms which have lost electrons; therefore they are now positively charged. The like charges repel and the leaf diverges from the rod.

*QUESTION 3:* If a charged polythene strip is held near the cap, the leaf diverges. What kind of charge is on the leaf?

### 17.4 Insulators or conductors?
Charge an electroscope by rubbing a negatively charged strip along the edge of the cap. Some of the charge on the strip will be transferred to the cap of the electroscope, free electrons will be repelled and the negative charge will spread all over the metal cap and down to the leaf (see Fig. 17.7). Now touch the cap with different things, e.g. a comb, your finger, a piece of paper, a piece of cotton and any other things that are available. Sometimes the leaf collapses quickly, sometimes slowly, and sometimes not at all.

When the leaf collapses the charge on the electroscope is passing through you to earth via the substance touching the cap. When the leaf does not collapse this means that the charge on the electroscope is not getting through the substance touching the cap.

Substances that do not allow electrons to move through them or along their surfaces are called *insulators*. Substances that allow electrons to pass easily through them are called *conductors*. All metals are very good conductors.

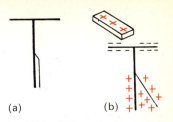

(a)　　　(b)

**Fig. 17.6** *When a positively charged strip is brought up to an uncharged electroscope the leaf diverges*

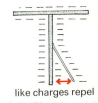

like charges repel

**Fig. 17.7** *A negatively charged electroscope*

When a metal is touched on the cap of a charged electroscope the electroscope is immediately discharged, but if a strip of polythene touches the cap of a charged electroscope the leaf does not collapse.

Without good insulating materials we could not make use of electric power, and where insulation is inadequate or faulty, electrical equipment is dangerous. If, for instance, the insulating material of the flex containing the wires for an electric fire is damaged then anyone touching the flex can get an electric shock.

### 17.5   Is air an insulator or a conductor?

The fact that the leaf of a charged electroscope remains diverged when charged means that the charge does not disappear through the air. But if you do the experiment on a damp day the leaf will probably collapse slowly. This is because moist air is not such a good insulator as dry air.

The presence of a flame makes air a better conductor. Charge an electroscope and then hold a burning match near the cap. The leaf immediately collapses. The experiment illustrated in Fig. 17.8 will help us to understand

the reason for this. A Bunsen burner or candle flame is placed between two metal plates. The plates are connected to an EHT (extra high tension) supply. When the EHT supply is switched on, the plate at one side of the flame becomes negatively charged and the plate at the other side positively charged. The flame is observed to widen out and divide into two parts; one part moves towards the negatively charged plate and the other towards the positively charged plate. This indicates that both negative and positive charges are present in the flame. The charges result from the action of the heat on the air molecules.

When air molecules are heated they become so excited that some of them lose an electron. When this happens we say the molecule is "ionized". A molecule that has lost an electron will have a net positive charge and is called a *positive ion*. If one of these detached electrons attaches itself to a neutral molecule then that molecule will have a net negative charge and is called a *negative ion*. If the electroscope is positively charged and a flame is held near it, negative ions will be attracted to the cap until the charge on it is neutralized. If the electroscope is negatively charged it will attract positive ions until the charge on it is neutralized.

### 17.6   Testing for charge

Charge an electroscope by rubbing a charged polythene strip along the edge of the cap. Rub your comb or fountain pen on your jacket and bring it close to the cap of the charged electroscope. The leaf diverges further.

*QUESTION 4:* What kind of charge is on the comb?

If the leaf of an electroscope diverges further when a body is brought up towards the cap, then the charge on that body is the same as the charge on the electroscope.

**Fig. 17.8** *The flame splits into two halves when the EHT supply is switched on. This is because the flame contains positive and negative ions.*

flame before p.d. applied

EHT supply

metal plates

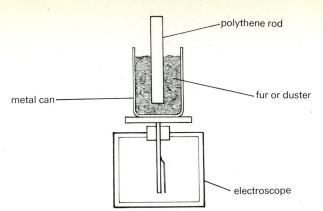

**Fig. 17.9** *When the polythene is rotated, no movement of the leaf is observed. When the polythene rod is withdrawn the leaf diverges but collapses again when the rod is returned. What can you conclude from this experiment?*

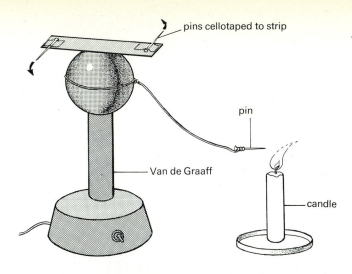

**Fig. 17.10** *An electric wind blows from the charged pin point*

## 17.7 Rubbing produces equal amounts of positive and negative charge

Wrap some fur or a duster round a polythene rod, and stuff the fur and rod into a metal can placed on the top of an electroscope (Fig. 17.9). Without touching the electroscope twist the polythene rod backwards and forwards a few times. Remove the rod from the fur and observe that the leaf diverges. Replace the rod in the fur and observe that the leaf collapses again. When the rod was first rotated both the fur and the rod became charged, one positively and one negatively. The fact that the leaf only diverges when the rod is removed shows that the negative charge on the polythene is exactly equal to the positive charge on the fur.

## 17.8 Discharge from points

A machine called a Van de Graaff generator (see Fig. 17.15) is frequently used in nuclear physics research to produce very high voltages. Most schools now have a small Van de Graaff generator capable of producing high voltages (you can see one in Fig. 17.13). When the machine is set running, positive charge builds up on the top sphere. The charge is produced by the friction of a moving belt running over rollers. If the sphere is connected to a pin (as in Fig. 17.10) an electric wind streams away from the pin point. The wind can be felt, and can be made to blow a candle flame. A strip of metal with two pins attached as shown and placed on top of the sphere will spin round as a result of the wind streaming away from the points.

The toy "windmill" in Fig. 17.11 consists of four wires arranged as spokes of a wheel with the pointed ends bent as shown. The wires are connected to a Van de Graaff generator. The electric wind streaming from the points causes the points to recoil and the wheel rotates.

The explanation of the wind produced in these two experiments is that electric charges tend to pile up on

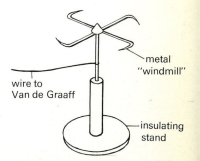

**Fig. 17.11** *The "windmill" rotates when connected to a Van de Graaff generator. Why?*

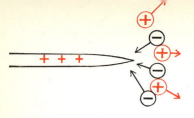

sharp points and exert large forces on the outermost electrons of the air atoms. Some of these electrons are removed from their atoms, and if the point is positively charged these electrons will be attracted to the point. The repulsion of the resulting positive ions produces the electric wind (Fig. 17.12).

American scientists have developed a rocket whose thrust is produced by the repulsion of charged ions. The force produced is very small but it acts over a long period of time, and can therefore produce large rocket velocities. Such rockets may be ideal for space travel between manned space stations. One of their advantages is that they use very little fuel.

**Fig. 17.12** *The high concentration of charge at the point ionizes the air. Positive ions are repelled and move away from the point.*

**Fig. 17.13** *The boy is charged to about 150 000 V. The soap bubbles he blows are charged and move away from him. What do you think is happening to the bubbles blown by the girl?*

An impressive experiment with a Van de Graaff generator is to place a wig of long hair on top of it. What do you think will happen? In Fig. 17.13 the boy with his hand on the top sphere of the generator is charged to a potential of about 150 000 volts. Notice his hair standing on end. (For the bubble experiment see "Things to do" at the end of the chapter.)

**Fig. 17.14** *A lightning flash*

## 17.9 Lightning

When a positively charged thundercloud passes overhead it attracts electrons to a point on the ground immediately below the cloud. If the build-up of charge is great enough, a lightning flash occurs as electrons jump from the ground to the cloud (Fig. 17.14).This is exactly like the spark that occurs when a Van de Graaff discharges (Fig. 17.15). The air in the path of the spark heats up, expands and then cools and contracts. This results in sound waves being produced and hence the resulting thunderclap. The electrons are most likely to jump from the highest points in the area over which the cloud is passing. If this happens we say that the point has been struck by lightning. The tremendous heat produced by the lightning discharge can cause considerable damage. In order to prevent this, lightning conductors are fitted to tall buildings (Fig. 17.16). The conductor is a long metal rod, one end of which is attached to a plate buried deep in the ground, and the other end, which is pointed, sticks up above the building. When, for example, a negatively charged thundercloud passes overhead negative electrons on the point are repelled to earth, leaving the point positively charged. An electric wind of positive ions moves away from the point. This has the effect of neutralizing some of the charge on the cloud. Any discharge from the cloud is then much less violent and if it does occur, the charge passes harmlessly to earth through the metal rod. If the cloud is positively charged a stream of electrons and negative ions flows from the point to the cloud, making the violent discharge of a lightning flash less likely.

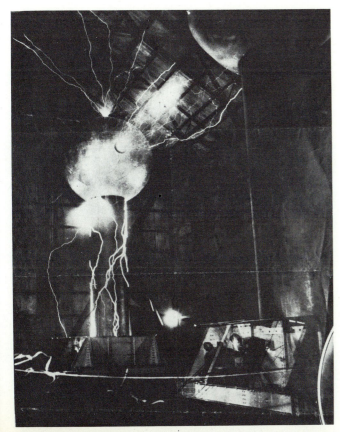

**Fig. 17.15**  *A Van de Graaff generator discharging*

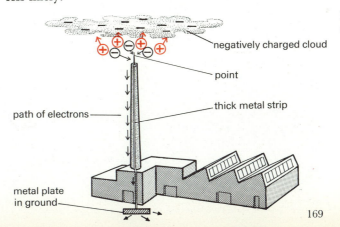

**Fig. 17.16**  *The action of a lightning conductor*

negatively charged cloud

point

thick metal strip

path of electrons

metal plate in ground

## 17.10 Why does a charged rod attract small pieces of paper?

If you rub your comb on your sleeve the comb will attract small pieces of paper from the table.

*QUESTION 5*: If the comb is negatively charged what sort of charges are on the parts of the paper nearest to the comb?

Negative electrons on the paper are repelled by the negative charge on the rod. This leaves the top of the paper positively charged. The paper is attracted to the comb because of the attraction of unlike charges.

Figure 17.17 shows a stream of water being deflected by a charged rod. Try it with your comb.

**Fig. 17.17** *A stream of water is deflected by a charged rod*

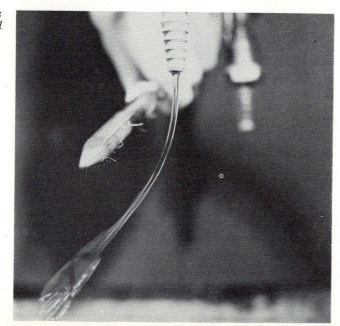

**Fig. 17.18**

A.  a. Stand on a strong, dry, upturned polythene bucket or washing-up bowl. Get someone to beat your back with a piece of fur (a duster sometimes works). After a few moments bring your pointed forefinger slowly towards your friend's ear. A spark will jump between your finger and his ear (Fig. 17.18). Repeat the experiment but this time when you are charged get your friend to hold his hand above your head. Some of your hair will stand on end.

b. Hold a charged comb near a soap bubble.

Both the above experiments are more startling if the charge is produced using a Van de Graaff generator.

c. Stand on the washing-up bowl used in the first experiment, place one hand on the top sphere of a Van de Graaff generator, and blow bubbles from a pipe or bubble-blower held in the other hand (see Fig. 17.13).

B.  a. Suspend two balloons by cotton thread, and charge them by rubbing with a dry duster. Bring them close to each other and observe what happens.

b. Rest a sheet of glass across two books (Fig. 17.19). Underneath the glass put a piece of kitchen foil and on this lay some tiny paper figures cut from tissue paper. Rub the glass with a dry duster. The figures will be seen to "dance".

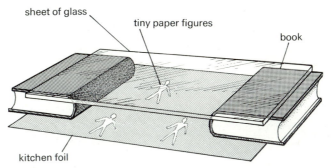

sheet of glass — tiny paper figures — book

kitchen foil

**Fig. 17.19**  *When the glass is rubbed the paper figures jump up and down. Why?*

c. Make a glider out of aluminium foil. Charge it with an ebonite rod. By repelling it with the charged rod it can be kept in the air for a long time.

C.  Wedge a stick in the top of a nylon stocking (or one leg of some tights) as shown in Fig. 17.20. Blow up a balloon and tie a piece of nylon or woollen thread to it. Holding the balloon by the thread draw the nylon stocking over the balloon. Repeat this until the stocking fills out as though it contains a phantom leg. Repeat this with another stocking (or the other leg of the tights). Bring the stockings close to one another. Explain what you observe. When the stockings are close to one another bring a balloon between them. Again explain what you observe.

D.  Investigate whether a gramophone record becomes charged during playing.

**Fig. 17.20**

E.    Design and make an electroscope. You could use a glass jar with a cork or polythene stopper. Pass a piece of wire through the centre of the stopper, and on the end of the wire fix two leaves of aluminium cooking foil (you will probably find it easier to fix two leaves to the bottom of the wire than to use the wire and one leaf). Test your electroscope by bringing up a rubbed comb to the wire protruding from the top of the stopper. The aluminium leaves should diverge.

F.    Find out all you can about Benjamin Franklin (1706–90). What did his famous kite experiment demonstrate?

G.    Find out (a) how buildings like the Empire State Building in New York and the Eiffel Tower in Paris are protected against lightning, (b) what special electrical precautions must be taken when wiring up a bathroom.

1. After you have brushed and groomed a cat you can get an electric shock if you put your fingers near to its ears. Why? Does the possibility of getting a shock depend on the time of year or the heating conditions?

2. Explain why a rubbed balloon will often adhere to the wall. Why does it only work sometimes?

3. When sifting some sugar for cake making, the sugar first fell into the bowl below the sieve, but as the sifting continued more and more sugar was thrown to one side of the bowl. Why?

4. Why is it important that blobs of solder on high voltage equipment should be smooth and spherical in shape?

5. Early movie films were made on nitrate film which was very inflammable. The films often caught fire during rewinding. Why?

6. When you take off nylon clothing sparks sometimes occur, and (if you're a man) the hairs on your chest may stand on end! Explain why these things happen.

7. The Brown family go off to have a picnic with the Joneses one sunny day. As they arrive at their destination, Tommy Jones rushes up to meet them. When he touches the car he gets an electric shock. Johnny Brown says it's because the passengers have been sliding on the seats and electricity has been generated by friction. Sarah says it's because the sun's rays produce electricity when they strike the car. Mr Brown says it's because of the sparks from the sparking plug, but Mrs Brown says it's the friction caused by air passing over the car. Discuss these explanations.

8. The temple at Jerusalem was built at a place where thunderstorms are common, yet there is no record of the temple ever having been struck by lightning as a sign of displeasure of the Almighty. Does II Chronicles Chapter 3 verse 7 offer any explanation?

9. Explain why a fountain pen rubbed on your clothes will pick up pieces of paper.

10. It is sometimes said that "lightning never strikes at the same place twice". Discuss this statement.

11. Two metal-coated balls X and Y are suspended from nylon threads. A nylon comb is rubbed on a piece of cloth and brought up to each ball in turn. If you find that both balls are repelled by the comb, what do you conclude? If you were to repeat the experiment with different balls X and Y, what would you conclude if (a) both balls were attracted by the comb, (b) ball X was attracted and ball Y was repelled?

12. Most gramophone records have a negative charge on them. Why? This charge attracts dust which results in considerable crackle when playing the record. In order to remove the charge an anti-static cloth is used. The *anti-static* used is *hydrophilic* and produces a film of moisture on the record.

   Explain the meaning of the words in italic, and why wiping the record with the cloth removes the electricity.

13. An electroscope is negatively charged. Describe what will happen when (a) a negatively charged rod, (b) a positively charged rod, (c) your hands, are brought up to the cap in turn.

14. Describe a gold-leaf electroscope.

   With the help of diagrams describe and explain what will happen to the leaf in the following circumstances. (a) A negatively charged rod is brought up to the cap and then removed. (b) A negatively charged rod is brought up close to the cap. While the rod is held near the cap, the cap is touched by the finger and then the finger is removed. The rod is then removed. (Think carefully — this is a difficult question.)

## CHAPTER 18

## 18.1 A simple circuit

Figure 18.1(a) is a photograph of part of a television set, and (b) is that of part of a microcircuit used in a computer. It would be difficult to draw diagrams of such circuits without using the agreed symbols illustrated in Fig. 18.2. In this chapter we shall explain some simple circuits and how they work.

**Fig. 18.1** (a) *Section of a TV circuit*

174

# ELECTRIC CIRCUITS AND THE HEATING EFFECT OF A CURRENT

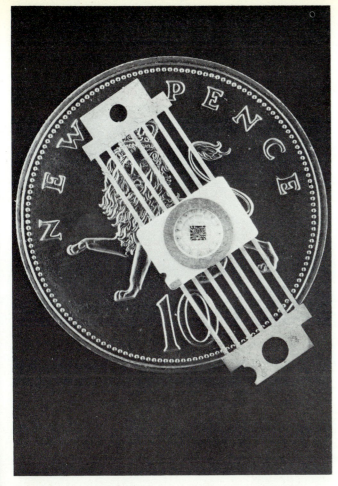

**Fig. 18.1** (*b*) *A microcircuit*

We all know that when we switch on a torch, the battery is connected to the bulb, and the bulb lights up. Figure 18.3(a) shows a bulb in a bulb-holder being lit by

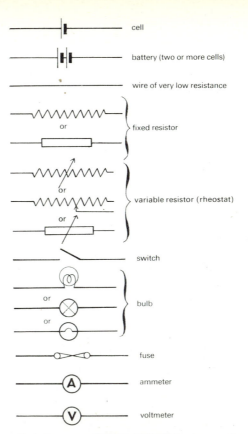

**Fig. 18.2** *Standard symbols*

cell

battery (two or more cells)

wire of very low resistance

or — fixed resistor

or — variable resistor (rheostat)

or

switch

or — or — bulb

fuse

ammeter

voltmeter

means of a battery. Figure 18.3(b) shows the circuit diagram of the arrangement using the correct electrical symbols. The bulb is only lit when the switch is closed. No current flows unless there is a complete circuit for the electrons to flow round.

Does it matter what materials are used to complete the circuit? To find an answer to this question replace the switch in Fig. 18.3 by several objects in turn, each of them made of different materials. A convenient way to

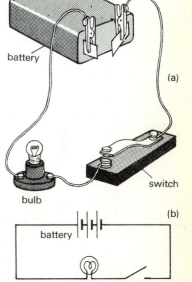

battery

(a)

switch

bulb

(b)

battery

bulb   switch

**Fig. 18.3** *When the circuit is complete the bulb lights up*

**Fig. 18.4** *A Worcester circuit board*

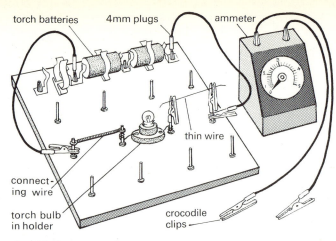

torch batteries     4mm plugs     ammeter

thin wire

connect-ing wire

torch bulb in holder

crocodile clips

do this is to use the special circuit board shown in Fig. 18.4. The different objects are connected between the crocodile clips. Add, in turn, a stick of wood, a strip of paper, a strip of copper and a pencil lead. Which are the best *conductors* of electricity? Substances that do not allow

**Fig. 18.5** (*a*) *Insulator about to be fixed in position on the national grid*

**Fig. 18.5** (*b*) *Insulator in position*

electricity to pass through them are called *insulators* (see also page 165). Figure 18.5(a) shows insulators which are about to be fitted to a 132kV grid line, and in Fig. 18.5(b) the insulators can be seen in position on a grid line crossing the Thames.

A continuous electric current can only flow in a circuit if there is a complete circuit of conductors.

## 18.2 An electric current is a flow of electric charge

In the experiment illustrated in Fig. 18.6 a Van de Graaff generator is connected, as shown in the diagram,

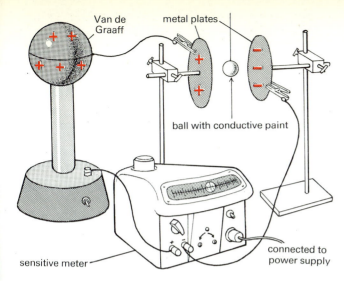

**Fig. 18.6** *Charge is transferred between the plates by the ball*

Labels in figure: Van de Graaff; metal plates; ball with conductive paint; sensitive meter; connected to power supply

to two metal plates via a sensitive meter that will detect the current (a galvanometer with an internal light beam is ideal). When the Van de Graaff is switched on, one of the plates has a positive charge on it and the other has a negative charge, but as the plates have air between them, the circuit is not complete and no current flows. The Van de Graaff is then switched off and a table-tennis ball coated with aquadag (a conductive paint) is suspended between the plates. When the generator is switched on again the table-tennis ball, by moving backwards and forwards, touching one plate and then the other, acts as a carrier of electric charge across the gap. We can detect that charge is moving round the circuit by the movement of the light spot on the meter. The movement of the table-tennis ball is caused by repulsion from one plate and attraction by the other plate, as a result of the ball becoming charged when it touches a plate.

## 18.3 Some electrical terms

Before going any further it will be helpful to define some electrical terms.

a. An *electric current* in a wire is a flow of electrons.

b. A *galvanometer* is an instrument for detecting an electric current.

c. An *ammeter* is an instrument that measures the actual size of an electric current.

d. The *ampere* (A) is a unit used in the measurement of current. With a current of 2 amperes (2A) the number of electrons per second passing any point is twice as great as with a current of 1A.

e. A *battery* (a collection of cells) is a device which, by chemical means, causes a current to flow in a circuit. By convention, one terminal is marked "positive" (+) and the other "negative" (−). The positive is sometimes painted red, and the negative black.

## 18.4 The effects of an electric current

a. *Heating and lighting effect* (see Fig. 18.3). These are really the same thing, as light is produced when the filament of a bulb gets very hot.

b. *Magnetic effect* (Fig. 18.7). Wind lots of turns of wire onto a large nail. Hold the end of the nail near some small pins and connect the ends of the wire to a battery. The nail becomes magnetized and attracts the pins.

c. *Chemical effect* (Fig. 18.8). Connect a metal coin to the negative terminal (marked −) of a battery, and a strip of copper to the other terminal. When the copper and coin are immersed in a solution of copper sulphate an electric current flows (this may be detected by putting a meter or bulb in the circuit). After a while you will see that a coating of copper has formed on the metal coin.

A battery provides another example of the chemical effect of a current. A battery converts chemical energy to electrical energy. The chemical action in the battery produces the energy needed to make the electrons "flow"

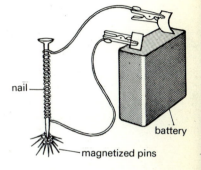

Labels in figure: nail; battery; magnetized pins

**Fig. 18.7** *An electric current has a magnetic effect*

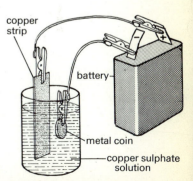

Labels in figure: copper strip; battery; metal coin; copper sulphate solution

**Fig. 18.8** *Copper-plating a coin*

round the circuit. We may picture the chemical action in the battery as transferring lots of electrons from the atoms of the positive terminal onto the negative terminal.

*QUESTION 1:* What will happen when the terminals are connected by a wire? Why?

### 18.5 Set up your own simple circuits

It is very difficult to understand electric circuits simply by reading a textbook. Every reader is encouraged to set up as many different circuits as possible. At school this can be done by using a Worcester circuit board (Fig. 18.4) which provides a simple means of joining different components together. But a special board is not necessary and the equipment may easily be purchased from an electrical shop and used as shown in Fig. 18.3.

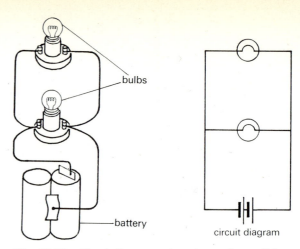

**Fig. 18.10**   *Two bulbs connected to a battery in parallel*

**Fig. 18.9**   *Two bulbs connected to a battery in series*

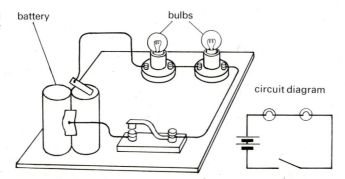

Connect one torch bulb to one torch battery, and observe the brightness of the bulb. Reverse the connections on the battery. Connect two bulbs to one battery as shown in Fig. 18.9. Use two batteries with two, three and four bulbs. Try reversing one of the batteries. In each case estimate the current from the brightness of the bulb. (When one bulb is connected to one battery refer to the

current as "one bulb's worth of current". An extra-brightly-lit bulb would have more than one bulb's worth of current flowing through it. A very dull bulb would have less than one bulb's worth of current.) When the bulbs are connected as shown in Fig. 18.9 we say they are connected *in series*. In Fig. 18.10 they are connected *in parallel*. Try connecting one, two and three bulbs across a battery in parallel. What do you notice about the brightness compared with the brightness achieved by using a similar number of bulbs connected across the battery in series?

### 18.6 Variable resistors

Now try putting a variable resistor in a circuit you have set up.

One type of resistor, a *rheostat*, is drawn in Fig. 18.11(a). Moving the slider changes the length of wire in the circuit. As the length of the wire is increased the bulb lights

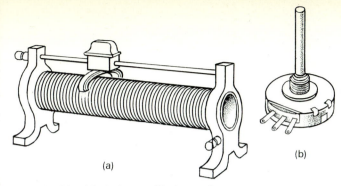

**Fig. 18.11** *(a) A rheostat (b) A potentiometer*

up less brightly. Another design of variable resistor (a *potentiometer*) is illustrated in Fig. 18.11 (b). The volume control of your radio looks like this. Inside the case, a slider (a piece of metal) moves over a coil of wire and changes the length of wire in the circuit.

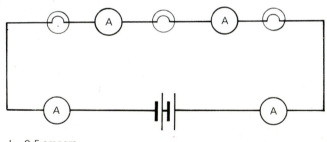

I = 0·5 ampere

**Fig. 18.12** *What will be the reading on the ammeters?*

### 18.7 The current in a series circuit

Connect up the series circuit shown in Fig. 18.12. If only one ammeter is available it may be connected at each of the points shown in turn.

*QUESTION 2:* The current ($I$) passing through the ammeter connected to the negative terminal of the battery is 0.5A. What current will pass through the other ammeters?

In a series circuit the current at every point is the same.

An analogy might help you to understand why this is so. Some central heating systems have a pump to pump the water round. The pump is similar to a battery in an electrical circuit: it provides the driving force needed to keep the water circulating. A battery provides the energy necessary to keep electrons "flowing" round a circuit. We call this energy the e.m.f. (*electromotive force*) of the battery.

*QUESTION 3:* If you had a flow meter to measure the rate of flow of water at various points in a central heating system would the rate of flow at every point in the water circuit be the same?

To continue the analogy further, we can compare the quantity of water flowing per second to the "flow" of the electric current.

An electric current is a "flow" of electric charge, and when an ammeter reads 1 ampere a quantity of charge called a *coulomb* (C) is passing through it every second.

Amperes are coulombs per second. A current of 2A means that 2C flow past any point every second.

*QUESTION 4:* How many coulombs flow past a point every second when the current is (a) 3A, (b) 4A?

### 18.8 The current in a parallel circuit

Study the circuit diagram in Fig. 18.13; it consists of two identical bulbs connected in parallel across a battery. The ammeter nearest the negative terminal reads 0.4A and the ammeter nearest to this, and in series with the first bulb, reads 0.2A.

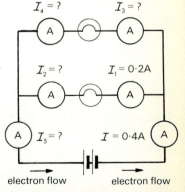

$I_4 = ?$  $I_3 = ?$

$I_2 = ?$  $I_1 = 0·2A$

$I_5 = ?$  $I = 0·4A$

electron flow  electron flow

**Fig. 18.13** *What will the ammeters read in a parallel circuit?*

*QUESTION 5:* What will the other ammeters read? It will help you to answer this question if you think back to the experiments you did when you connected bulbs in parallel.

Try the experiment shown in Fig. 18.13 with different numbers of bulbs in each of the branches. You will discover the following fact:

The total current is equal to the sum of the currents in the separate branches.

**Fig. 18.14** *The same current flows through each bulb, but one is much brighter than the other. Why?*

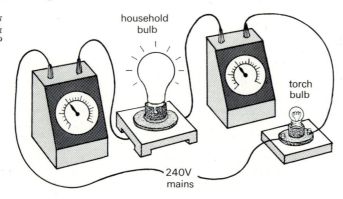

household bulb

torch bulb

240V mains

### 18.9  Potential difference

In the experiment illustrated in Fig. 18.14 a household bulb and a torch bulb are connected in series. As in all series circuits the same current flows through each bulb, but the household bulb emits much more light energy per second than the torch bulb. A measure of the current in coulombs per second is not enough to tell us how much light to expect from the bulb. We must also take into account the fact that coulombs are carriers of energy. As a very crude picture of what takes place, let us imagine coulombs of charge moving along with bundles of energy on their backs (Fig. 18.15). As they go round the circuit they give up their energy. The greater the bundle of

**Fig. 18.15**  *Mr Coulomb gives up his energy as he goes round the circuit*

energy given up by each coulomb the greater is the light energy produced. To measure how much energy we get from the bulb every second we need a device that measures how much energy is given up by each coulomb.

An instrument that measures the joules of energy given up by each coulomb is called a *voltmeter*. Volt (V) is just a shorthand way of writing the number of joules of energy transformed from the electrical form to some other form when one coulomb has passed. The quantity of energy each coulomb carries is determined by the battery providing the energy in a circuit. For instance a 3V battery gives a coulomb passing through it 3 joules of energy.

*QUESTION 6:* If two 3V batteries are connected in series what will a voltmeter read when it is connected across them?

As a coulomb passes through a 6V battery you can imagine the battery giving it 6J of energy and sending it off at the negative terminal with the instruction, "Give up all your energy as you pass round the circuit and when you return I'll give you 6 more joules of energy". If we put that into scientific terms we say that 1 coulomb on the negative terminal of a 6V battery is given 6 joules of potential energy as a result of chemical energy in the battery, and the potential difference (p.d.) between the terminals is 6 volts.

1 volt is the potential difference existing between two points when 1 coulomb of charge passing between them does 1 joule of work.

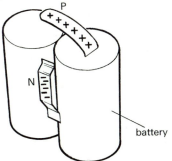

**Fig. 18.16** *Chemical action in a battery moves electrons onto the negative terminal*

We can picture the chemical action in a battery as forcing negative charge onto the negative terminal (Fig. 18.16). Work has to be done to get it there and, as we have already stated, the charge on the negative terminal possesses potential energy. If the terminals are joined, electrons flow from the negative terminal to the positive terminal.

The energy transferred per coulomb from chemical form to electrical inside the battery is the electromotive force (e.m.f.) of the battery. It is measured in volts.

## 18.10 Resistance

When a potential difference (p.d.) is applied across a conductor a current flows. If there is no p.d. then there is no current. If a p.d. of 2V is applied across a thick piece of wire a current of 10A may flow, but if the same p.d. is applied across a very thin piece of wire of the same material there may only be a current of 0.1A. (Use a Worcester circuit board to investigate the effect of using wires of different lengths and different cross-sectional areas.) The magnitude of the current is determined by the ease with which electrons can pass through the wire. For a given p.d. the current will be small for a long thin wire and large for a short thick wire. Another simple analogy may help you to understand this. Imagine a vast crowd of people (representing electrons) leaving a football ground, some through a long thin passage and some through a short wide passage. In which passage will the rate of flow of people be greatest?

The opposition a conductor offers to the flow of electricity through it is known as its *resistance*. If for a given p.d. a large current flows we say there is a small resistance, but if for the same p.d. a very small current flows we say there is a large resistance. We measure the resistance of a conductor in ohms* ($\Omega$) and it is defined by the equation

$$\text{Resistance (ohms)} = \frac{\text{p.d. across conductor (volts)}}{\text{current through conductor (amps)}}$$
(equation 1)

*QUESTION 7:* What is the resistance of a conductor if a p.d. across it of 10V causes a current of 2A to flow in it?

$$R = \frac{10 V}{2 A} = 5 \Omega$$

* The unit is named after George Simon Ohm (1789–1854), a teacher of physics and mathematics at Cologne.

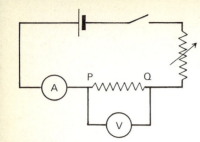

**Fig. 18.17** *Circuit to measure p.d. across a resistor and the current flowing through it*

## 18.11 Ohm's law

In Fig. 18.17 the current flowing *through* a resistor PQ and the p.d. *across* it are measured. Notice that the voltmeter is connected *across* the resistor. (You will be less likely to make mistakes when connecting a voltmeter if you always make the voltmeter the last component to be added to the circuit and remember that it goes across the resistor.) If the value of the variable resistor is altered, then the readings on both the voltmeter and the ammeter change. The table opposite shows values of the p.d. across PQ and corresponding values of the current through PQ. The readings are given in scale divisions because some ammeters and voltmeters are calibrated using the results of this experiment, i.e. Ohm's law. In the third column the value of the ratio p.d./current is calculated. Notice that for this particular piece of wire the ratio is a constant. Figure 18.18(a) shows a graph of current against p.d. for this piece of wire. The graph is a straight line through the origin. (Mathematical reminder: a straight line through the origin always results when two quantities are proportional; that is, the ratio is a constant.)

### READINGS OF POTENTIAL DIFFERENCE AND CURRENT FOR A WIRE

| p.d. (scale divs) | Current (scale divs) | Resistance $= \dfrac{\text{p.d.}}{\text{current}}$ |
|---|---|---|
| 1.00 | 0.62 | 1.6 |
| 1.25 | 0.80 | 1.6 |
| 1.50 | 0.95 | 1.6 |
| 1.75 | 1.10 | 1.6 |
| 2.00 | 1.25 | 1.6 |

The resistor PQ may be replaced by other components and the experiment repeated. In Fig. 18.18, (b) shows the results for a torch bulb, (c) for copper sulphate with copper electrodes (copper strips in the solution where the current enters and leaves), (d) a rectifier, (e) a neon lamp. Notice that a rectifier allows the passage of a large current in one direction but only a very small current when the p.d. is reversed.

The ratio of the p.d. to the current for a torch bulb is constant when the p.d. is small, but as the bulb warms up the graph begins to curve. The next table shows a series of readings for a torch bulb. Notice that the resistance increases as the temperature of the bulb increases. When a substance is heated, the atoms vibrate with greater energy and it is harder for the electrons to pass through the substance, i.e. the resistance has increased.

### READINGS OF POTENTIAL DIFFERENCE AND CURRENT FOR A TORCH BULB

| p.d. (scale divs) | Current (scale divs) | Resistance $= \dfrac{\text{p.d.}}{\text{current}}$ |
|---|---|---|
| 1.00 | 2.00 | 0.50 |
| 1.63 | 2.50 | 0.65 |
| 2.52 | 3.00 | 0.84 |
| 3.19 | 3.50 | 0.91 |
| 4.00 | 4.00 | 1.00 |

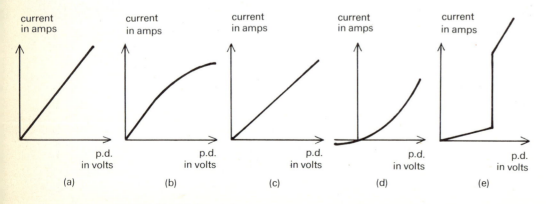

**Fig. 18.18** *Graphs showing current/p.d.*

Ohm's law may be stated as follows:

Provided the temperature of a conductor is constant, the current flowing through it is proportional to the potential difference across it.

Conductors that obey Ohm's law are known as *ohmic conductors*.

QUESTION 8: Which of the conductors whose p.d. current graphs are drawn in Fig. 18.18 are ohmic conductors?

## 18.12   Fuses

Connect a piece of fuse wire (or a thin strand of copper wire) in series with an ammeter, battery and variable resistor (a Worcester circuit board like the one in Fig. 18.4 with a variable resistor and an ammeter in place of the bulb is a convenient arrangement). As the current is increased the wire becomes hot and eventually melts.

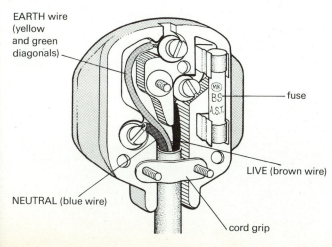

EARTH wire (yellow and green diagonals)

fuse

LIVE (brown wire)

NEUTRAL (blue wire)

cord grip

**Fig. 18.19**   *A 13A plug*

Household plugs like the one illustrated in Fig. 18.19 contain a fuse which melts and breaks the circuit when wires are overloaded. Obtain a number of different cartridge fuses and pieces of fuse wire. Use the above circuit to determine the strength of the current through them when the wire melts. The most common types of cartridge fuse are labelled 3A (blue cartridges), 5A (grey cartridges) and 13A (brown cartridges). This means that the wire should melt and break the circuit when the current exceeds 3A, 5A and 13A respectively.

In any mains circuit a fuse with its low melting point is a necessary safeguard against fire. Without a fuse a surge in current resulting from a fault in an appliance (i.e. a short circuit) could make wires dangerously hot.

## 18.13   Household wiring

Electricity is usually brought to our homes by underground cables. When the supply cable enters the house it passes through the company's fuse which is in a sealed box. It then passes through the meter which tells the electricity company how much electrical energy you have used (see page 184). The main types of circuit are shown in Fig. 18.20 (a) and (b). Most lighting circuits are wired with a 5A fuse, the lamps being wired in parallel, so that when each switch is closed the full mains voltage is applied to each lamp. If a fault develops and the fuse blows, all the lights on that circuit will go out. Points for fires are usually wired in the form of a ring main. The ring is fused at 30A, so that if there are 10 points in use each can take 3A before the fuse blows. Each of the plugs has its own fuse so that if a fault develops on one appliance and the fuse goes, other appliances connected to the ring are not affected.

The two small pins in the electric fire plug are the neutral and live pins which supply electrical energy to the appliance. The third pin on a plug is connected to

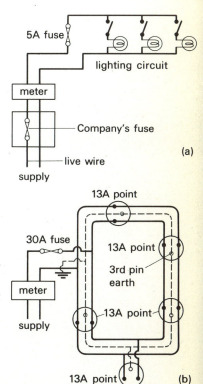

5A fuse

lighting circuit

meter

Company's fuse

live wire

supply

(a)

13A point

30A fuse

13A point

3rd pin earth

meter

13A point

supply

13A point

(b)

**Fig. 18.20**   (a)   *A   lighting   circuit*
(b)   *A ring main*

183

earth as a safety precaution. If, for example, the element of an electric fire were to break, and the end of the wire touched the metal case of the fire, anyone touching the metal case would get an electric shock because an electric current would flow through him. But if the case is connected to the third pin the current flows in a path that has virtually zero resistance. The resulting heavy current in the earth wire causes the fuse to blow.

## 18.14  Power in electrical circuits

We saw in Chapter 12 that power is the rate of doing work. In the circuit shown in Fig. 18.21 electrical energy is being converted into heat and light energy by the bulb. If the p.d. across the bulb is 10V then for every coulomb of charge that passes through the bulb 10 joules of work are converted into heat and light energy (see page 181). If the current is 2A (2 coulombs per second), the rate of conversion of energy is $10 \times 2 = 20\text{J/s} = 20\text{W}$ (see page 118).

In general, if a current of $I$ amps flows through a p.d. of $V$ volts then the work done per second (power) is given by the following equation:

$$\text{Power} = VI \text{ watts} \qquad \text{(equation 2)}$$

If the current $I$ flows for $t$ seconds then Electrical energy $= VIt$ joules.

In passing through the bulb the moving electrons collide with the atoms of the metal filament and give up some of their energy to them, and hence the temperature of the filament rises.

Using equation 1 on page 181 and substituting in equation 2 we get:

$$\text{Power} = I^2R \text{ watts}$$

i.e. the power dissipated as heat in a resistor is proportional to the square of the current.

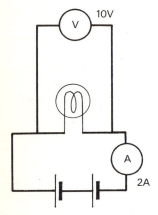

**Fig. 18.21** *Electrical energy is being converted into heat and light energy at a rate of $10 \times 2 = 20$ J/s*

## 18.15  Buying electricity

If you look at your electricity meter you will discover that it is calibrated in a unit called a kilowatt-hour (kWh).

A kilowatt-hour is the electrical energy consumed in 1 hour when it is being used at the rate of 1000W; or

$$\text{Energy consumed} = \frac{\text{power in watts}}{1000} \times \text{time in hours}$$

**Fig. 18.22** *An electricity meter*

184

QUESTION 9: How many joules is a kilowatt-hour? To discover this write down (a) the energy in joules consumed per second when electricity is being used at the rate of 1kW, (b) the energy consumed in joules per hour when the rate at which it is being used is maintained at 1kW.

The dials of an electricity meter are shown in Fig. 18.22. The reading is 87939. Notice that when the pointer is between two figures the lower figure is read. (This does not apply when the pointer is between 9 and 0. Can you say why?). If you look at your household electricity bill you will find that two readings appear on it. These are the meter readings at the beginning and end of the quarter. By subtraction the electricity board knows the number of units (kilowatt-hours) consumed in the quarter.

QUESTION 10: What is the cost of running five 60W bulbs for 10 hours if the cost of a unit is 3p?

## 18.16   Electric filament lamps
In this chapter we frequently referred to the lighting effect of an electric current as seen in a bulb. How do such bulbs function? Most modern lamps consist of a coil of tungsten wire inside a glass bulb which contains argon or nitrogen at reduced pressure (Fig. 18.23). Tungsten is used because it has a high melting point (3400°C). The wire filament is coiled to reduce heat loss and thus increases the temperature of the filament. The gas helps to prevent the filament from evaporating. Why do you think the filament will tend to evaporate?

## 18.17   The heating effect of a current
This effect is used in many appliances in the home. Fig. 18.24 shows three of these. They each contain a *heating*

*element* which gets hot when an electric current flows through it. When you turn on an electric fire you can see the element glowing red. The heat energy comes from the electrons colliding with the metal atoms in the heating element. The electrons give up some of their energy to the atoms which vibrate with greater energy and the temperature of the heating element rises.

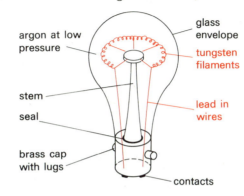

**Fig. 18.23**

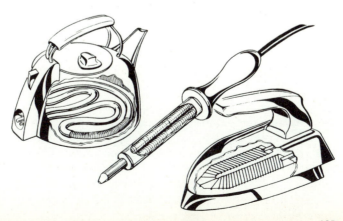

**Fig. 18.24**  *Electric irons, kettles and soldering irons all have heating elements*

**THINGS TO DO**

A. Make an electric quiz game (Fig. 18.25). Cardboard and drawing pins may be used but it is better to use wood and nuts and bolts. The bolt heads protrude above the wood and have the questions and answers written beside them. The dotted lines show the connections underneath. The ends A and B are bare bits of wire. Many more questions and answers may be added. When the end A is put on the bolt that says "flock of geese", the bulb will only light if the end B is placed on the bolt by the correct answer.

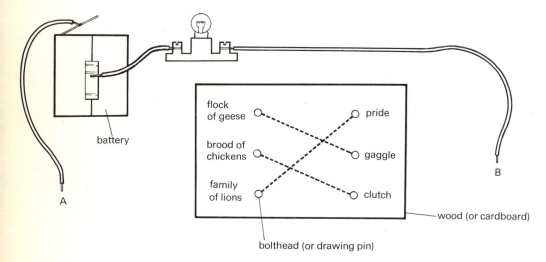

**Fig. 18.25** *An electric quiz game*

B. Read your electricity meter now and at weekly intervals. Draw a graph showing how the number of units consumed varies from week to week. Find out the cost of a unit and draw another graph showing how the cost of electricity varies from week to week. Show the graphs to your teacher and your parents.

C. Find out what fuses (2A, 5A, 13A or 30A) are used in plugs connected to the following appliances: a two-bar electric fire, a cooker, a washing machine, an electric fan heater, an electric clock, an electric vacuum cleaner.

D. Find out where the fuse box is in your home, where the fuse wire is kept and how to change a fuse. The operation is dangerous unless proper precautions are taken; ask your parents' permission before you start, and ensure that the mains switch is off before you open the box. Find out also how to change a cartridge fuse.

E. Find out what type of switch is used in a bathroom. Why is a special type of switch needed?

F. Digital computers and logic-machines make use of a number of basic circuits which are illustrated in Fig. 18.26. You can make these circuits for yourself using tapping keys and a battery and bulbs. In (i) the bulb will light if both A and B are depressed. In (ii) the bulb will light if either A or B is depressed. In (iii) either A or B must be depressed but not both in order to light the bulb.

See if you can design and build a nor circuit. This is a circuit where the bulb lights when neither A nor B is depressed but goes out if either A or B is depressed.

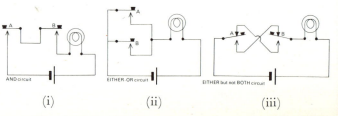

**Fig. 18.26**

1.  (a) You switch your torch on and it fails to light. List the possible causes of failure. How would you set about finding out and correcting the fault(s)?
    (b) Describe what tests and experiments you would carry out to decide which was the best buy out of three different makes of torch battery.

2.  Figure 18.27(a) shows a 6V bulb connected to a 6V cell. The bulb lights with normal brightness. In (b) to (e) the bulbs and cells are all 6V. Say in each case whether the bulbs will glow with about the same brightness, less brightness or more brightness than the bulb in (a).

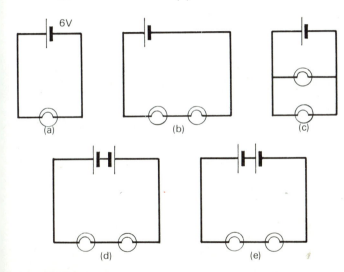

**Fig. 18.27**

3.  The lamp in Fig. 18.28 is fully lit but the ammeter shows no reading. Suggest a reason for this.

4.  What can you say about the reading on the ammeter and the brightness of the bulbs in each of the circuits in Fig. 18.29. The ammeter would read 0.2A if the bulbs are at normal brightness. One bulb connected across one battery lights to normal brightness.

**Fig. 18.28**

5.  What happens to the reading on the ammeter A in Fig. 18.30 if the slider on the variable resistor is moved from the left-hand end to the right-hand end.

6.  What can you say about the brightness of the bulbs in Fig. 18.31? What happens if the points X and Y are joined by a wire?

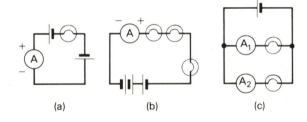

(a)          (b)          (c)

**Fig. 18.29**

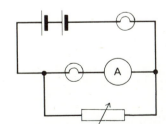

**Fig. 18.30**

7.  Three people carry out a project to see how many cars are using a certain section of a suburban traffic system every hour. They station themselves

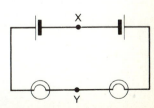

**Fig. 18.31**

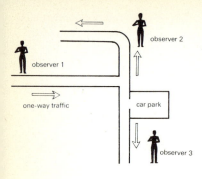

observer 2

observer 1

one-way traffic

car park

observer 3

**Fig. 18.32**

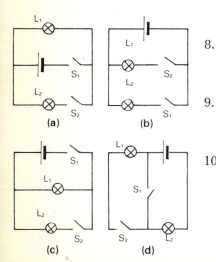

$L_1$

$S_1$

$L_2$

$S_2$

**(a)**

$L_1$

$L_2$

$S_2$

$S_1$

**(b)**

$S_1$

$L_1$

$L_2$

$S_2$

**(c)**

$L_1$

$S_1$

$S_2$

$L_2$

**(d)**

**Fig. 18.33**

as shown in Fig. 18.32. All the roads are one-way and the direction of the traffic is shown by the arrows. The following table gives their observations for each hour from 9am to 12 noon.

Plot these results on a graph. From your graph deduce the resistance of the coil. Draw a diagram of the circuit used to obtain the readings.

|  | Cars between 9.0 and 10.0 | Cars between 10.0 and 11.0 | Cars between 11.0 and 12.0 |
|---|---|---|---|
| Observer 1 | 100 | 50 | 60 |
| Observer 2 | 70 | 20 | 30 |
| Observer 3 | 10 | 30 | 30 |

How does the first column differ from the other two? Give a possible explanation for the difference. Which columns would best describe the behaviour of water flow and electron flow? In complete water circuits or electrical circuits do we ever get situations like that shown in column 1? Why?

8. Figure 18.33 shows four circuits. Describe for each circuit which bulbs will light when (i) $S_1$ is closed, (ii) $S_2$ is closed, (iii) both $S_1$ and $S_2$ are closed at the same time.

9. Johnny Brown says that the two headlamps of a car are connected in parallel, but Sarah says they are connected in series. Who is right? What test would you carry out to show your answer is correct?

10. The following table of readings shows the current flowing through a coil when various potential differences are applied across it.

p.d. (volts)      0.5    1.0    1.5    2.0
current (amps)  0.22  0.43  0.65  0.88

Plot these results on a graph. From your graph deduce the resistance of the coil. Draw a diagram of the circuit used to obtain the readings.

11. (a) A p.d. of 10V is maintained across a resistor of $5\Omega$. What current flows in the resistor? (b) A 240V mains drives a current of 4A through the element of an electric fire. What is the resistance of the element? (c) If 2A is flowing through a resistor of resistance $4\Omega$, what is the p.d. across the resistor?

12. When 1 coulomb passes through a 6V cell it gains 6 joules of energy. If two such cells are arranged (a) in series, (b) in parallel in a circuit, how much energy does a coulomb gain when it passes the cells? Give reasons for your answers. What is the connection between your answers and the p.d. across the terminals of the cells?

13. When two resistors ($R_1$ and $R_2$) are joined in series the combined resistance ($R$) is given by $R = R_1 + R_2$. Use this equation to calculate the combined resistance of $2\Omega$ and $4\Omega$ connected in series. A 12V battery is now connected across them. What current flows? Draw a diagram of the circuit.

14. Bulbs marked 20V 0.2A are to be used to light a Christmas tree. If the mains is 200V how many bulbs must be used and how should they be connected?

If the filament of one of the bulbs burns out, what will happen to the other bulbs? Give a reason for your answer.

15. Figure 18.34 illustrates the use of two-way switches. Explain what happens to the bulb when first $S_1$ is switched to the dotted position and then $S_2$.

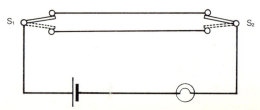

$S_1$                          $S_2$

**Fig. 18.34**

16. 120 joules of heat is produced in a resistor by the passage of 12 coulombs. What is the p.d. (voltage) across the resistor?

17. 18kJ of heat are produced in a heater when 2A flow for 300 seconds.
    a. How many coulombs flow every second?
    b. How many coulombs flow in 300 seconds?
    c. What is the p.d. across the heater?

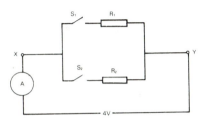

**Fig. 18.35**

18. In Fig. 18.35 $R_1$ and $R_2$ are $2\Omega$ and $4\Omega$ respectively. How much current flows in the circuit when $S_1$ is closed? What is the current when $S_2$ is closed? What will the ammeter read when both switches are closed together? What single resistor (R) must be placed between X and Y to get this current? Repeat the question when $R_1$ is $1\Omega$ and $R_2$ is $4\Omega$. Are R, $R_1$ and $R_2$ related by the following equation?

$$\frac{1}{R} = \frac{1}{R_1} + \frac{1}{R_2}$$

Notice that connecting resistors in parallel means that the resistance of the circuit is less than if either resistor were there on its own. Does the analogy of pupils streaming out of a building with two doors open instead of one help you to understand why putting resistances in parallel reduces the total resistance? Explain.

19. What are the ammeter readings in the circuits shown in Fig. 18.36? To answer this question you need to know the equations in questions 13 and 18.

**Fig. 18.36**

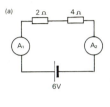

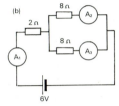

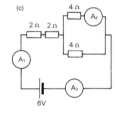

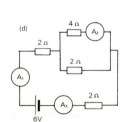

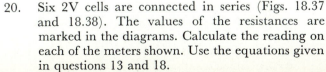

20. Six 2V cells are connected in series (Figs. 18.37 and 18.38). The values of the resistances are marked in the diagrams. Calculate the reading on each of the meters shown. Use the equations given in questions 13 and 18.

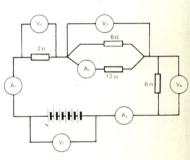

**Fig. 18.37**

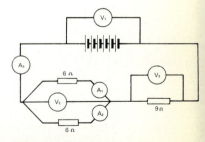

**Fig. 18.38**

189

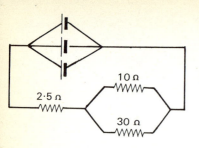

**Fig. 18.39**

21. Each of the cells in Fig. 18.39 are of e.m.f. 1.5V. Calculate the current flowing through each resistor. Use the equations given in questions 13 and 18 (neglect the resistance of the batteries).

22. If $R$ is the resistance of a piece of wire of length $l$ and cross-sectional area $A$ then

$$R = \rho \frac{l}{A}$$

where $\rho$ is a constant known as the resistivity of the material.
a. Complete the following sentences: (i) $R$ is proportional to ... (ii) $R$ is inversely proportional to ...
b. Solve the equation for $\rho$ and show that the units of $\rho$ are ohm $\times$ m.
c. If a wire of length 1m and cross-sectional area 0.5mm² has a resistance of 1.2Ω, what is the resistivity of the material?
d. If a material has a resistivity of $1.8 \times 10^{-8}$ ohm m, what is the resistance of 2m of wire of cross-sectional area 0.4 mm²?

23. A current of 0.05A through the heart can cause death. The size of the current flowing through the body for a given potential difference depends on the resistance of the body, and this in turn depends on the dampness of the skin. For a 240V supply and body resistance 3000Ω, is the current likely to be fatal? It is likely to prove fatal if the body resistance is 30 000Ω? If the body resistance is 50 000Ω what voltage would prove fatal? Why is it safer to work model railways off 12V rather than 240V?

24. At the beginning and end of a quarter the readings on a household electricity meter are 60 960 and 61 870. How much is the bill if electricity costs 3p a unit?

25. Suggest a reason why some of the dials on an electricity meter are labelled clockwise and some are labelled anticlockwise.

26. An electric fire is labelled 250V 1kW. Use equation 2 on page 184 to calculate the current it takes. Then apply Ohm's law to calculate its resistance.

27. What is the cost of running (a) a 1kW fire for 3h, (b) a 3kW immersion heater for 20h, (c) a 100W electric light bulb for 10h, (d) a 60W light bulb for 10h, (e) a 100W light bulb and a 60W light bulb together, for 10h, (f) six 100W light bulbs and ten 60W light bulbs all on together for 6h each day for a week? The cost of a unit of electricity is 3p.

28. A car sidelight bulb is labelled 12V 6W. What experiments would you carry out to determine its true wattage? Show how the wattage is calculated from the readings.

29. At one time power points were connected in parallel like the lighting circuit in Fig. 18.20(a). Why is a ring main preferable to this? What is the advantage of having each plug fused? What is the purpose of the main fuse in the live wiring connected to the ring?

30. (a) Why is an earth wire always used in a power circuit but not necessarily in a lamp circuit? (b) The main supply to your house is 250V. What fuse would you put in the plug of (i) a 2kW electric fire, (ii) a bedside lamp containing a 60W bulb? The standard fuses available are 2A, 5A and 13A. Give reasons for your choice.

31. A 12V car battery needs recharging. To do this is it necessary to drive a current through it in the opposite direction to that in which it discharges. A 240V d.c. mains is available. Draw the circuit you would use if the charging current was to be 2A. What is the value of the resistor in your circuit? (Neglect the resistance of the battery.)

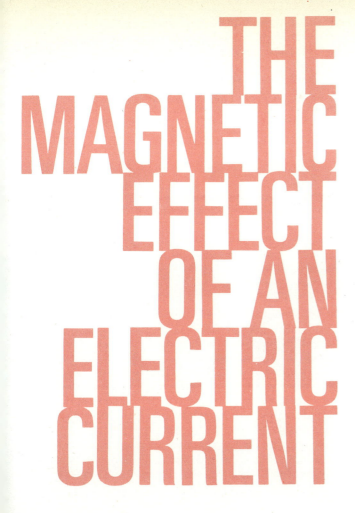

# THE MAGNETIC EFFECT OF AN ELECTRIC CURRENT

## 19.1  An early experiment

At the beginning of the nineteenth century when scientists first began an extensive study of magnetism and electricity, certain similarities between the two were noticed. We have already done experiments which lead us to believe that there are only two kinds of charge and only two kinds of magnetic pole and, moreover, that like charges repel each other and like poles repel each other.

Scientists tried to detect a magnetic field near stationary electric charge, but they were unsuccessful. It was a chance discovery made by Hans Christian Oersted in 1819 that led to an understanding of the real connection between magnetism and electricity. At the end of a lecture on a simple cell, Oersted was using a cell to drive

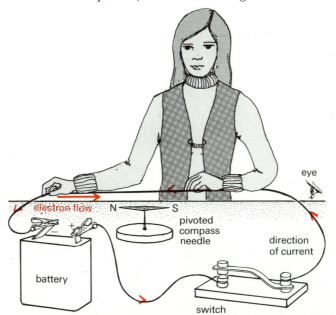

**Fig. 19.1**  *Which way does the compass needle move?*

electron flow

battery

N      S

pivoted
compass
needle

direction
of current

eye

switch

a current through a wire. The wire happened to be lying parallel to a compass needle. To the surprise of everyone, including himself, the compass needle moved when the current was switched on and off.

Stationary electric charge does not produce a magnetic effect but moving charge does.

You can easily repeat Oersted's experiment. Hold a wire over a compass needle which is pointing north-south (Fig. 19.1). When the current is switched on the compass needle moves. When the direction of the current is reversed, the needle moves in the opposite direction. (Do not hold the switch down too long or you will damage the battery. Why?)

**Fig. 19.2** *Lines of flux resulting from a straight wire carrying a current*

electric current is passed through the wire. Iron filings sprinkled on the card set in concentric circles, as in Fig. 19.2(a), when the card is tapped. We conclude therefore that a straight piece of wire carrying a current has a magnetic field round it represented by concentric magnetic lines of flux. When plotting compasses are placed on the card they set along the lines of flux, as in Fig. 19.2(b). The compass needles point in the opposite direction when the direction of the current is reversed, as in (c), thus showing that reversal of the current reverses the direction of the magnetic field. In Chapter 16 we defined the direction of a magnetic field as the direction in which a north pole points when placed in the field. You can determine the direction of the magnetic field resulting from an electric current by looking along the wire; when the electrons are coming towards you, i.e. from the negative terminal, the direction of the magnetic field is clockwise.

*Note on direction of current.* Electrical engineers talk about the movement of positive electricity from the red (positive) terminal of a battery to the black (negative) terminal. This direction is conventionally taken as the direction of the electric current. This was decided long before anybody knew that an electric current was a flow of electrons which of course is in the opposite direction, from negative to positive. We talk about current in this old way because we would find it difficult to change. (We still talk about the sun rising and setting though we know perfectly well it does no such thing.) It is common practice, therefore, to put arrows on wires in diagrams showing the direction of the conventional current from the positive terminal to the negative terminal, but it is always wise to check in any book on electricity whether arrows indicate "electron flow" or "conventional current direction". In this book arrows *on* wires in circuit diagrams indicate conventional current. To indicate electron flow an arrow *by the side of* the wire is used (see Figs

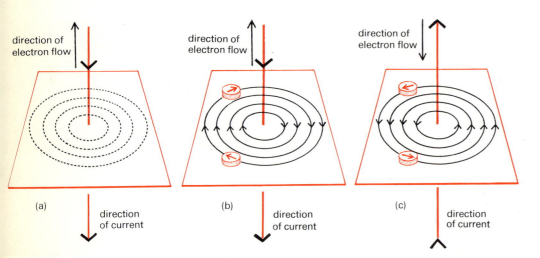

**19.2 Investigation of the magnetic field resulting from an electric current flowing in a wire**

We can continue our investigation into the magnetic field around a wire carrying a current by the experiment illustrated in Fig. 19.2. A straight wire passes through the centre of a piece of cardboard held horizontally, and an

19.1 and 19.2). In the text the term "**current**" refers to conventional current.

The following *clockwise rule* may now be stated in terms of conventional current:

If you imagine yourself to be looking along the wire with the current flowing away from you, then the direction of the magnetic field is clockwise.

*QUESTION 1:* (a) In Fig. 19.1 which way will the north pole move when the current is switched on? (b) Which way will the needle move if the wire is held below it?

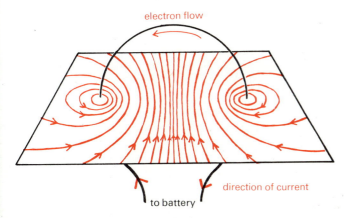

Fig. 19.3   *Lines of flux resulting from a circular coil carrying a current*

## 19.3   The magnetic field resulting from a circular coil

The lines of flux resulting from a circular coil are illustrated in Fig. 19.3. Near the points where the wire passes through the cardboard the lines of flux are nearly concentric circles, because the rest of the coil is too far away to have much effect. For a small area near the centre the field is very nearly uniform. Apply the clockwise rule to Fig. 19.3 to check the direction of the lines of flux.

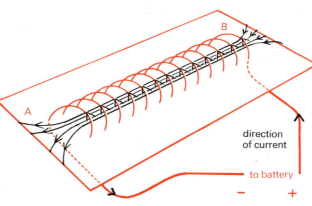

Fig. 19.4   *The lines of flux resulting from a solenoid carrying a current*

## 19.4   Field resulting from a solenoid

A solenoid is a spiral or helix of wire (Fig. 19.4). A strong magnetic field may be produced if there are many turns of wire forming the solenoid. The lines of flux run parallel down the centre of the solenoid and the fields at either end resemble those of a bar magnet (see Fig. 19.5).

*QUESTION 2:* If the north pole of a compass needle is brought up to the end of the solenoid marked A (Fig. 19.4), will it be attracted or repelled? (Hint: compare the lines of flux to those of a bar magnet.)

*QUESTION 3:* An unmagnetized piece of magnetic material is placed inside a solenoid. What will happen when the current is switched on?

Strong permanent magnets are made by allowing the molten metal to solidify while it is in the magnetic field of a solenoid.

**Fig. 19.5** *Lines of flux (a) in and (b) around a current-carrying solenoid*

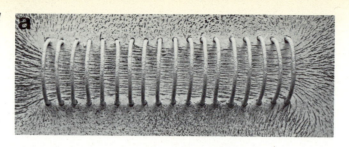

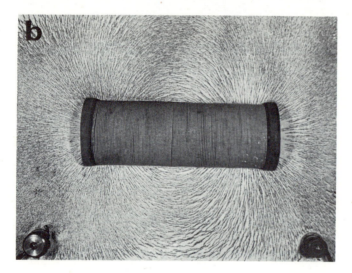

## 19.5  Electromagnets

A simple electromagnet may be constructed by winding wire round a nail or soft iron cylinder, i.e. wind a solenoid on a soft iron cylinder (Fig. 19.6). When a current flows in the wire the soft iron becomes magnetized, and will pick up tacks. When the current is switched off the tacks fall off.

**Fig. 19.7**  *A 165cm diameter magnet picking up pig iron. Where is the field strongest?*

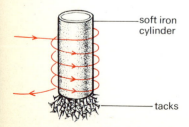

soft iron cylinder

tacks

**Fig. 19.6**  *A simple electromagnet*

*QUESTION 4:* A magnet is inside a solenoid through which an alternating current flows. This means that the current flows in one direction for a short time and then in the opposite direction. The a.c. mains alternate 50 times every second (see page 318). If the magnet is slowly withdrawn along the axis of the solenoid until it is a long way away from the solenoid, what do you think will happen to its magnetism? Could this method be used to demagnetize a watch?

194

**Fig. 19.8**  *Nimrod, a giant accelerator of charged atomic particles*

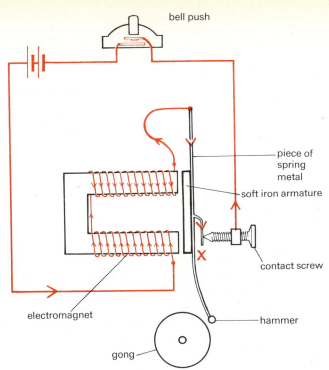

**Fig. 19.9**  *A trembler bel*

There are many uses for electromagnets. Figure 19.7 shows a very powerful electromagnet at work in a steel mill. In Fig. 19.8 you can see part of a giant electromagnet used to accelerate charged atomic particles to high velocities. On page 151, at the beginning of this section of the book, there is a photograph of an electromagnet being used to extract a metal splinter from a patient's eye — naturally this will only work if the splinter is of a magnetic material.

Let us look now at three everyday uses of electromagnets.

## 19.6  The trembler bell and door chimes

Figure 19.9 is a diagram of a trembler bell. The path of the current when the bell push is depressed is shown by the red arrows. The electromagnet attracts the soft iron armature and the circuit is broken at X. The electromagnet then no longer attracts the armature, and the springy piece of metal pulls the armature back so that contact is again made at X. The hammer moves backwards and forwards striking the gong.

In door chimes the hammer is drawn in when the bell push is depressed, and held against the chime until the bell push is released. The circuit is such that a current flows through a solenoid (which attracts the hammer) and continues to flow all the time the bell push is kept depressed.

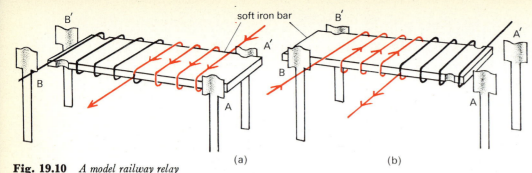

**Fig. 19.10** *A model railway relay*

(a)          (b)

### 19.7   A relay

There are a number of different kinds of relay that use
electromagnets. The relay shown in Fig. 19.10 is for
automatic train and signal control on model railways. It
is an electrically operated switch. To understand how it
works, hold a piece of soft iron so that one end is just
inside a solenoid. When the current in the solenoid is
switched on you find that soft iron is pulled into the
solenoid, because the soft iron becomes an induced
magnet. When the train on a model railway goes over a
certain "trip" on the track, a circuit is completed and
current flows in one part of a solenoid — this is marked
with red arrows in Fig. 19.10(a). The soft iron is drawn
into this part of the solenoid and joins the contacts AA'.
AA' is connected to a circuit arranged in such a way that
when AA' is joined the signal goes green. When the train
has passed the signal it goes over another trip, which
causes a current to flow in the end of the solenoid marked
red in Fig. 19.10(b). The soft iron is now drawn into the
left-hand end of the solenoid and completes the circuit of
which BB' is part. This is the circuit containing the red
bulb, and the signal goes to red. The relay may also be
used to make one train start or to stop another train. If
the electricity supply to the track is fed via a circuit
containing AA', then when the relay is in the position
shown in Fig. 19.10(b) no current will get to the track.

196

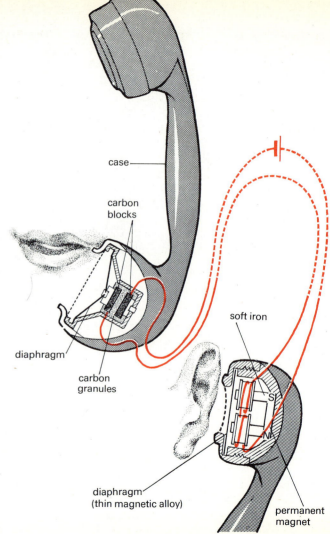

**Fig. 19.11** *Telephone earpiece and receiver*

When the relay is in the position shown in Fig. 19.10(a) then the track will be "live". A train going over a trip anywhere else on the layout can be used to actuate (i.e. set in motion) the relay.

### 19.8   Telephones

One end of a household telephone contains a microphone and the other end, which you put to your ear, contains a small loudspeaker. When you speak into the mouthpiece the sound waves impinge on a diaphragm which is in contact with carbon granules in a box. The pressure changes resulting from the sound waves move the diaphragm backwards and forwards. When the carbon granules are compressed their resistance decreases. The variations in pressure in the sound wave are thus converted into varying electric currents by the microphone. These varying electric currents arrive at the earpiece. In the earpiece the currents pass through the coils of an electromagnet. The electromagnet pulls the diaphragm (Fig. 19.11) towards it, the amount of movement depending on the current in the coils. As the diaphragm in the earpiece moves in and out sound waves are produced. Thus in the microphone sound energy is turned into electrical energy, and in the earpiece the electrical energy is turned back into sound energy.

197

**THINGS TO DO**   A.   Make a signal from odd pieces of wood and cardboard like the one illustrated in Fig. 19.12. Wind a solenoid on a cardboard cylinder. When the circuit is complete the signal will go up.

**Fig. 19.12**   *A home-made signal*

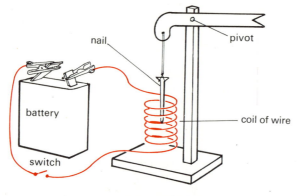

B.   Make a vibrator like the one illustrated in Fig. 19.13. A hacksaw blade is fixed to a block of wood. Wind lots of turns of wire on a piece of soft iron (a thick nail will do), and arrange the electrical circuit as shown. It may work better with a lump of plasticine on the end of the blade. If you want to make a din let the end of the blade hit against a gong.

C.   Construct a simple electromagnet from an iron bolt which has a nut screwed on at one end. Wind layers of insulated copper wire onto the bolt, leaving enough wire at the beginning and the end to connect to the batteries. Wind insulated tape over the wire to prevent it unwinding from the bolt.

D.   Find out all you can about some of the pioneers in systems of communication, e.g. Wheatstone, Bell, Marconi, Edison.

E.   Make an illustrated chart showing all the appliances which function by means of an electromagnet within their construction.

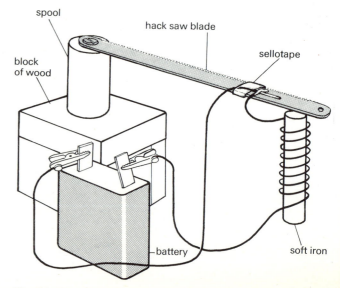

**Fig. 19.13**   *A home-made buzzer*

1. A compass needle is placed due east of a vertical wire. Would it be possible by observing the compass needle to know when the current in the wire was switched on? In which direction is the current flowing if the needle points south? If the compass needle is placed due north of the wire, in which direction is the current flowing when the needle points due east?

2. How would you investigate the magnetic field around a straight wire carrying a current? What difference would be observed in the pattern of the lines of flux if at a point a few centimetres from the wire the field caused by the current flowing in the wire is the same strength as the earth's magnetic field at that point?

3. What are the main differences between weak and powerful electromagnets?

4. Draw a diagram of an electric bell-push system with at least two bell pushes and explain how it works.

5. You have constructed an electromagnet which proves to be very weak. Suggest how it could be strengthened.

6. Describe with diagrams how you would magnetize a steel bar by electrical means. Mark clearly the direction of the current and the induced poles (show clearly whether your arrows indicate electron flow or conventional current).

7. It is possible for a piece of steel to be magnetized but to have no poles. What shape must the steel be, and how are its domains aligned? How could such a piece of steel be magnetized?

8. Figure 19.14 is a diagram of a relay. Study it carefully and then describe how it works. Suggest some possible uses for such a relay.

9. Figure 19.15 is a diagram of a modern underdome type of trembler bell. Study the diagram and describe how it works.

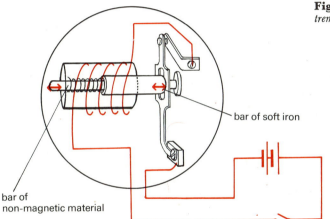

**Fig. 19.15** *An underdome-type trembler bell*

bar of soft iron

bar of non-magnetic material

10. Figure 19.16 is a simple form of current balance: (a) What do you think it is used to measure? (b) Why does the straw move when a current is passed through the coil? (c) Would it be better if the pin were in the middle of the straw? Give reasons for your answer.

11. You are provided with steel and iron bars of similar shape. Describe what experiments you would perform to decide which is the better for making (a) an electromagnet, (b) a permanent magnet.

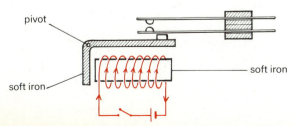

pivot

soft iron

soft iron

**Fig. 19.14** *Diagram of a relay*

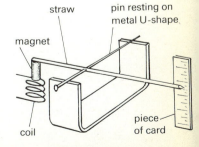

straw

pin resting on metal U-shape

magnet

coil

piece of card

**Fig. 19.16** *A simple current balance*

**CHAPTER 20**

## 20.1 Liquids as conductors

On page 177 we described an experiment showing that an electric current can pass through copper sulphate. Are all liquids conductors of electricity?

**Fig. 20.1** *Pure water is a bad conductor, but when salt is added the bulb lights up*

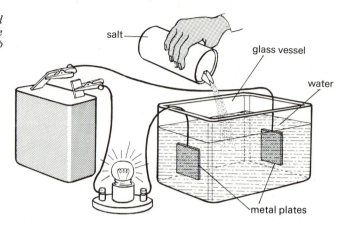

salt

glass vessel

water

metal plates

We can test liquids to see if they are conductors by using the apparatus shown in Fig. 20.1. A battery and a torch bulb (or galvanometer) are joined in series with two metal plates. The plates are dipped into various liquids. If the bulb lights, or the galvanometer shows a reading, then the liquid is conducting electricity. We will find that some liquids are good conductors (e.g. acids) whilst others are poor conductors (e.g. oils). Pure distilled water is a very poor conductor, and the bulb will not light if the plates are put in pure water. But if salt is added to the water it then becomes a good conductor. Even the impurity obtained by rubbing your hands together in the distilled water is enough to decrease considerably the resistance of the water.

The study of changes that occur when electricity flows through liquids is called *electrolysis*. The liquid through which electricity is passed is called the *electrolyte*. The wires or plates where the current enters and leaves the

# THE CHEMICAL EFFECT OF AN ELECTRIC CURRENT

liquid are called the *electrodes*. The electrode connected to the positive terminal of the battery is called the *anode*, and the electrode connected to the negative terminal of the battery is called the *cathode*.

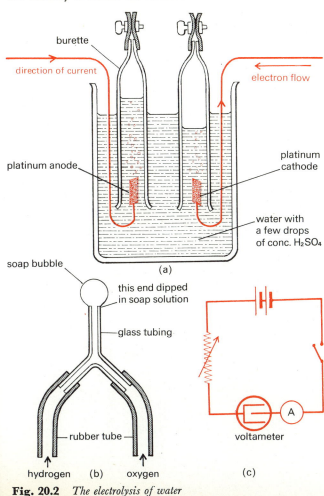

**Fig. 20.2**   *The electrolysis of water*

## 20.2   Two experiments

a.   *The electrolysis of water*

The apparatus used for studying the passage of electricity through an electrolyte is called a *voltameter*. A voltameter suitable for passing electricity through water is shown in Fig. 20.2. A few drops of concentrated sulphuric acid are added to the water to decrease its resistance and make it a better conductor. While the electric current is flowing, bubbles appear both at the cathode and at the anode. When the gases are tested it is found that oxygen is given off at the anode and hydrogen at the cathode. You will probably know from your chemistry that a mixture of hydrogen and oxygen is explosive. If the single stem of a Y-branch is dipped into soap solution and the two arms are attached to the burettes by means of rubber tubing, as in Fig. 20.2(b), a soap bubble containing a mixture of the two gases is formed. The bubble explodes if a lighted taper is held near it.

b.   *The electrolysis of copper sulphate*

Use the same electrical circuit for this experiment as that shown in Fig. 20.2(c), but substitute the voltameter illustrated in Fig. 20.1. If copper electrodes are used no bubbles will be observed either at the anode or the cathode. If, after some time the cathode and anode are examined it will be seen that the cathode is covered with a bright fresh deposit of copper and the anode appears dull. If the electrodes are weighed before and after the passage of the current it will be found that the cathode has gained in weight, and the anode has lost weight. Moreover, if the anode is pure copper, the gain in weight of the cathode will be the same as the loss in weight of the anode. The concentration of the copper sulphate solution is the same at the end of the experiment as it was at the beginning. It would appear that copper has been transferred from the anode to the cathode.

*QUESTION 1:* What sort of charge is carried by the copper as it travels towards the cathode?

### 20.3 The ionic theory

The passage of electricity through some chemicals in solution led to the belief that charged carriers of electricity exist within such solutions. These charged carriers are called *ions* (see also page 166) and are formed by the dissociation (splitting up) of molecules. For example, sodium chloride dissociates into positive sodium ions ($Na^+$) and negative chlorine ions ($Cl^-$). A sodium ion is a sodium atom which has lost an electron, and a chlorine ion is a chlorine atom which has gained an electron.

Copper ions are copper atoms which have lost two electrons ($Cu^{++}$). In electrolysis the positive copper ions travel to the negative cathode and, on arrival, get two electrons from the cathode (which in turn accepts the electrons from the negative terminal of the battery), and become copper atoms again. The copper atoms are deposited on the cathode and hence a deposit of copper gradually builds up. Negative sulphate ions travel towards the anode but they are not discharged there. Instead copper atoms from the anode go into solution.

*QUESTION 2:* Hydrogen ions are positively charged because they are hydrogen atoms which have lost an electron. (a) Which electrode will the hydrogen ions travel towards? (b) What do you think happens to the hydrogen ions when they arrive at the cathode?

The ions of metals and hydrogen are positively charged and hence always travel towards the cathode.

### 20.4 Faraday's laws of electrolysis

How does the mass of a substance deposited in electrolysis depend on the quantity of electricity which has flowed? This investigation may be carried out using the circuit of

Fig. 20.2(c). The mass of substance liberated at the cathode is determined by weighing the cathode before and after the passage of the current. (How could you determine the mass if the substance liberated is a gas?) Repeat the experiment for the same length of time with twice the current and then three times the current. Then keep the current constant and repeat the experiment doubling and trebling the time. From your results it may be concluded that

$$\text{mass deposited} \propto \text{current} \times \text{time}.$$

These results bear out our theory that the passage of electricity within the solution is due to charged particles. Referring to page 179 it follows that charge (coulombs) = current (amperes) × time (seconds). Thus if twice the charge flows then twice the mass is liberated. Faraday's first law of electrolysis may be stated as follows:

The mass of a substance liberated in electrolysis is proportional to the quantity of electricity (current × time) which has flowed.

### 20.5 Electroplating

Electroplating is a common application of electrolysis. The iron bumpers of a motor car are covered in this way, first with a nickel coating and then with a chromium coating, to prevent rust forming on them.

A silver-plated spoon is simply cheap metal coated with a surface of silver. Since metals are always deposited at the cathode the object to be coated with silver is made the cathode and the electrolyte is a silver salt. For nickel plating, a nickel salt is used. The process is illustrated in Fig. 20.3 (a) and (b).

### 20.6 Obtaining metals from their ores, i.e. refining

The deposit formed on the cathode during electrolysis is a very pure substance. Metals such as copper are often refined by electrolysis. In refining copper, the cathode is

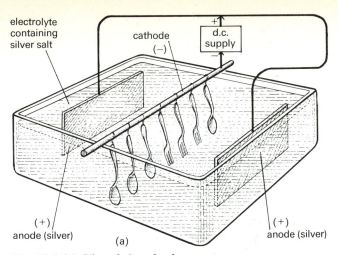

electrolyte containing silver salt

cathode (−)

+
d.c.
supply

(+) anode (silver)

(a)

(+) anode (silver)

**Fig. 20.3** (*a*) *Silver-plating of cutlery*

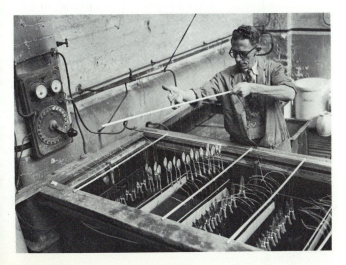

**Fig. 20.3** (*b*)   *The bars from which the cutlery is suspended are connected to the negative terminal of the supply*

a thin copper strip, the electrolyte a copper salt, and the anode impure copper. As electrolysis proceeds, the cathode becomes a thick piece of copper and the anode gradually disappears, any impurities fall to the bottom of the vessel containing the electrolyte.

## 20.7   Cells

The chemical effect of an electric current is used in a cell to convert chemical energy to electrical energy. One of the earliest cells, known as the *simple cell* consisted of a zinc plate and a copper plate in dilute sulphuric acid. A common form of cell, the Leclanché type cell, is illustrated in Fig. 20.4.

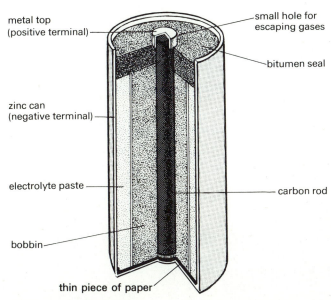

metal top (positive terminal)

small hole for escaping gases

bitumen seal

zinc can (negative terminal)

electrolyte paste

carbon rod

bobbin

thin piece of paper

**Fig. 20.4**   *A torch battery*

The positive terminal is the central carbon rod and the negative terminal is the zinc can which is the outside of the cell and forms the container for the other ingredients. The bobbin, which is the positive element, consists of manganese dioxide, carbon and ammonium chloride. These substances are moistened with water and formed under pressure around the carbon rod. The electrolyte is made of ammonium chloride and zinc chloride mixed with starch, flour and water to form a thick paste. This electrolytic layer is in contact both with the positive element (the bobbin) and the negative terminal (the zinc can). The energy is derived from the oxidation of the zinc by the manganese dioxide.

In the absence of manganese dioxide, hydrogen gas would form on the carbon and so prevent the cell from functioning. The manganese dioxide (which is called a depolarizer) oxidizes the hydrogen to water, thus preventing the formation of bubbles.

Figure 20.5 shows a cell composed of a fork, a knife and a lemon. The cell is being used to drive a motor. The fork and knife are the electrodes and the acid of the lemon is the electrolyte.

**Fig. 20.5** *A lemon motor* (Crown copyright, Science Museum, London)

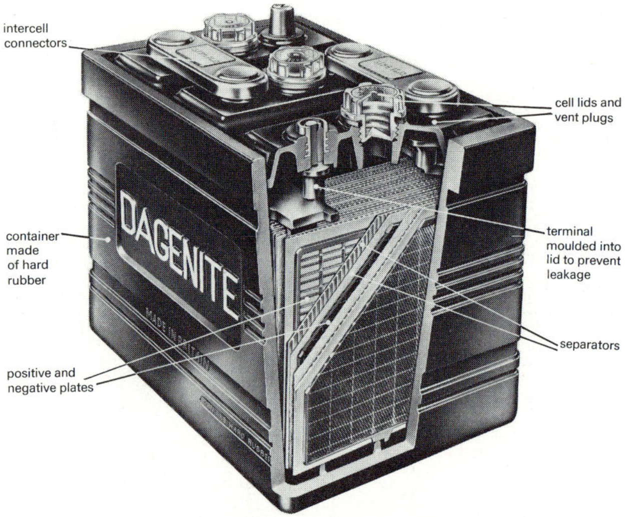

intercell connectors

cell lids and vent plugs

container made of hard rubber

terminal moulded into lid to prevent leakage

separators

positive and negative plates

**Fig. 20.6** *A car battery*

More familiar is the lead-acid accumulator (Fig. 20.6) used in motor cars. The negative terminal is a lead plate, and the positive terminal a lead peroxide plate. The electrolyte is dilute sulphuric acid.

This type of cell, known as a secondary cell, may be recharged by passing a current through it in the opposite

direction, and so can be used over and over again. Every time the self-starter of a car is used energy is taken from the battery (very large currents of the order of 100 amps are needed to drive the starter motor). When the car is running the dynamo is recharging the battery. On the dashboard of a motor car a red light shows up when the ignition is switched on. When the engine is running this light goes out, indicating that the dynamo is charging the battery. If the dynamo fails to function, as a result, for example, of a broken fan belt, the red light comes on. If this happens at night time it is advisable not to go too far before having it seen to. Why?

When a lead-acid accumulator is recharged, the current passed through it in the reverse direction reverses the chemical reaction that takes place during discharge. In charging the battery, electrical energy is converted into chemical energy and the relative density (see page 82) of the acid rises. When the accumulator is discharging, chemical energy is converted into electrical energy, and the relative density of the acid falls as lead sulphate is deposited on both plates. Thus, the state of charge or discharge can be determined by using a hydrometer (see page 107). When not in use, the electrical energy is stored in the form of chemical energy. It is important to keep the accumulator "topped up" with distilled water to make good the loss by evaporation.

Figure 20.7 shows the *Ariel III* space research satellite in the space chamber of the Royal Aircraft Establishment at Farnborough during the final stages of testing. The masses of tiny rectangles are *solar cells*. In this type of cell it is the light energy from the sun which is turned into electrical energy. The source of energy is the release of electrons as a result of this light energy falling on substances known as semiconductors.

**Fig. 20.7** *Research satellite* Ariel III. *The small rectangles are solar cells which convert the light energy from the sun into electrical energy for operating the equipment in the satellite*

**THINGS TO DO**

A. Connect two wires to a battery and hold them fairly close together in a saturated salt solution. Try and discover what is happening at each electrode. (Hint: salt is sodium chloride; there are negative chlorine ions in the solution.)

B. Nickel-plate a metal article, using an electrolyte made up from about 240g of nickel sulphate, 55g of nickel chloride and 30g of boric acid dissolved in 1000cm³ of water. Use a piece of nickel for the anode.

C. Use a paper clip to fasten one end of a piece of wire to a metal plate. Connect the other end to the negative terminal of a battery of about 12V (or three 4.5V flat batteries connected in series). Soak some white blotting paper in a solution of potassium iodide which has had a little starch added. Put the blotting paper on top of the metal plate. Write on the blotting paper with a copper wire connected to the positive end of the battery. The iodine liberated at the anode forms a dark compound with starch.

D. Find out how electroplating is used in the construction of gramophone records.

1. Suggest how the chemical effect of an electric current may be used to measure the magnitude of an electric current in amps.

2. "Metals or hydrogen are always deposited at the cathode." What does this statement indicate about the charge on metallic ions and hydrogen ions?

3. Why is pure water a bad conductor of electricity? Why do a few drops of acid turn it into a good conductor?

4. Some liquids conduct electricity, others do not. What is present in electrolytes that is not present in non-conducting liquids?

5. In Fig. 20.8 a voltameter is connected between P and Q.

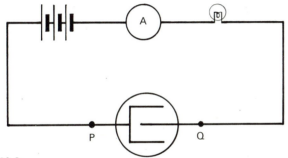

**Fig. 20.8**

a. In the electrolysis of copper sulphate using copper electrodes, is the copper deposited on the electrode connected to P or the electrode connected to Q? What change takes place at the other electrode? What will happen to the reading on the ammeter and the brightness of the bulb if the plates are moved closer together? What will happen if the plates are allowed to touch?

b. Draw the apparatus that you would connect between P and Q for the electrolysis of water. Show clearly which gas is given off at P and which gas at Q. Are the gases given off at the same rate? Were the gas atoms part of a charged particle before being liberated? What charge would you associate with each gas? Show on your diagram the direction of movement of the positive and negative ions in the electrolyte, and of the electrons in the connecting wires.

c. Positive charge is flowing in the electrolyte so we would expect positive charge to be flowing in the wires. Comment on this statement.

6. The mass of substance liberated by the passage of 1 coulomb of electricity is called the electrochemical equivalent (e.c.e.).

a. Describe an experiment you would do to determine the e.c.e. of copper.

b. If a current of 1.5A deposited 0.89g of copper in 30min, calculate the e.c.e. of copper.

c. A steady current is passed for 1 hour through a copper voltameter with copper electrodes. If 1.00g of copper is deposited, what is the magnitude of the current?

7. Use the information given in question 6b to draw a graph of mass of copper deposited against time over a period of 2 hours. State clearly any assumptions you make in drawing your graph. From your graph determine (a) the time for 1.00g of copper to be deposited, (b) the mass deposited in 75 minutes.

8. Fig. 20.8 shows a voltameter in series with a battery, an ammeter and a bulb. The voltameter has a piece of steel as one electrode which is to be electroplated.

(a) Redraw the diagram and mark clearly (i) the anode and cathode (ii) the piece of steel.

(b) What difference would there be in the rate of deposit of copper on the steel if the bulb were replaced by a bulb of much higher resistance?

(c) In what way is the current flow in the electrolyte different from that in the connecting wires?

(d) Explain why water is not an electrolyte but water with a little acid added to it is an electrolyte.

9. In problems where we have calculated currents flowing in circuits we have assumed that the cell in the circuit has no resistance. Is this justified? Give reasons for your answer.

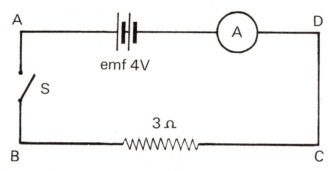

**Fig. 20.9**

10. This is a hard question. Think carefully!

The ammeter in the circuit in Fig. 20.9 reads 1A when the switch is closed.

a. What is the p.d. between B and C?
b. What is the p.d. between A and C?
c. What is the p.d. between A and D?
d. When the switch is open the p.d. between A and D (known as the terminal p.d.) is 4V, but when S is closed it is less than 4V. Can you explain why? (Hint: think about the answer to question 9.)
e. Fill in the gaps in the following sentences:
(i) The e.m.f. of a cell is the potential difference (voltage) across its terminals when no .............. is flowing.
(ii) The potential difference across the terminals of a cell falls when a .............. is flowing through it, because of its internal ....................

11. A cell has a potential difference of 6V across its terminals when it is not connected in a circuit. When it is connected in a circuit and delivers a current the potential difference across its terminals falls because of the internal resistance of the cell.

a. What do you think is meant by the internal resistance of the cell?
b. Why does a cell have internal resistance?
c. If the potential difference across its terminals falls by 1V when it delivers a current of 2A, what is the internal resistance?

12. A high resistance voltmeter connected across the terminals of a cell reads 1.5V. When a resistor of resistance $2\Omega$ is connected across the cell the voltmeter reading falls to 1.0V.

a. What is the e.m.f. of the cell?
b. What current flows in the circuit when a $2\Omega$ resistor is connected across the cell?
c. How many joules of energy does each coulomb give up as it goes round the circuit?
d. How much energy does a coulomb lose as it passes through the $2\Omega$ resistor?
e. What is the internal resistance of the cell?

13. A cell has an internal resistance of $1\Omega$ and an e.m.f. of 6V. A resistor of $5\Omega$ is connected across the terminals of the cell. What will a high resistance voltmeter connected across the terminals of the cell read?

# HEAT ENERGY

## Part Four

*Solar furnace in Massachusetts USA. Heliostat (right) collects sunlight. Concentrator (left) focuses it onto the test chamber (centre).*

two panes of glass with space between

**Fig. 21.1** *(a) Double glazing prevents the loss of heat through glass (b) Overall warmth from a room heater*

# THE TRANSMISSION OF HEAT

## 21.1  How does heat travel?

In Fig. 21.1 we see two typical examples of advertisements that confront us in newspapers and on television most days of our lives. One is concerned with the selling of appliances to provide heat in our homes, and the other with a means of keeping that heat in the home.

The technologists who developed both of these appliances had to consider the same question — how does heat travel? Only by knowing the answer to this question was it possible for them to find ways of spreading heat efficiently all over a room or house, and ensuring that as little heat as possible is lost.

The transfer of heat always takes place when bodies are at different temperatures. Heat energy is transferred from objects at high temperatures to objects at lower temperatures. During this transfer it may have to travel through solids, liquids, gases or empty space.

To find out how heat travels through different mediums try the following experiments:

a.  Hold one end of a metal spoon in boiling water and the other end between the fingers.

b.  Place a glass tube (or empty ball point pen casing) upright in the centre of a glass beaker containing cold water. Drop a crystal of potassium permanganate (a colouring agent) into the water through this tube, and then carefully remove the tube so that very little colour is imparted to the water. Gently heat the beaker from underneath.

c.  Get someone to stand to one side of an electric fire (the type that has an element and a reflector) so that he cannot see or feel it directly. Another person then takes a large mirror and holds it in such a position that the first person can see the fire in the mirror.

In the first experiment the end of the metal spoon soon becomes too hot to hold. The heat has been transferred along the metal by the method known as *conduction*.

In the second experiment the water that is nearest to the heat, at the bottom of the beaker, is seen to rise, and

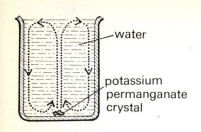

**Fig. 21.2** *Convection currents in water*

cold water from the top flows downwards to take its place (see Fig. 21.2). This process is continuous, setting up a constant circulation of water clearly shown by the colouring of the currents by the dye. Eventually the circulating water reaches boiling point. The heat has been transferred through the water by the method known as *convection*.

In the third experiment the person standing to the side of the fire can feel its heat as soon as the mirror is directed towards him. Previously he had felt no heat. The heat has been transferred by the method known as *radiation*.

These three methods of heat travel are illustrated in the following problem. Imagine the parcel to be heat.

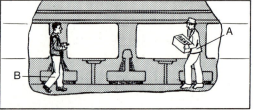
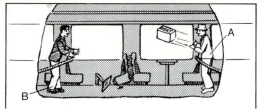

**Fig. 21.3** *Drawings to illustrate the three methods of heat travel*

QUESTION 1: (Fig. 21.3) Two friends A and B get into a railway carriage by two different doors, and are half a carriage apart. The carriage has a central gangway. Before B gets off at the next station A needs to give him a parcel. How do you think A will have to pass the parcel to B under the following conditions: (a) when the carriage is packed with people standing side by side along the central gangway, (b) when the carriage is empty, (c) when the carriage is empty but an area of the gangway between them has been roped off so that they cannot reach one another? How are these methods similar to the transfer of heat?

## 21.2 Explanation of heat transfer

Conduction is the flow of heat energy through a body which is not at a uniform temperature, the heat flowing from places of higher to places of lower temperature without the body as a whole moving.

The transfer of energy by conduction results from energy being passed first to the molecules nearest the source of heat, and then to other molecules along the length of the body. There is no movement of the molecules from their average positions.

This is the method by which heat travels through solids, but it can also occur in liquids and gases.

In the case of metals, conduction mainly involves the movement of free electrons (see Chapter 17). Whilst moving, these electrons bump against the metal molecules and cause them to vibrate. By this means the energy is transferred much more rapidly.

Convection is the transfer of heat by the circulation of the matter which is experiencing the temperature differences.

The molecules nearest the source of heat increase their vibrations and therefore spread further apart. Thus the material in this region is less dense and it rises, being replaced by cooler and denser material which flows downwards and inwards. This circulation of the material is called a convection current. It is the usual method by which heat travels through liquids and gases, and it involves an actual movement of the molecules.

Radiation is the transfer of heat by electromagnetic waves.

Radiant heat can only be detected as heat when the

energy strikes and is absorbed by an object in its path. It does not heat the air through which it travels. It requires no molecules for its transfer and is the method by which the sun's heat reaches the earth. Radiant heat is greatest when the temperature of the heat source is very high, white- or red-hot, although even objects at relatively low temperatures always radiate a little heat.

QUESTION 2: For which method of heat travel are the following heating appliances (or parts of appliances) designed, (a) an electric cooker plate, (b) the polished back plate of an electric fire, (c) the vents at the top of a solid fuel room heater, (d) the radiator in a central heating system?

## 21.3   Good and bad conductors

QUESTION 3: If you were blindfolded how would you distinguish between a metal surface and a planed wood surface?

The apparatus in Fig. 21.4 can be used in conjunction with a Bunsen burner to show the difference between

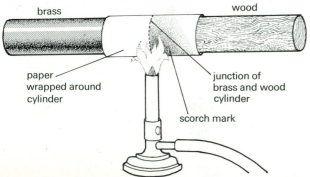

**Fig. 21.4**  *Experiment to determine which is the better conductor of heat, brass or wood*

wood and metal in their conduction of heat. The paper is scorched where it covers the wood, but not where it covers the brass.

The paper over the brass does not burn because the heat is conducted away quickly by the brass. It is only conducted slowly away from the paper by the wood and therefore the heat energy is built up at this spot. Such a build-up may cause the material to burn when the temperature reaches ignition point.

Generally, metals are better conductors of heat than other materials, and some metals are better conductors than others.

Materials that resist the passage of heat along them are *bad conductors* or *heat insulators*. Materials that transfer the heat quickly are termed *good conductors*.

## 21.4   Experiment to compare conductivity in different materials

This experiment is illustrated in Fig. 21.5. Rods of different materials, e.g. copper, iron, brass, lead and glass, are bound together at one end with copper wire. Melted paraffin wax (candle wax) is brushed onto each rod and balls of lead shot are placed at regular intervals along the underside of each rod; when the wax solidifies the lead shot is firmly held. Heat is then applied to the bound ends of the rods.

As the heat is conducted along each rod, so the wax melts, releasing the lead shot at that point. Thus we can compare the conductivity of each material by the extent to which the wax has melted along each rod. Of the materials listed copper is found to be the best conductor. After copper come brass, iron, lead and glass, in that order.

QUESTION 4: Why were the ends of the rods bound with *copper* wire?

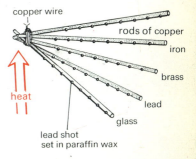

**Fig. 21.5**  *Experiment to compare the conductivity of different materials*

F.O.P.—P

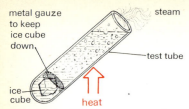

**Fig. 21.6** *Water is a poor conductor of heat*

In comparing the conductivity of materials, air is sometimes taken as the standard unit and other materials compared with it. A good conductor has a high conductivity.

The poor conductivity of water is demonstrated clearly by the experiment shown in Fig. 21.6. The water can be boiled for a considerable time before the ice even starts to melt.

### 21.5 Everyday uses of good and poor conductors

It is quite impossible to list all the uses made of materials because of their high and low conductivity. You will find a few common uses in the table opposite, and illustrations of some of these in Figs. 21.7 to 21.10. Try to extend this list yourself.

**Fig. 21.7** *Cooking utensils are made of good conductors, such as aluminium*

**Fig. 21.8** *Metal fins on a 2-stroke engine provide a large surface for cooling*

**Fig. 21.9** *In winter animals develop layers of fat and grow thick coats*

QUESTION 5: (a) Would you be warmer if you wore your fur coat inside out? If so, why? (b) Gardeners worry about frost killing off young shoots. Why is it that they do not worry about plants buried in snow?

**Fig. 21.10** *Vermiculite, a natural mineral used for insulation, contains small pockets of still air*

214

# APPLICATIONS OF THE METHODS OF HEAT TRAVEL

| | MATERIAL | APPLICATIONS |
|---|---|---|
| **CONDUCTION**<br>Good conductors | Copper | Heating appliances, boilers, (Davy) safety lamps, fire tubes in steam engines, tip of soldering iron |
| | Aluminium | Cooking utensils |
| | Iron | Shaped into fins to give a large surface area for cooling in air-cooled engines |
| Bad conductors<br>(i.e. heat insu-<br>lators) | Felt, glass fibre | Lining of roofs, lagging to pipes and boilers |
| | Asbestos | Lining of roofs, lagging to pipes and boilers, protective clothing |
| | Paper | Wrapping |
| | Plastic | Handles to cooking utensils |
| | Wood | Handles to cooking utensils, wooden spoons, table mats |
| | Cork | Table mats |
| | Straw | Roofing, hay-box used in camping |
| | Glass, porcelain | Crockery (little heat transfer to hands and lips) |
| | Fat | In the skin and around organs of animals and humans |
| | Bricks, plaster | Housing |
| | Air | Woollen clothing (fibres of wool have many air spaces), feathers, fur, eiderdowns, string vests, cellular blankets, cavity walls, double glazing |
| | Vermiculite | Insulation |
| **CONVECTION** | In water | Hot water systems<br>Central heating systems<br>Car radiators |
| | In air | Heating<br>Ventilation in homes, mines, tunnels<br>Gliding on thermals<br>Hot air balloons<br>Refrigeration systems |
| **RADIATION**<br>Good absorbers<br>Bad absorbers<br>(i.e. good reflectors) | Soot<br>White clothing<br>Whitewash<br>Polished metal<br>Aluminium foil | Mixed with garden soil to give warmth<br>Tropical countries<br>Houses in hot countries, on glass roofs of buildings to maintain constant temperature<br>Electric and gas fires, solar furnaces<br>Cooking, food storage |
| Good radiators<br><br>Bad radiators | Black painted metal<br>Shiny metal surfaces<br>Whitewash | Car radiator, fins on refrigerator condenser unit<br><br>Teapots, kettles (to keep contents hot)<br>On glass roofs to maintain constant temperature |

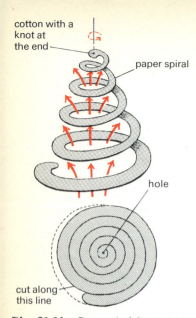

cotton with a knot at the end

paper spiral

hole

cut along this line

**Fig. 21.11** *Paper spiral detects rising convection currents*

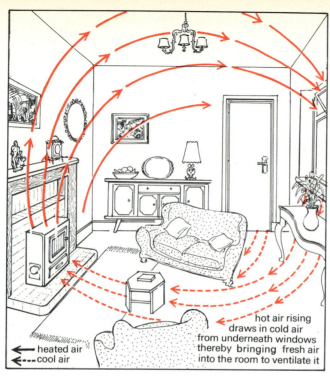

hot air rising draws in cold air from underneath windows thereby bringing fresh air into the room to ventilate it

← heated air
← --- cool air

**Fig. 21.12** *Convection currents help to ventilate a room*

### 21.6 Convection in air

We have already seen an experiment that shows how convection takes place in water. Convection currents can be detected in air by observing their effect on smoke from a candle or taper, or on a circular piece of paper cut to form a spiral and suspended from a thread (Fig. 21.11).

Convection currents are very important in the heating and ventilation of the home (see table on previous page and Fig. 21.12). Try using the paper spiral to trace convection currents in your rooms at home.

216

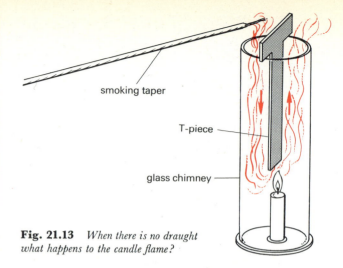

smoking taper

T-piece

glass chimney

**Fig. 21.13** *When there is no draught what happens to the candle flame?*

The significance of convection currents is most strikingly shown in the experiment illustrated in Fig. 21.13. The candle is placed at the bottom of a gas jar or a glass chimney (which is sealed at the bottom by a layer of grease). All attempts at keeping the candle alight fail until the T-shaped piece of card is placed in the cylinder, the best position for this being just to one side of the candle. The candle then continues to burn, and if a smoking taper is held at the top of the cylinder the convection current which has been set up is clearly seen. This convection current brings in the necessary "fresh" air, containing the oxygen, which is required for the combustion of the candle.

*QUESTION 6:* If a room were to be made completely draughtproof what would be the effect on (a) a coal fire, (b) the people in the room when all windows and doors were shut?

On an altogether larger scale, convection currents in

**Fig. 21.14** *Convective cells in cloud structure over the ocean—photographed from space*

**Fig. 21.15** *Onshore breezes in daytime; offshore breezes at night*

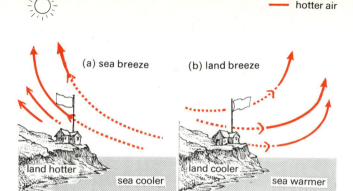

····· cooler air
——— hotter air

(a) sea breeze    (b) land breeze

land hotter    sea cooler    land cooler    sea warmer

air are responsible for cloud formation, coastal breezes and the earth's wind systems (see Figs 21.14 and 21.15). Gliders rely upon convection currents of warm air which they call thermals. These help them to rise (Fig. 21.16).

**Fig. 21.16** *Glider rising on thermals*

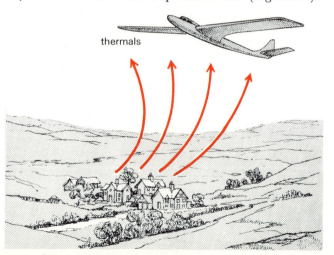

thermals

## 21.7 Convection in water

In the home, convection forms the basis for the design of all hot water supplies and central heating systems (see Fig. 21.17).

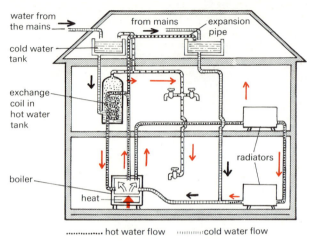

water from the mains    from mains    expansion pipe

cold water tank

exchange coil in hot water tank

boiler    radiators

heat

············· hot water flow    ······ cold water flow

**Fig. 21.17** *Hot water and central heating system of a house*

QUESTION 7: Why is it necessary to have (a) the hot water cylinder higher than the boiler, (b) the pipes carrying cold or cooler water entering at the bottom of the cylinder and boiler?

In central heating systems the water in the system is kept entirely separate from the normal hot water supply. The exchange of heat takes place in the hot water cylinder by a means of an exchange coil at the centre of the cylinder. This ensures that when hot water is drawn off this will not greatly affect the temperature of the central heating system.

Look at Fig. 21.18 to see how convection currents are set up within a car radiator.

On a large scale, convection currents in water account

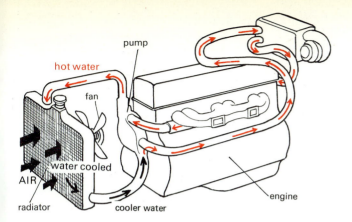

**Fig. 21.18** *Water removes heat from car engine, rises and circulates within radiator system*

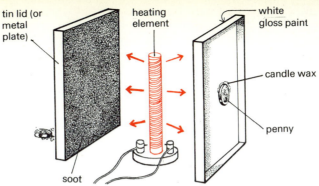

**Fig. 21.19** *Experiment to determine what surfaces are good absorbers of radiant heat. The penny on the sooted surface falls off first.*

for some of our ocean currents — the Gulf stream, for example, which warms our northern shores. Such currents have an important effect on climate in various regions.

## 21.8 Radiation: good and poor absorbers

Our first experiment (page 212) showed us that radiated heat can be reflected from a shiny surface in the same way as light. The effect of other surfaces on radiated heat can be explored using the apparatus illustrated in Fig. 21.19.

Flat pieces of metal, such as tin lids, are coated on one side only with different surfaces: matt black paint, black gloss paint, whitewash, white gloss paint, for example. Coins are then stuck by means of candle wax to the other side of the metal (or heat-sensitive paper may be placed against this side). The pieces of metal are then placed at equal distances from an electric heating element.

By the melting of the wax and the releasing of the coins, we can detect which surface is the best absorber of heat.

Illustrated in Fig. 21.20 is another experiment which will enable you to investigate the absorption of radiation.

First, hold your hand in the position shown and feel the heat from the heating element coming through the hole.

Next, place a very thin sheet of aluminium foil over the back of your hand and again put your hand in the position shown. (The foil will stick if you lick it first.)

Finally, paint a mixture of lampblack and methylated spirit on the foil, which is still stuck to your hand, and hold your hand near the hole again.

From the above experiments we reach the following conclusions:

Black or dark surfaces are better absorbers of radiant heat than white or light surfaces.

Rough surfaces are better absorbers of radiant heat than polished, smooth surfaces of the same colour.

*QUESTION 8:* (a) What type of surface would you need for a container of petrol in a hot country, and why? (b) Why does snow under fallen leaves melt more quickly than snow exposed to sunlight?

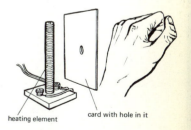

**Fig. 21.20** *The heat from the heating element passes through the hole and can be felt on the back of the hand*

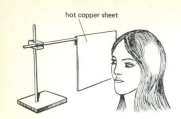

Fig. 21.21 *When the cheek is near the blackened surface of the hot copper plate, more heat can be felt than when it is near the shiny surface*

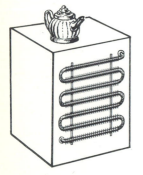

Fig. 21.22 *Silver teapot keeps contents hot. Cooling fins on refrigerators are painted black.*

## 21.9 Good and poor radiators

In the experiment illustrated in Fig. 21.21 a copper sheet which is polished on one side and blackened on the other is heated and positioned in a stand, as shown. Bring your cheek up to each side of the sheet in turn. This and similar experiments lead to the general conclusion that:

Dark rough surfaces are good radiators of radiant heat. Light smooth surfaces are poor radiators of radiant heat.

Some everyday uses of good and poor absorbers and good and poor radiators of heat are listed in the table on page 215 and illustrated in Figs 21.22 and 21.23.

*QUESTION 9:* If light surfaces are poor radiators of heat why is it that radiators in central heating systems are not painted black rather than white?

## 21.10 Infra-red rays

A *thermopile*, an instrument made up of a series of thermocouples (see page 236), is used to detect radiant heat. When this instrument is moved along in the visible spectrum of light towards the red end it gives a reading on a scale. This reading is greatest in the region beyond the red end of the spectrum, known as the infra-red region. The range of electromagnetic waves (see page 50) from this region make up heat radiation as we know it.

You have probably seen or read about infra-red grills for cooking, and infra-red lamps which are used to relieve muscular pain. Because infra-red lamps give out invisible rays they can be used in darkness as signalling lamps; their signals cannot be observed unless the correct receiver is used. Aeroplanes used such lamps during the last war. Infra-red rays can also be used for "seeing" objects. Look at the print from an evapograph camera on page 251. These rays are not easily scattered and therefore photographs taken on plates sensitive to infra-red

Fig. 21.23 *Houses in hot countries are painted white: less heat is absorbed and interiors remain cool*

radiation often reveal details that the human eye cannot detect. Such cameras respond to heat radiation emitted by any object that is at a temperature above absolute zero (page 238). They can even detect the passage of a nuclear submarine 24 hours after it has travelled through a particular section of water.

## 21.11    Reflection of heat radiation
Bad absorbers of heat radiation are good reflectors (see the experiment described on page 219). Nowadays it is possible to reflect heat radiation from the sun and to make direct use of this solar heat. The frontispiece to this section shows one such solar furnace where the heat is concentrated to a point by a large concave reflecting surface. Solar energy (see page 401) can be used to cook food, heat the water supply and provide central heating in a house. In the future the sun's rays may provide the peoples of the world with much of the energy they require. This is discussed more fully in Chapter 32. Figure 21.24 shows an apparatus that demonstrates in a spectacular manner the result of the reflection of heat rays. The match lights when the heat rays are concentrated onto it by means of the curved reflectors.

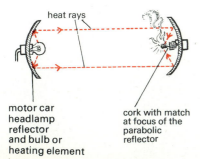

heat rays

motor car
headlamp
reflector
and bulb or
heating element

cork with match
at focus of the
parabolic
reflector

**Fig. 21.24**  *The reflection and focusing of heat rays causes the match to light when the bulb is switched on*

## 21.12    Radiant heat from the sun
The sun's heat reaches the earth purely by heat radiation. Some is absorbed by the earth's atmosphere, but most of it is absorbed at the earth's surface, causing a rise in temperature. This temperature rise occurs as long as the earth is not losing radiant heat at a faster rate than it is receiving it. Naturally at night the earth's temperature drops as it receives no heat from the sun but continues to radiate the heat that it has already received.

*QUESTION 10:* Are late frosts more likely to occur when it is a cloudy or cloudless night? Explain your answer.

Perhaps you have noticed that when people buy a house they always prefer one where the main windows face south. The sunlight comes into the rooms for most of the day and the house is warmed. The following experiment shows us why this happens.

## 21.13    Experiment to find out if glass transmits radiant heat
First, stand in front of a closed window when the sunlight is directly upon it. Then stand in front of a red hot fire holding a piece of glass between yourself and the fire. In the first case you felt the warmth of the sun through the glass, in the second very little heat reached you from the fire.

It appears that glass allows through the radiant heat from very hot bodies such as the sun, but does not allow through the heat that comes from less hot objects such as the fire. The infra-red rays of shortest wave-length are transmitted by the glass. But objects in the room, or plants in the greenhouse, which absorb this radiant heat give off infra-red rays of longer wave-length. These rays of longer wave-length do not get through the glass. Thus most of the heat is kept within the room or greenhouse and the temperature rises quite considerably.

Obviously in factories and offices with large windows, such a build-up of temperature can be unbearable (particularly in hot countries). Window glass is now being produced which does not noticeably cut down visible light, but which absorbs or reflects a larger proportion of infra-red rays than normal glass. The absorbent power of such glass in increased by the introduction of metallic particles, approximately 0.0001cm in diameter, below the surface during the manufacturing process. Types of heat-rejecting glass tend to be tinted, but even the shade of the tint can be controlled to suit the purpose of the building.

## 21.14 Vacuum flask (invented in 1892 by Sir James Dewar)

We all know that a vacuum flask keeps hot drinks hot and cold drinks cold. It does this by drastically reducing the heat exchange between its contents and their surroundings. Its design is such that all three methods of heat travel are reduced to a minimum (Fig. 21.25).

**Fig. 21.25**  *A vacuum flask*

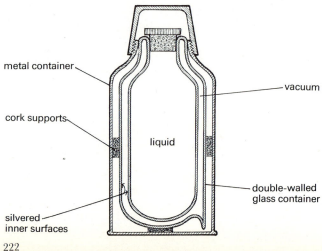

metal container

vacuum

cork supports

liquid

double-walled glass container

silvered inner surfaces

## 21.15 Problems of heat transfer in space probes and fast-moving aircraft

a. *Space vehicles*
On leaving and re-entering the earth's atmosphere a space capsule is subject to intense heat because of the frictional forces involved. The nose of the capsule is therefore made of a bad conductor, and has a heat shield made of a honeycombed coating of glass resin. On re-entry from space, the command module is turned so that its heat shield points towards the earth. The resin melts away taking the heat energy with it away from the capsule. The angle of re-entry of a spacecraft must be accurately calculated so that the frictional heat does not exceed that for which the shield is designed. *Apollo 11*'s heat shield was subjected to a temperature of 5000°C. The burning effect on such a shield is seen in Fig. 21.26.

b. *Spacesuits*
Astronauts also are subjected to extremes of temperature and their spacesuits are designed and tested to make sure that their bodies will be protected from the heat and cold (Fig. 21.27). Spacesuits designed for use on the moon have to protect the astronaut from temperatures ranging from 100°C during the lunar day to −150°C during the lunar night.

c. *Supersonic aircraft*
The surfaces of supersonic aircraft become heated as a result of friction with the atmosphere. At twice the speed of sound the temperature of the body surface

Fig. 21.26 *The scorched heat shield on a space capsule*

of the aircraft is approximately 120°C. In order to keep the passengers cool some method of heat transfer is required. In the case of *Concorde* (Fig. 21.28) the fuel in the fuel system is at a lower temperature than the air in the air-conditioning system. The fuel is used to remove excess heat from the air and thus the cabin of the aircraft

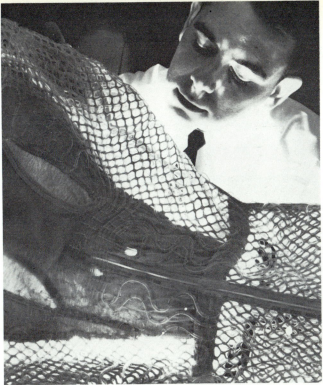

Fig. 21.27 *The water-cooled under-garment for astronauts. Water in the tubing absorbs heat from the body*

is kept cool. It also removes the heat from the hydraulic fluid and the lubricating oil for the engine.

Another precaution against frictional heating is taken at the front of the aircraft where a visor acts as a heat shield. This is raised in front of the windscreen during supersonic flight.

*QUESTION 12:* "It keeps the heat in and the cold out." This is a statement made by a pupil about a material used to keep an object at an even temperature. What is wrong with this statement?

**Fig. 21.28** Concorde 002

A. Use a paper saucepan to heat water. Obtain either an unwaxed paper cup, or make a paper box out of a square of paper. Put water into the cup until it is half full. Heat the bottom of the cup over the flame of a burner. The water boils but the cup does not burn.

B. Construct your own form of container which will keep its contents either hot or cold. The sort of materials you can use are a jam jar, a tin can, aluminium foil, cork mats on which to stand the jar inside the tin can, fibre glass or any other form of insulation to pack between the jar and the sides of the can, and cork or cotton wool plugs to act as lids.

Put a hot liquid in your container and compare its cooling rate with that of an equal quantity of liquid in a similar non-insulated jam jar. This will give you some idea of the effectiveness of your container.

C. Design an experiment to test the insulative properties of different fabrics.

D. Make your own schlieren effect. Place a light source on one side of a darkened room. Direct the light towards a plain wall, and between the light and the wall position any form of flame. Observe the patterns of the convection currents on the wall. Schlieren is a German word meaning "streaks". The schlieren technique is based on the fact that the refractivity of liquids and gases is altered if their density varies. Variations in density can occur, for instance, with temperature changes or pressure changes.

E. Construct your own convection toys as illustrated in 21.11, and form some of them into mobiles.

F. Find out what is meant by the "U" factor of different building materials? Give some examples. (These can be found in leaflets advertising different industrial products.)

G. Find out the best method of lagging a hot water tank, and what effect the filling of cavity walls with plastic foam would have on fuel bills.

H. Trace the hot water pipes around your own house, and map out the complete hot water system.

I. Either make up an illustrated catalogue of materials you would use in the design of a house that was to be well insulated in winter, or design a year's wardrobe of clothes that would keep you cool in summer and warm in winter (include illustrations or samples of materials).

J. Find out where the largest solar furnace is established. Solar energy is likely to become a major source of energy in the future. Describe some of the plans that scientists have in mind for this form of energy.

K. Find out about the structure of a sunshine recorder and how it works.

1. List the advantages and disadvantages of night storage heaters. What type of heating would you suggest should be installed to meet the demands and income of (a) a young married couple, both of whom work, living in a small house, (b) an elderly retired lady living in a bungalow, (c) a family living in a six-bedroomed house built 70 years ago?

2. Explain the following statements:
   a. In winter some animals develop thicker coats and a layer of fat under the skin.
   b. String vests, although they appear to be full of holes, keep the body very warm.
   c. A shiny metal teapot keeps the tea hotter than a matt brown teapot.
   d. The outsides of houses in hot countries are painted white in order to keep the interiors cool.
   e. A paper cup containing water will not catch fire when heated from below by a naked flame.

3. Draw a simple diagram of a vacuum flask. State the purpose of the following parts: (a) silvered surfaces, (b) vacuum, (c) double-walled container, (d) cork supports and cap.

4. Write down the names of three good conductors of heat and three poor conductors of heat. In each case state an everyday use for the material.

5. Describe experiments you would use to demonstrate convection currents (a) in water, (b) in air.

6. A garden chair, made from nylon cloth on an aluminium frame, was left on a beach all morning in very hot sun. A person lying a little way from the chair observed that there was a heat haze (such a haze indicates that hot air is rising) over the black painted arms. This haze was not visible over any other part of the chair. Explain this phenomenon.

7. Figure 21.29 illustrates a thermoscope, an instrument used to detect radiant heat. Explain how it works.

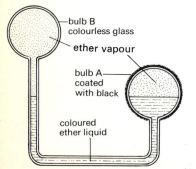

bulb B colourless glass

ether vapour

bulb A coated with black

coloured ether liquid

**Fig. 21.29** *A thermoscope*

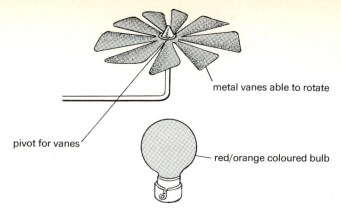

metal vanes able to rotate

pivot for vanes

red/orange coloured bulb

**Fig. 21.30**

8. Figure 21.30 illustrates the apparatus which is used to create the flickering effect behind artificial coal. Explain how it works.

9. By what methods do the following transmit their heat, (a) an electric blanket, (b) the electric element in a kettle, (c) the sun, (d) a radiator, (e) an electric fire?

10. By means of a diagram show the positioning of a reflector in an electric wall heater which is to be placed near the ceiling in a bathroom. Draw in the heat rays.

11. Why is aluminium foil used in cooking joints of meat? Why will a joint cooked by this method be more tender and moist?

12. Explain the following statements:
    a. A radiator will blacken the wall to which it is attached unless a shelf is built over it.
    b. An advertisement claiming that a certain product will keep out the cold is scientifically incorrect.
    c. A frost kills off young shoots, but a covering of snow helps them to survive the winter.

d. Projectionists who use a hand to stop a fast rotating plastic reel usually get it burnt. If the reel were metal there would be less chance of a burn.
e. The double glazing of windows helps towards reducing fuel bills.

13. (a) What is heat? (b) What is meant by the kinetic theory of heat? (c) Explain why it is impossible for heat transmission by convection to occur in a solid. (d) Describe an experiment to demonstrate that four metal rods made from aluminium, iron, copper and lead have different conductivities.

14. Explain the following statements:
a. A glass garden frame may be used to trap heat.
b. Dew and frost are more likely to form during a cloudless night than during a cloudy one.
c. Heating elements are always placed at the bottom of electric kettles.
d. During a total eclipse of the sun the temperature of the air drops.

15. Describe an experiment to show that water is a very poor conductor of heat.

16. Draw the convection currents that will be set up in a room with an open fire, one door and one window, all positioned on different walls.

17. In what ways does heat travel from an electric light bulb filament to the glass envelope (assume that there is a vacuum inside)? How is the heat transferred from the glass envelope to the room?

18. Describe (a) the best radiating surface, (b) the worst heat-absorbing surface.

19. Describe an experiment which can be used in the laboratory to illustrate the action of a solar furnace.

20. Draw a diagram showing how the transmission of heat by convection is applied to (a) the hot water system and (b) the central heating system of a house.

21. When the immersion heater (Fig. 21.31) is switched on the paddle wheel rotates. (a) Why does the paddle wheel rotate? (b) Why is there a short delay between switching on the immersion heater and the rotating of the paddle wheel? (c) In which direction will the paddle wheel rotate? (d) Without touching the paddle wheel, how could you make it rotate in the opposite direction?

22. When the wire protruding from a delicate electronic component needs to be soldered it is often held by a pair of pliers, and the pliers grip the wire between the component and the solder joint. Why are the pliers used? Would a wooden clip be as effective?

23. Figure 21.32 is a photograph of Skylab, which is a laboratory orbiting clear of the earth's atmosphere. It will continue orbiting for about 9 years. A solar panel matching the one on the right was lost during the launching. At the far end there is a telescope. (a) What is the purpose of the solar panel? What energy changes take place in the solar panel? (b) What are the advantages of having a telescope above the earth's atmosphere?

**Fig. 21.31**

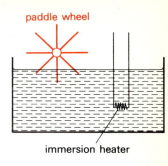

paddle wheel

immersion heater

**Fig. 21.32** *Skylab. The telescope is at the far end (with paddle wheel power panels). Astronomers hope to find out about the origins and make-up of stars, nebulae and interstellar space*

## CHAPTER 22

# EXPANSION

**Fig. 22.1** *Weights and pulleys on the overhead electric cables of a railway line*

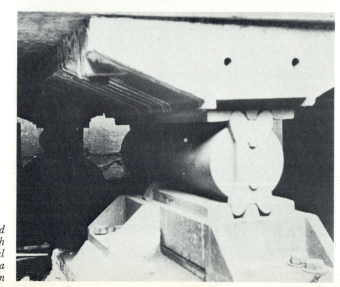

**Fig. 22.2** *Roller bearings at the end of the 22.2m span of the Hammersmith flyover. These rollers and a central joint in the span allow for a movement of 35cm*

## 22.1 How does heat affect structures?

In Fig. 22.1 we can see weights hung on the end of the electric cables of an electrified railway, and in Fig. 22.2 rollers under the end of a bridge. What is the purpose of the weights and the rollers?

We can find the answer to that question by doing an experiment in the laboratory (Fig. 22.3).

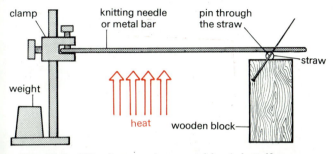

**Fig. 22.3** *What happens when a metal bar is heated?*

In this experiment the knitting needle represents a bridge clamped at one end, with the other end on a drinking-straw roller. A pin is stuck vertically downwards through the straw to serve as a pointer. The needle is then subjected to a change in temperature by applying heat from a Bunsen burner at its centre (a candle will do).

It is quite surprising to see how little heat is required before the pointer moves over to the right (when viewed as seen in the diagram). Any rolling movement of the pointer can only result from a change in length of the needle. As the needle is clamped at one end, movement can only occur at the pointer end. In this case a movement of the pointer to the right shows that the needle is increasing in length. This is confirmed when the needle is allowed to cool. The pointer is then seen to move back in the opposite direction, showing that the needle is decreasing in length, i.e. going back to its original length.

*QUESTION 1:* (a) What would happen to a bridge on a very hot day if it was firmly fixed at both ends? (b) What movement of the weights at the end of the electrical cables takes place (i) on a hot day, (ii) on a cold day?

The increase in size of a substance when it is heated is called *expansion* and the decrease in size when it is cooled (heat is removed) is called *contraction*.

## 22.2 When a substance is heated, does it expand in length only?

The apparatus illustrated in Fig. 22.4 (the ball and ring) helps us to answer this question. At normal room temperatures the ball easily slips through the ring. After being heated it will no longer go through the ring, no matter in what direction it is rotated. On cooling it will again go through the ring.

At home you can use a nut and bolt to give the same results.

Expansion and contraction take place in all directions.

## 22.3 Making allowances for forces of expansion and contraction

If we place a car jack lengthways between two concrete pillars and start to lengthen it, terrific forces and strains are set up. Eventually two things can happen: (1) the jack will buckle and break as it fails to move the pillars, or (2) the pillars will start to crack and will possibly collapse.

Similar forces are set up if a heated object is not given enough room to expand. This can result in damage to the structure of the object or its surroundings and, in a mechanical device, its failure to work. You can see an example of this in Fig. 22.5.

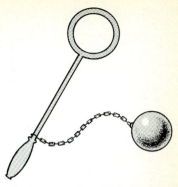

**Fig. 22.4** *Ball and ring experiment*

F.O.P.—Q

**Fig. 22.5** *In a heat wave on 1 July 1969 firemen had to spray the locks of Tower Bridge so that the bridge could be closed*

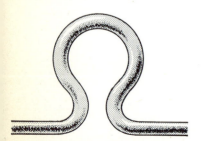

**Fig. 22.6** *A horseshoe loop used in steam pipes*

If a machine or structure is likely to be used in conditions that cause expansion and contraction, then it must be designed to allow for these changes.

Two examples of the provisions made for expansion and contraction have already been seen in Figs. 22.1 and 22.2. Figure 22.6 shows a horseshoe loop widely used in steam pipes. When the pipe expands, the extra length is taken up in the loop, so the pipe remains as it was laid. Figure 22.7 shows the insertion of cellular fibre board in between concrete sections of road. This board is easily compressible and makes an ideal expansion joint filler. It is placed in position, wet concrete is poured in around it, and finally a molten sealing compound is poured in on top of the fibre board to a depth of 2.5cm.

**Fig. 22.7** *An expansion joint being constructed for a concrete road*

Figure 22.8 shows the construction, on site, of a section of continuous welded railway line. Using hydraulic machinery, the rail is "de-stressed" for some temperature between 21°C and 27°C (i.e. stretched so that when the temperature rises to, say 25°C, there is no stress in the rail). It is then welded and anchored to sleepers which are fixed to massive pyramids of ballast.

**Fig. 22.8** *Continuous welded rail is made "stress free"*

*QUESTION 2:* What provisions are made for the expansion of the moving parts inside a cooker?

The force of contraction can be convincingly demonstrated by using the bar-breaking apparatus illustrated

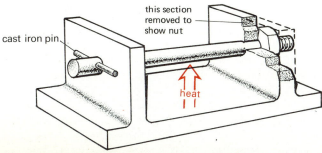

**Fig. 22.9** *Experiment to show the force of contraction*

in Fig. 22.9. A cast iron pin is put through the hole in the bar, which is then heated. After a few minutes the nut at the other end is screwed up so that the cast iron pin is pulled as closely as possible against end A of the apparatus. This screwing up is done time and time again as the bar is heated. When the nut cannot be screwed up any further the bar is allowed to cool down. The tremendous force exerted when the bar contracts easily snaps the cast iron pin neatly in two!

*QUESTION 3:* How can this apparatus be adapted to show the force of expansion?

## 22.4 Uses of the forces of expansion and contraction

Engineers have long recognized the usefulness of these forces in cases where they wish to make parts of machinery or structures fit very closely together. Tools cannot exert the forces that are required.

It is not easy, for instance, to take measurements and then cut a complete ring of paper so that it fits tightly around a metal tin. This type of problem in industry can be solved by using a method known as "sweating". For example a toothed rim can be "sweated" onto the flywheel which is used to engage the starter motor in a car. The rim is made slightly too small; it is heated so that it expands and then it easily slips onto the flywheel. When it cools and contracts the fit between the two parts is tight and secure. A tight fit is very necessary because the flywheel is subject to tremendous forces when it engages and starts the engine from rest (i.e. overcomes its inertia). If the rim were bolted on it could soon become loose, and if the whole flywheel and rim were cast in one piece it would be very expensive to replace when teeth were damaged. Similar uses of these forces are illustrated in Figs. 22.10, 22.11 and 22.12.

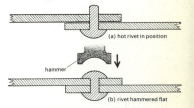

**Fig. 22.10** *Riveting: when the rivet cools it draws the metal plates closely together*

231

**Fig. 22.11** *A cold diesel wheel is lowered into its hot metal tyre. The tyre cools and contracts and thus a very tight fit is achieved.*

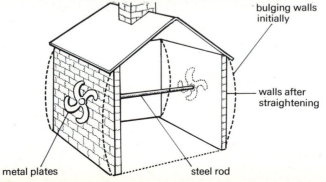

**Fig. 22.12** *A way of dealing with bulging walls. The steel rod is heated and the metal plates are screwed up. When the rod cools it contracts, pulling the walls inwards and straightening them.*

bulging walls initially

walls after straightening

metal plates

steel rod

## 22.5  What happens to bad conductors of heat when they undergo expansion and contraction?

In your geography lessons you have probably studied the weathering of rocks and know that the effect of heat on rocks such as granite causes thin sheets, or spalls, to split off.

Granite is a bad conductor of heat. When it is exposed to the sun during the day, it is only the outer layers that rise in temperature and expand. Similarly, during the night it is only the outer layers that cool down and contract. The forces set up between the outer and inner layers of rock by continual expansion and contraction eventually cause the outer layers to fall off. Similar weathering processes on the spalls reduce them to even smaller particles, and eventually soil is formed.

*QUESTION 4:* Why is it that (a) milk bottles crack when boiling water is poured into them, (b) thick paper covers on books curl when they are placed on a radiator?

## 22.6  Do different materials expand by different amounts?

Have you ever been presented with the problem of removing a very tight metal cap from a glass bottle? Someone may have suggested that you place the top of the bottle in very hot water, and perhaps you were surprised to discover how easily the cap came off after that. You were heating two different materials, metal and glass. What happened?

Let us find out what happens when two different materials of the same length are subjected to the same change in temperature (Fig. 22.13). The two metals, aluminium and iron, are riveted together in what is called a bimetallic strip, so that they cannot move independently. On heating, the strip becomes curved as in Fig. 22.13(b); this shows that the metals must have expanded by different amounts. If they had expanded by

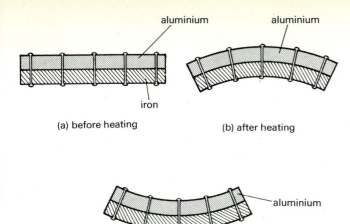

aluminium      aluminium

iron

(a) before heating      (b) after heating

aluminium

(c) on cooling

**Fig. 22.13** *The heating and cooling of a bimetallic strip*

equal amounts the strip would have remained straight. In this bimetallic strip the aluminium is on the outside of the heated curve.

*QUESTION 5:* Which has expanded by the greater amount, the aluminium or the iron? Which material expanded by the greater amount when you removed the tight metal cap from the glass bottle?

## 22.7   Uses of differences in expansion of different materials

In some instruments, such as timing devices, any increase in the size of the moving parts would alter the delicate balance and lead to inaccuracies. An increase in the length of a pendulum of a clock, for example, causes it to lose time. If this were to happen to Big Ben, it might create quite a disturbance! There has to be, therefore, a built-in corrective to expansion. One form of compensated pendulum is shown in Fig. 22.14. Zinc expands approximately twice as much as iron When the temperature increases the iron rod expands downwards and the zinc upwards to exactly the correct amount. By this means the centre of gravity of the pendulum is kept at the same distance from the suspension point of the pendulum, the swing is unaltered and the clock keeps perfect time. The balance wheel of a watch is constructed from a bimetallic strip (Fig. 22.15) to give similar compensation.

In thermostat devices which keep temperatures constant in, for instance, rooms or appliances, the general principle is that one end of a bimetallic strip is fixed and the other end is left free to move (Fig. 22.16). Thus in position (1) it allows the electricity to flow, and in position (2) it prevents it from flowing. A control knob is set to the required temperature. At this temperature the bimetallic strip goes to position (2). When the bimetallic strip cools it returns to position (1), thus a constant temperature is maintained. In shutting off a flow of energy when it reaches a dangerous level, the thermostat also can act as a safety switch.

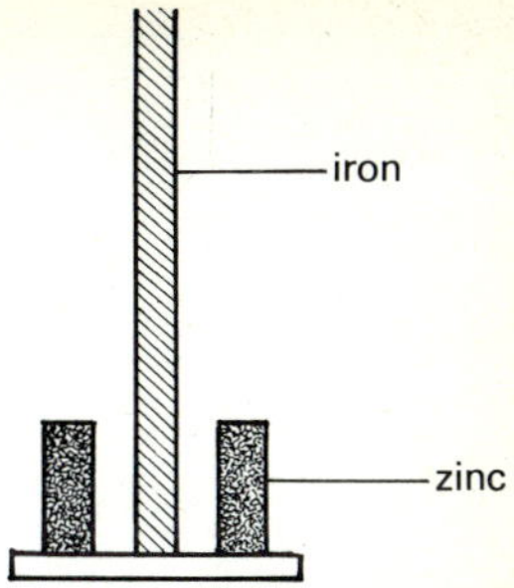

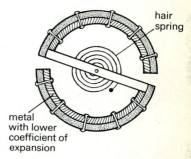

iron

zinc

**Fig. 22.14** *A compensated pendulum*

hair spring

metal with lower coefficient of expansion

**Fig. 22.15** *Compensated balance wheel of a watch. Any increase in radius resulting from a temperature rise is compensated for by the inward curving of the bimetallic strip.*

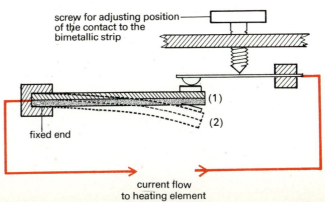

screw for adjusting position of the contact to the bimetallic strip

(1)

(2)

fixed end

current flow to heating element

**Fig. 22.16** *The principle of a thermostat*

**Fig. 22.17** *A fire alarm system: heat from the fire expands the bimetallic strip, closing the electrical circuit*

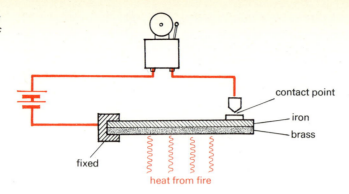

Some other uses of bimetallic strips are illustrated in Figs. 22.17, 22.18 and 22.19.

**Fig. 22.18** *Gas oven thermostat: on expansion the brass rod draws the invar rod with it, thereby cutting down the supply of gas to the burners*

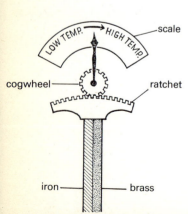

**Fig. 22.19** *A bimetallic thermometer*

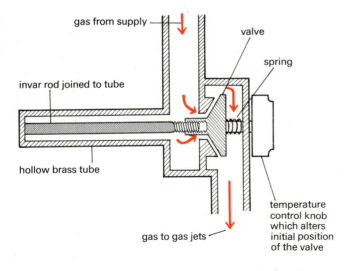

QUESTION 6: How does the bimetallic thermometer in Fig. 22.19 work?

234

## 22.8 Do liquids expand and contract?

This can be answered quite simply by using the apparatus illustrated in Fig. 22.20.

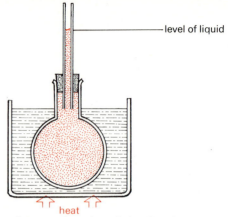

**Fig. 22.20** *Experiment to determine whether liquids expand when heated*

The flask full of liquid, with no air bubbles, is heated either directly or by placing it in a water bath. The level of the liquid in the tube can be observed easily. At first it drops and then is seen to rise rapidly. On cooling, the level of the liquid goes down again. This clearly shows that a liquid does expand when heated. Probably you have already realized this on occasions when you have filled a kettle right up to its lid and then heated it.

QUESTION 7: (a) Why does the level of the liquid in the tube drop when it is first heated? (b) Does this experiment tell you anything about the relative expansion of liquid and solids?

We can compare the expansion of different liquids by filling several flasks with different liquids and putting them in a common water bath. Their expansion can then be compared over the same rise in temperature.

## 22.9 Thermometers

A thermometer is an instrument used to measure the hotness, or temperature, of a body.

It does *not* measure the quantity of heat energy. We will discuss this again in Chapter 24.

Our skin, particularly at the elbows, is very sensitive to temperature, but is not always a reliable guide. Try putting your elbow first into a bowl of warm water, then under running cold water, and then again into the bowl of warm water. Does the temperature of the warm water feel the same the second time?

The expansion of liquids offers a good general means of measuring temperature reliably. It is necessary to choose a liquid that is readily visible and a good conductor, that expands by a suitable amount over as wide a range of temperatures as possible, and that moves freely within a glass container.

Two liquids, mercury and coloured alcohol, fulfil all these requirements, each compensating for deficiencies in the other: mercury boils at 357°C but freezes at −39°C; alcohol boils at 78°C and freezes at −115°C. If you were going to the south pole you would take an alcohol thermometer.

## 22.10 Temperature scales

To measure and record temperature we must have a scale and a unit.

A scale can be obtained by taking two temperatures as *fixed points* and dividing the distance between them into a number of equal divisions which are called *degrees*.

On the Celsius (Centigrade) scale the lower fixed point is the temperature at which pure ice melts under normal atmospheric pressure, and is taken as 0°C. The upper fixed point is the temperature at which pure water boils under normal atmospheric pressure, and is taken as 100°C. There are 100 equal divisions between these

points on the thermometer, and when the mercury or alcohol moves through one division this corresponds to a change in temperature of 1°C.

The kelvin scale will be discussed later (see page 238).

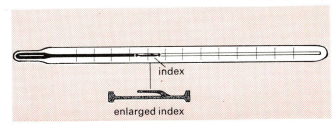

**Fig. 22.21** *A maximum thermometer*

## 22.11 Different types of thermometer

1. *Maximum thermometer* (see Fig. 22.21)
This type of thermometer usually consists of a thin glass capillary tube with a uniform internal diameter (bore), and a bulb containing the liquid at one end. The other end is sealed. Mercury is used, and above the mercury in the capillary tube is a small glass *index*. This is kept in position by a very fine thread of glass that acts as a spring against the side of the tube. The spring allows the expanding mercury to push the index up the tube, but prevents it from moving back again. Thus the index records the maximum temperature reached in any period of time. The thermometer is reset by means of a magnet which attracts the piece of metal in the index back to the surface of the mercury.

2. *Clinical thermometer* (see Fig. 22.22)
This is a special form of maximum thermometer used for taking body temperature. It has no glass index, but there is instead a constriction in the capillary just above the bulb. When the mercury expands it has enough force to get past the constriction, but when the thermometer is removed from the patient's mouth, the mercury does not return to the bulb. The thread of mercury is therefore

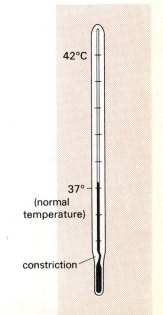

**Fig. 22.22** *A clinical thermometer*

broken at the constriction and it remains in the stem of the thermometer so that the temperature can be read off at any time. After use the mercury can be shaken back into the bulb — a technique of wrist flicking at which only doctors and nurses seem to be expert!

### 3. *Minimum thermometer*
This contains alcohol and has an index similar to that found in a maximum thermometer, but in a different position — below the surface of the alcohol in the capillary tube.

*QUESTION 8:* What happens to the index in a minimum thermometer when the thermometer is (a) heated, (b) cooled, and how would you read off the minimum temperature?

### 4. *Non-liquid thermometers*
These generally cover a wider range of temperature than do the liquid thermometers.

A *resistance thermometer* uses the effect of an increase in temperature to increase the resistance of a metal (e.g. platinum) to the flow of electricity (see page 181). It has a range from approximately $-200°C$ to $1200°C$.

A *thermocouple* consists of two wires of different materials (e.g. copper and iron) joined together at the ends, one junction being at a constant temperature. When the temperature of the second junction differs from the fixed temperature an electric current flows and is measured by a galvanometer inserted in the circuit as shown in Fig. 22.23. This is a very sensitive instrument. A *thermopile* consists of several linked thermocouples.

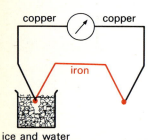

Fig. 22.23   *A thermocouple*

### 5. *A gas thermometer*
This thermometer can give readings from approximately $-260°C$ to $2500°C$. The temperature is measured by the amount the gas expands. Why does it have such a wide range?

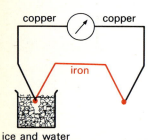

236

## 22.12   Expansion of gases
We can see how gases expand by a very simple experiment (Fig. 22.24). Only the warmth of the hands around the flask is needed to cause the expansion of the

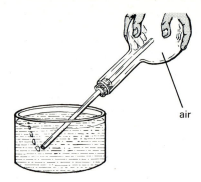

Fig. 22.24   *Experiment to determine whether gases expand*

air. If the flask is heated by a burner and then allowed to cool, the remaining air contracts, and the water rushes in to form a fountain within the flask.

Fig. 22.25   *An early hot-air balloon*

**Fig. 22.26** *Sealed containers of volatile liquids must not be placed near direct heat*

This experiment clearly demonstrates that gases expand much more than solids or liquids for the same temperature rise (in fact about twenty times more than mercury). Figures 22.25 and 22.27 show two uses for gas expansion, and Fig. 22.26 shows the danger of not making allowances for gas expansion (and of not following instructions on labels!).

Experiment shows that all gases expand by the same amount when heated through the same temperature rise.

**Fig. 22.27** *The gases produced by the explosion are heated and expand; their expansion causes the ejection of the missile*

### 22.13   The kinetic theory of matter

We have already stated in Chapter 10 that matter consists of molecules that are constantly vibrating. Heat is a form of energy and when matter is heated the molecules move with increased kinetic energy. They therefore move faster, collide with one another and gradually force themselves further apart, in spite of molecular forces that try to keep them together. This causes an increase in size, i.e. expansion. As more heat energy is given to a substance, the vibration of its molecules increases and its temperature goes up. As the heat is lost, kinetic energy decreases and the molecules return to their former state.

Our experiments have also shown us that liquids and gases expand more than solids. This is because the molecular structure of these three states of matter are different from one another. The molecules of a gas are moving faster, have more space between them and greater freedom of movement than when in the solid state. This fits in with the kinetic theory of matter which explains the three states of matter as follows (Fig. 22.28).

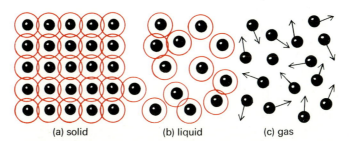

(a) solid          (b) liquid          (c) gas

**Fig. 22.28** *Molecular arrangements in the three states of matter*

a.   In a solid a molecule vibrates backwards and forwards about a mean position. The molecules remain close together under strong forces of attraction, but there are also close-range forces of repulsion which prevent a solid from being easily compressed.

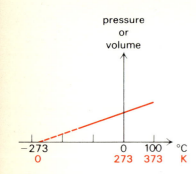

**Fig. 22.29** *Graph of pressure or volume against temperature for a fixed mass of gas*

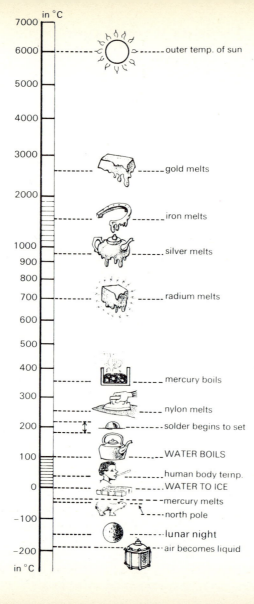

in °C

- 6000 — outer temp. of sun
- 2600 — gold melts
- 1500 — iron melts
- 960 — silver melts
- 700 — radium melts
- 357 — mercury boils
- 250 — nylon melts
- 200 — solder begins to set
- 100 — WATER BOILS
- 37 — human body temp.
- 0 — WATER TO ICE
- — mercury melts
- — north pole
- — lunar night
- — air becomes liquid

in °C

**Fig. 22.30** *A selection of temperatures measured in °C*

238

b. When the solid matter melts it becomes a liquid. The molecules are moving faster, but they still remain close together under forces of attraction. However, they are able to move past one another and they often collide.

c. When the liquid changes to a gas the molecules move even faster and are spread well apart, thus they can move quite freely although they still collide. There is very little attraction between molecules (a gas always fills the vessel in which it is contained).

## 22.14  Absolute zero

We have stated that during the process of cooling the vibration of molecules becomes slower and slower. Therefore, at some point we should reach a temperature where the molecules have the minimum possible kinetic energy. This temperature, *absolute zero*, is estimated to be −273°C. Fig. 22.29 shows a graph of pressure (or volume) of a gas plotted against temperature. Why do you think part of the graph is dotted? At absolute zero the molecules no longer exert a pressure on the container. Such a temperature has never been reached, although scientists have come within a few millionths of a degree of it. At such very low temperatures gases become lique-fied (oxygen, for example, liquefies at −183°C) and objects lose all their normal properties — a rubber ball, for instance, becomes just like glass and shatters if dropped.

The kelvin scale of temperature has its zero at absolute zero. Thus it follows that −273°C = 0K, 0°C = 273K and 100°C = 373K.

The size of the divisions on the Celsius scale and the kelvin scale are the same (1°C = 1K), therefore intervals of temperature on the two scales are identical.

QUESTION 9: Look at the range of temperatures in °C shown in Fig. 22.30. (a) What is the temperature of molten iron? (b) What is the range of temperature over which solder begins to set?

## 22.15   The peculiar expansion of water

QUESTION 10: You have all, no doubt, broken the ice on a puddle at some time or other. Bearing this in mind, why do you think that skaters are warned about skating on deep ponds at the first appearance of ice?

Most liquids, when they solidify, do so from the bottom upwards (for example, molten metal in the workshops, and molten fat in the kitchen). Why should water do otherwise?

In the following experiment (Fig. 22.31) we examine the effect of heating water from 0°C (273K) to 12°C (285K). The apparatus is set up as shown in Fig. 22.31(a), and the flask is left surrounded by iced water for several hours. In this way we ensure that all the water in the flask reaches 0°C. The level of the water in the capillary tube at this temperature of 0°C is marked. The ice melts meantime and when all of it has gone the temperature of the water in the flask is seen to rise. At 1°C, 2°C, 3°C, 4°C and 5°C and so on, the level of the water in the capillary tube is observed.

The observed level shows clearly that as the temperature rises from 0°C (273K) to 4°C (277K) the water contracts. Above 4°C the water expands. In fact, water occupies its smallest volume, and is therefore densest at a temperature of 4°C. As the graph in Fig. 22.31(b) of volume against temperature indicates, at 4°C (277K) 1kg of water occupies 1000cm³.

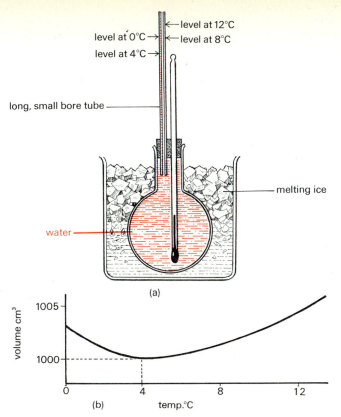

Fig. 22.31   Experiment to examine the effect on the volume of 1 kg of water when the temperature is raised from 0°C to 12°C

Now we can explain why water freezes from the surface downwards. As the temperature drops so the water at the surface of a pond is cooled. When its temperature reaches 4°C it is at its greatest density and this surface water sinks, but when the temperature of the surface water falls below 4°C (277K) it is then less dense and remains at the surface. If the temperature falls to 0°C (273K) it freezes

0°C ice →
0°C water →
1°C →
2°C →
3°C →
denser water at 4°C →

**Fig. 22.32** *Pond temperatures in winter*

and the temperatures in the pond are as shown in Fig. 24.32. Ice is less dense than water and therefore remains at the surface. Many days of very low temperatures are required before the layer becomes really thick.

This peculiar expansion of water can also be explained in terms of the molecular theory. Water molecules exist in groups and a rearrangement of these groups takes place when water is heated from 0°C to 4°C (273K to 277K). Although there is an increase in the kinetic energy of the molecules, their rearrangement causes an overall decrease in volume.

## 22.16 How does an engineer know how much allowance to make for the expansion of a bridge?

The engineer cannot just build the bridge and then heat it up to find out! He must have a way of calculating the amount of expansion at the design stage. His greatest problem is the great length involved, but we have already seen how the expansion of a piece of metal can be detected by experiment. He can therefore solve his problem if he finds out experimentally by what amount a unit length of metal expands. Once he knows this, he can calculate the expansion of much greater lengths of the metal. In fact, he must find out what is called the

*coefficient of linear expansion* for the material (*linear* means measured along its length).

The coefficient of linear expansion is the increase in length of unit length when the temperature is raised by 1 degree.

## 22.17 Experiment to measure the coefficient of linear expansion of a material

The apparatus used in this experiment is illustrated in Fig. 22.33. The length of the rod of material is measured by means of a metre rule before it is placed in the steam jacket. When it is in position in the jacket the temperature is read. A special instrument designed for the measurement of small lengths, a *micrometer screw*, is used. This is screwed up until it just makes contact with the bar (the electric bell rings when the circuit between bar and micrometer is completed) and a reading is taken. The micrometer is then unscrewed. Steam is sent through the

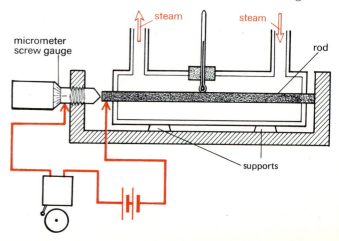

micrometer screw gauge

steam    steam

rod

supports

**Fig. 22.33** *Apparatus used in determining the coefficient of linear expansion of a material*

steam jacket and the micrometer is screwed up at various intervals until it gives the same reading over a period of time. (The micrometer is unscrewed after each reading.) The temperature is then read.

QUESTION 11: Explain (a) why the micrometer screw is never kept in contact with the bar for any long period of time, (b) why the final reading is only taken after the micrometer reading has been constant for several minutes.

The coefficient of linear expansion is calculated from the results as follows:

Initial length of tube at
   room temperature    $= x$mm
Increase in length of rod  $=$ (final micrometer screw reading) —
                          (first micrometer screw reading)
                      $= y$mm
Increase in temperature    $=$ (final temperature) — (initial
                           temperature)
                      $= t°$C
Then $x$mm of material expands in length by $y$mm
    when the temperature is raised by $t°$C

Therefore 1mm of material expands in length by $\frac{y}{x}$ mm

    when the temperature is raised by $t°$C

Therefore 1mm of material expands in length by $\frac{y}{x \times t}$ mm

    when the temperature is raised by 1°C

This expression is, from the definition on page 240, the coefficient of linear expansion.

$$\text{Coefficient of linear expansion} = \frac{\text{increase in length}}{\text{original length} \times \text{temperature rise}}$$

Units are: per degree Celsius (1/°C), per kelvin (1/K). On the kelvin scale an interval of one unit is exactly the same as an interval of one degree on the Celsius scale (see page 238).

Thus if a bar 50cm long is heated from 20°C to 100°C and the expansion is 0.068cm, the coefficient of linear

expansion will be:

$$\begin{aligned}\text{Coefficient of linear expansion} &= \frac{\text{increase in length}}{\text{original length} \times \text{temperature rise}}\\ &= \frac{0.068}{50 \times (100 - 20)}\\ &= \frac{0.068}{50 \times 80}\\ &= 0.000017 \text{ per degree Celsius}\end{aligned}$$

The coefficients of linear expansion for some common substances are given in the following table:

| COEFFICIENTS OF LINEAR EXPANSION | |
|---|---|
| **(fractional increase in length per °C temperature rise)** | |
| Zinc | 0.000 029 |
| Lead | 0.000 028 |
| Aluminium | 0.000 026 |
| Brass | 0.000 019 |
| Copper | 0.000 017 |
| Iron or steel | 0.000 012 |
| Glass | 0.000 0086 |
| Pyrex/glass | 0.000 0032 |
| Invar | 0.000 0009 |

## 22.18 Calculations using the coefficient of linear expansion

An engineer can use these coefficients to calculate the increase in length of his bridge as a result of expansion. For example, suppose he is going to build a steel bridge 150m long which may be subjected to temperatures ranging from −25°C to +40°C.
The table tells us that

1m of steel heated through 1°C expands by 0.000 012m
Therefore 150m of steel heated through (−25° to 40°C) expands
      by $0.000\ 012 \times 150 \times 65$m
      $= 0.117$m or 11.7 cm

QUESTION 12: If this same bridge were to be built on the moon, where the surface temperature (at a depth of 1mm) ranges from −150°C to +100°C, what expansion would take place?

## THINGS TO DO

A. Use a mercury thermometer to record the air temperature day by day over a period of months. Make a graph of your results. Is there any general pattern over each period of 24 hours? If so what is it?

B. Compile an illustrated catalogue showing the uses of specially prepared materials such as oven glassware and invar, which undergo little or no expansion. Find out how these materials are made.

C. Imagine you are a salesman for a firm producing thermostats. Make up a brochure for your customers, to illustrate the many uses of thermostats.

## THINGS TO WORK OUT

1. Describe the structure and functioning of (a) a bimetallic strip, (b) a thermostat, (c) a thermometer graduated on the Celsius scale.

2. Explain the following statements:
   a. A metal screw cap can be removed from a glass bottle by placing it under very hot running water. The same method cannot be used to get a metal cap off a metal container.
   b. A clinical thermometer has a constriction in the capillary just above the bulb.
   c. Baking powder in a moist cake mixture gives off carbon dioxide gas. It is disastrous to open the oven door before the cake mixture is thoroughly cooked (solidified).

3. State three everyday applications of the forces of expansion and contraction, and three examples of allowances that have to be made for the damaging forces of expansion. Draw diagrams for each.

4. Describe an experiment you could carry out to show the difference in expansion between a gas and a liquid. What conclusion would you expect to draw from this experiment?

5. How is expansion allowed for in (a) the balance wheel of a watch, (b) a pendulum clock?

6. What is meant by the "fixed points" of a thermometer scale? Draw a diagram of a clinical thermometer and describe how it works.

7. A householder in bed at night is woken up by the sounds of creaking boards and rafters in his loft. It has been a very hot day. Explain why he need not investigate.

8. In servicing his car Mr Brown removed the tightly fitting bearing from inside its seating in the wheel hub. When he comes to slide the bearing back into position he finds he cannot do it because the fit is too tight. Why does a garage mechanic suggest to him that he makes use of his oven and refrigerator?

9. To peel tomatoes easily, a housewife places them first in boiling water, then immediately into cold water. Explain.

10. If a mercury thermometer is plunged into a hot liquid the mercury level is sometimes seen to fall for a moment before it rises. Explain.

11. A tennis ball bounces better when it has a high internal pressure. Explain why, at first-class tournaments, balls are kept in constant temperature containers.

12. Describe an experiment to show that when water is heated from 0°C to 10°C it behaves differently from other liquids heated in the same way.

13. Figure 22.34 shows some air in a capillary tube trapped by some concentrated sulphuric acid. The length '$l$', of the trapped air was measured at different temperatures. The readings are shown below.

| Temperature (°C) | 20 | 40 | 60 | 80 | 100 |
|---|---|---|---|---|---|
| Length, $l$ (cm) | 15.6 | 16.6 | 17.8 | 18.9 | 20.0 |

a. Draw a diagram showing how you would heat the air.
b. Why is concentrated sulphuric acid used rather than water?
c. Plot a graph of $l$ against temperature.
d. What is the value of $l$ at 0°C?
e. Where would you expect the graph to cut the temperature axis when the length of $l$ is 0cm? Where does this graph cut the axis when $l = 0$cm? You may need to redraw your graph with the temperature axis labelled like the one in Fig. 22.29 on page 238.

14. Figure 22.35 shows a round-bottomed flask connected by a rubber tube to a Bourdon pressure gauge. The flask is immersed in water. The water is heated; the bunsen burner is removed and the water is stirred. The temperature of the water and the reading on the gauge are recorded. A series of readings obtained in this way are shown in the table below.

| $t$ (°C) | 0 | 15 | 60 | 80 | 100 |
|---|---|---|---|---|---|
| $p$ (N/m² $\div$ 10²) | 0.94 | 1.00 | 1.18 | 1.25 | 1.32 |

a. Plot a graph of p against t. Label your t-axis so that it goes from −300°C to 100°C (see Fig. 22.29, page 238.)
b. Use your graph to estimate the absolute zero of temperature.
c. How does the kinetic theory of gases explain the increase in pressure as the temperature is increased?
d. Is it best to use a long length of rubber tube for this experiment or a short length? Give reasons for your answer.
e. What assumption have you made about the behaviour of gases at low temperatures in using your graph to estimate the absolute zero?

For questions 14 to 16 use the values for coefficients of linear expansion as given on page 241.

15. Calculate the amount by which the following materials expand.
a. 100m of copper pipe, heated through 60°C (60K)
b. 50m of lead pipe, heated through 20°C (20K)
c. 20cm of iron pipe, heated through 75°C (75K)
d. 95cm of brass pipe, heated through 10°C (10K)

16. A copper rod is 100cm long at 12°C (285K). What is the length of the rod at 65°C (338K)?

17. Find the coefficient of linear expansion of a glass rod which measures 100cm at 15°C (288K) and increases by 0.024cm at 80°C (353K)

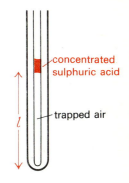

Fig. 22.34

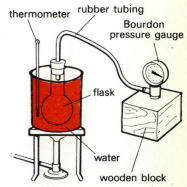

Fig. 22.35

243

## CHAPTER 23

# CHANGE OF STATE

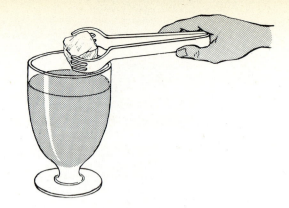

**Fig. 23.1** *Ice is added to orange squash to make a cool drink*

## 23.1 Three states of matter

In Fig. 23.1 orange squash is being made colder by the addition of ice. In Fig. 23.2 a woman is cooking fish by steaming it. The fish is wrapped in buttered paper

**Fig. 23.2** *Steam is used to cook food*

244

and placed in between two dishes over steaming water in a saucepan. Both the cooling effect of the ice and the heating effect of the steam result from changes in the state of the water. In the first instance the water is in the form of ice; in the second instance the water is in the form of steam. As we saw on page 237 the differing arrangements of molecules within matter give three kinds (states) of matter.

The three states of matter are solid, liquid and vapour.

In the case of water these three states are called ice, water and water vapour (steam).

QUESTION 1: (a) What state of matter does the ice change to when it is placed in the orange squash? (b) What state of matter does the steam change to when it comes into contact with the lower dish?

How can water on changing state both heat and cool other substances? The following experiment will help us to answer this question.

## 23.2    Experiment to observe the temperature of water during its change in state

Half-a-dozen ice cubes are removed from the refrigerator and placed in a beaker. A Celsius thermometer is placed between them. They are stirred, their temperature is recorded constantly until all the ice has melted, and then for several minutes after this time. At all times, the thermometer is in contact with the substance and not with the container.

The water formed from the ice, still in its container, is then heated slowly over a gas burner or electric ring. The temperature is again recorded constantly until the water boils, and then for several minutes after this. The following observations and recordings were made.

The ice cubes were at a temperature of 0°C (273K).

This was a lower temperature than their surroundings, and therefore they took in heat. Gradually the ice melted, becoming water. The mixture of water and ice did not rise above 0°C (273K) until all the ice had melted. At that point its temperature began to rise.

The rise in temperature increased rapidly when it was placed over a source of heat. At 100°C (373K) the water boiled and bubbles formed within it. The temperature did not go above 100°C (373K) no matter how long or how hard the water was boiled. The temperature of the steam given off was also found to be 100°C (373K).

## 23.3    Kinetic theory of matter, and change of state

In the above experiment heat was being supplied constantly to the water. But neither at the point when it changed from solid to liquid, nor at the point when it changed from liquid to vapour did the heat energy which was being supplied cause a rise in temperature.

What was the heat energy being used for if it was not increasing the temperature of the substance? It was being used to change the state of the molecules within the matter. In fact it was being absorbed by the molecules in the form of an increase in their energy.

To understand this we must think again about the kinetic theory of matter which was discussed in Chapter 10. The models in Fig. 23.3 will help you.

In (a) the balls represent molecules and the springs the forces holding them together within the solid. Each ball vibrates about a mean position. When the molecules receive energy in the form of heat they convert it into their own vibrational energy; this is seen as an expansion of the solid. If they are provided with sufficient energy they can overcome the forces holding them together. They are then free to move past one another, and they do not form a regular pattern. This is the liquid state and the behaviour of the molecules is represented, though rather inadequately, in (b).

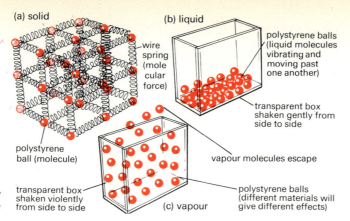

(a) solid

wire spring (mole cular force)

polystyrene ball (molecule)

transparent box shaken violently from side to side

(c) vapour

(b) liquid

polystyrene balls (liquid molecules vibrating and moving past one another)

transparent box shaken gently from side to side

vapour molecules escape

polystyrene balls (different materials will give different effects)

**Fig. 23.3** *Models to illustrate molecular structure*

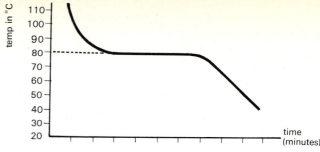

**Fig. 23.4** *Graph showing the cooling curve of naphthalene*

When given even more energy the molecules move faster and some have enough energy to escape from the liquid by overcoming (i) the forces of molecular attraction in the liquid (ii) the pressure of air on the surface of the liquid. They spread out into the air and move in all directions. In this condition, which is the vapour state represented in (c), the forces of attraction are small and the molecules can move at great speed and over large distances.

Thus, when a substance is changing state, all the heat energy being supplied is used to change the arrangement of the molecules. There is no energy available to cause a rise in temperature until the change of state is complete.

Because the heat energy seems to disappear without trace it is called *latent* (i.e. hidden) heat. Liquids and gases have this latent heat stored within them in a form of potential energy (page 116). This potential energy is released if there is a change of state from vapour to liquid or liquid to solid.

*QUESTION 2:* Some naphthalene (moth balls) is put into a test tube and heated in a water bath until it all melts. The heat is removed and, as the liquid cools, its

temperature is recorded every minute. Figure 23.4 is the graph drawn on the basis of the readings.

(a) What is happening to the naphthalene during the period shown as a horizontal path on the graph?
(b) What is the temperature at which naphthalene melts (i.e. its melting point)?

### 23.4 Types of latent heat

Figure 23.5 gives the terms used for the various changes of state and shows how the *specific latent heat of fusion* and the *specific latent heat of vaporization* is involved in each case.

Our first experiment (page 245) showed that the newly formed state of matter is at first always at the same temperature as the old state of matter. That is, ice melts to form water at 0°C (273K). Water vaporizes, forming steam, at 100°C (373K).

Specific latent heats are defined in the following way:

The specific latent heat of vaporization is the amount of heat energy needed to change unit mass of liquid into vapour at the same temperature.

The specific latent heat of fusion is the amount of heat required to change unit mass of solid into liquid at the same temperature.

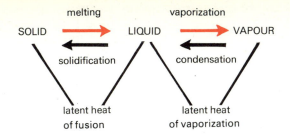

Fig. 23.5  *Terms associated with changes in state*

Heat is measured in joules (see page 114). 1 kg of water needs 2 260 000J of heat to change it to steam at 100°C (373K). Therefore the specific latent heat of vaporization of water is 2 260 000J/kg. Similarly the specific latent heat of fusion of ice is 336 000J/kg. Laboratory experiments used to calculate these values are discussed in Chapter 24.

*QUESTION 3:* (a) Why is ice used to cool drinks? (b) Why can steam be used to cook food?

Obviously, different materials have different specific latent heats. You might like to compare quite simply how much heat is needed to liquefy other materials, such as lead, solder, etc, by referring to the list of specific latent heats shown here.

**SPECIFIC LATENT HEATS OF VARIOUS MATERIALS (J/kg)**

| Fusion | | Vaporization | |
|---|---|---|---|
| Mercury | 11 760 | Ether | 369 600 |
| Lead | 21 000 | Mercury | 285 600 |
| Copper | 180 600 | Ammonia | 1 373 400 |
| Ice | 336 000 | Sulphur | 1 512 000 |
| Soft solder | 50 000 to 84 000 | Steam | 2 260 000 |

*QUESTION 4:* Steam causes a far more serious burn on the human skin than boiling water. Can you explain why?

Fig. 23.6

### 23.5  Evaporation and boiling

When you line up to have an injection (Fig. 23.6) you feel a number of sensations. Fear is perhaps one! Anticipation of pain is perhaps another. But when the doctor dabs on the surgical spirit what you feel is coldness. Why?

In the following experiment we study the evaporation of various liquids.

The apparatus is illustrated in Fig. 23.7. A small square piece of lint is soaked in surgical spirit and then tied round the bulb of a thermometer. Any changes in the reading of the thermometer and the dampness of the lint are observed over a period of time.

The experiment can be repeated, soaking a piece of lint each time in different liquids: water, ether, perfume, etc. As some of these liquids are inflammable it is essential not to do the experiments anywhere near a naked flame.

In each case the rate at which the temperature falls and the time taken for the lint to become dry is recorded.

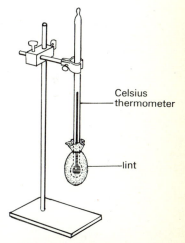

Celsius thermometer

lint

Fig. 23.7  *Experiment to study the evaporation of liquids*

Both changes are more rapid for surgical spirit, ether and perfume, than in the case of water.

In all cases the liquids change to vapours, but at different rates. Those that evaporate easily are said to be volatile. Volatile liquids are used as solvents in perfumes and in dry-cleaning fluids where quick drying is required. In all cases as the liquids evaporate they get the needed latent heat of vaporization from their surroundings. In both evaporation and boiling latent heat of vaporization is needed.

*An evaporating liquid cools its surroundings.*

This can be seen quite spectacularly if you place ether in a metal crucible and place this on top of a thin layer of water covering a wooden block. When air is blown through the ether to evaporate it quickly, the crucible freezes to the block.

*QUESTION 5:* One fibre-tip pen is labelled "water-based", a second is marked "spirit-based". Which would you buy if you needed a quick drying ink?

## 23.6  Differences between evaporation and boiling

In comparing the boiling of water in a beaker with the evaporation of water from a saucer, the following differences are observed.

| BOILING | EVAPORATION |
|---|---|
| 1. Takes place at a specific temperature, called boiling point | Takes place at any temperature |
| 2. Takes place within the liquid (bubbles can be seen within the body of the liquid) | Takes place at the liquid's surface (no bubbles form in the liquid) |

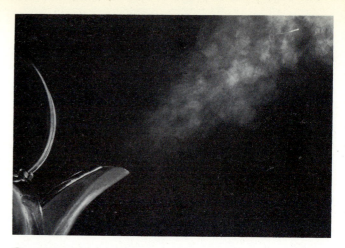

**Fig. 23.8**  *Where is the true steam?*

## 23.7  Can we see water vapour in the air?

Figure 23.8 shows the spout of a boiling kettle. Point to the area of the photograph which you think shows water vapour (dry steam). If you pointed to the "cloud" above the spout, then you are wrong. This is not water vapour. To find out what it is, place a candle or burner under the cloud. Immediately the cloud disappears, showing that in fact it was condensed water vapour, a cloud of water droplets.

Water vapour (dry steam) is invisible. Look carefully at the small area between the kettle spout and the cloud of condensed water vapour. In that gap there is water vapour. You cannot see it, but you can see where it is.

## 23.8  Can the rate of evaporation be increased?

A stranger from another planet might think that the women in Fig. 23.9 were undergoing some form of brain test! What is happening to them?

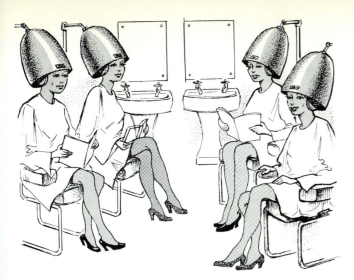

**Fig. 23.9** *Hair dryers*

When we dry hair with a hair dryer we are speeding up the rate of evaporation of the water. We can investigate the factors that affect the rate of evaporation by carrying out the following experiments (Fig. 23.10).

Several strands of wool are combed (teased) out in order to separate the fibres of wool. This teased-out wool represents the hair of the clients in the hairdresser's salon. Tests are then carried out to determine the effects of four factors on the rate of drying. These factors are (1) air movement, (2) heat, (3) amount of water in the air (humidity), (4) amount of surface area exposed. The time taken for the hair (wool) to dry is recorded in each case.

In each test the more effective method for quick drying is found to be as follows: 1(a), 2(b), 3(b), 4(b).

This experiment shows us the conditions necessary for speeding up the rate of evaporation.

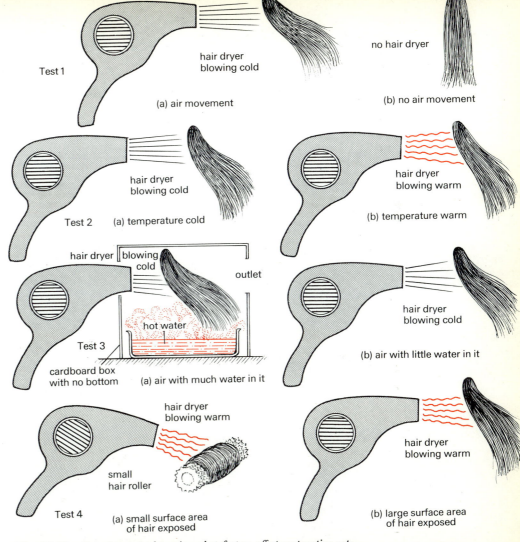

Test 1

(a) air movement

hair dryer blowing cold

no hair dryer

(b) no air movement

Test 2 (a) temperature cold

hair dryer blowing cold

hair dryer blowing warm

(b) temperature warm

hair dryer blowing cold

outlet

hot water

Test 3

cardboard box with no bottom (a) air with much water in it

hair dryer blowing cold

(b) air with little water in it

hair dryer blowing warm

small hair roller

Test 4 (a) small surface area of hair exposed

hair dryer blowing warm

(b) large surface area of hair exposed

**Fig. 23.10** *Experiments to determine what factors affect evaporation rate*

QUESTION 6: Which one of the following days would prove to be the best "drying day" for a housewife?
(a) warm, wet and no wind, (b) cold, wet and windy, (c) warm, dry and windy, (d) warm, dry and no wind, (e) cold, dry and windy.

**Fig. 23.11** *Cross-sectional diagram of a refrigerator*

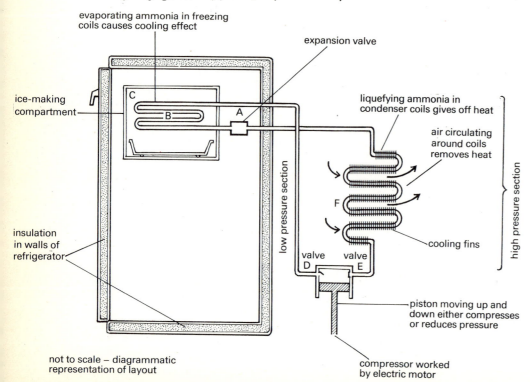

evaporating ammonia in freezing coils causes cooling effect

expansion valve

ice-making compartment

liquefying ammonia in condenser coils gives off heat

air circulating around coils removes heat

low pressure section

high pressure section

cooling fins

insulation in walls of refrigerator

valve D    valve E

F

piston moving up and down either compresses or reduces pressure

not to scale – diagrammatic representation of layout

compressor worked by electric motor

## 23.9  Uses of evaporation

### 1. *In the human body*

If you stand around in wet clothing your body feels chilled because of the evaporation of the water (see page 248). A similar cooling process occurs naturally through evaporation of water when the body sweats. Sweat is mostly water, containing a few body salts, which leaves the body by the pores in the skin. As it rests on the surface of the skin it evaporates and cools the body. The salts are left on the skin: after strenuous exercise you can taste them! Evaporation by sweating prevents overheating of the body, which would have a destructive effect on body tissues. (In fact, if evaporation and loss of heat by radiation and convection were stopped altogether by coating the body with gold paint, for example, death would quickly occur. This was the fate of a victim in a famous crime novel.)

In countries where temperatures are high and there is great humidity, the body becomes very uncomfortable because the rate of evaporation is poor. In such countries air conditioning is essential in homes and places of work.

### 2. *In refrigeration*

Figure 23.11 shows the working parts of one type of electric refrigerator. To study its action follow the sequence through by reference to the diagram, starting at A, the expansion valve. The volatile liquid (e.g. ammonia) is released through the expansion valve A into the coils B in the ice-making compartment C. The compressor, with its piston moving downwards and valve D open, reduces the pressure in this region and the liquid expands and evaporates and takes latent heat from its surroundings. Ice is formed in the ice compartment round the tubes and the contents of the refrigerator are cooled. The gas now passes through valve D. Valve D closes and E opens. The piston moves upwards and the compressed gas passes into the high-pressure section, and from there into the condenser coils, where it is cooled by surrounding air and liquefies. The heat must be convected away from the condenser coils if the refrigerator is to work properly. The liquid now passes through the expansion valve A and the whole process is repeated.

*QUESTION 7:* A man builds a working surface in his kitchen. In doing this he completely surrounds the top, back and sides of the refrigerator with wood. Why will the refrigerator not work properly and what danger is there in enclosing it in this way?

### 3. *In fire fighting*

When there is danger of a fire spreading to nearby buildings, firemen keep these buildings sprayed with water. The water evaporates on the heated surfaces and cools them down so that they do not reach their ignition point. When water is sprayed onto material that is already burning, the evaporation of the water removes so much heat that the burning is extinguished.

### 4. *In photography*

An evaporograph camera can take pictures in the dark. Radiant heat from the object to be photographed is focused on a thin oil film. The oil evaporates according to the temperatures received on different parts of the film. An example of the resulting photograph can be seen in Fig. 23.12.

## 23.10  Moisture in the air

Water from seas, rivers, lakes, land and living things evaporates into the atmosphere. When the amount of water vapour in the air is large we say that the humidity is high. Can moisture be removed from the air? The following experiment helps us to find out.

Air is bubbled through the ether (a volatile liquid) placed in the apparatus shown in Fig. 23.13. Immediately

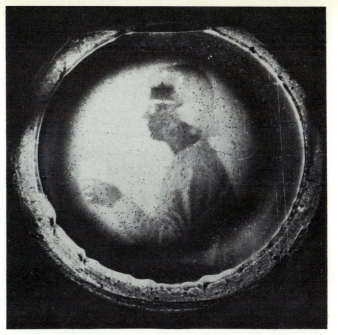

**Fig. 23.12**  *An evaporograph*

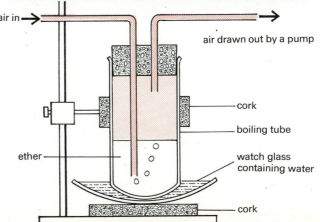

**Fig. 23.13**  *What happens when moist air is cooled?*

air in →

air drawn out by a pump

cork

boiling tube

ether

watch glass containing water

cork

droplets of moisture (misting) are seen on the outside of the tube. Gradually these freeze over and look like frost. The water in the watch glass turns to ice.

This shows us that when air is cooled sufficiently it gives up the water it contains.

Cold air cannot hold as much water vapour as warm air.

### 23.11 Weather

The condensing of water when air cools explains many of the weather situations we experience all the year round. Dew and ground frost are formed when water vapour in the lower atmosphere (much of it given off by plants) is cooled. The water vapour changes to a liquid, falls to the ground and freezes. Hoarfrost is formed when there is such a rapid fall in temperature that water vapour in the air freezes very rapidly.

Mist is formed when air containing a lot of water vapour comes into contact with cold ground or cold air.

When hot air laden with water vapour rises it meets cold air, and is cooled; the water vapour condenses, and forms a large mass of very fine water droplets. These masses appear as clouds.

*QUESTION 8:* (a) What is fog? (Hint: the use of smokeless zones is one of the ways of preventing severe fogs.) (b) Why do car windows mist up on the inside in cold weather?

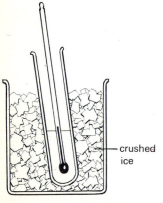

Fig. 23.14 *Ice and salt form a freezing mixture*

### 23.12 Experiment to investigate what is meant by a freezing mixture

Crushed ice is placed in a beaker and in the centre of it is placed a test tube containing a little water and a thermo-meter (Fig. 23.14). A second beaker is prepared in the same way but this time the crushed ice is mixed with an equal quantity of salt. (Additional beakers with varying amounts of salt can be used to extend the scope of the experiment.) Record the length of time it takes for the water in each test tube to freeze.

The water surrounded by the ice and salt freezes, but the water surrounded by the ice does not freeze. The ice and salt together must therefore be at a lower tempera-ture than ice alone (0°C, 273K).

Ice and salt form a freezing mixture.

If, on the other hand, water contains a dissolved im-purity such as salt, its freezing point is lowered below 0°C (273K). This is why a motorist puts anti-freeze into his car radiator and why road maintenance men spread salt and sand on icy roads.

### 23.13 Can anything else affect the freezing point of water?

Look at Fig. 23.15. These are familiar scenes of boys enjoying the snow, and the inevitable snowball fight. But

(a)  (b)

Fig. 23.15

look at the difference. In (a) the boy is just scooping up the snow and throwing it, and it leaves his hand as flakes. In (b) a ball of snow is thrown.

QUESTION 9: What have the boys done to the snow to make snowballs? Scientists would say that they have subjected the snow to ————. What is the word? (Hint: look at page 84.)

**Fig. 23.16**  *A stage magician cuts a woman in half*

The magician in Fig. 23.16 is cutting the woman in half, but at the end of the trick she appears unharmed. We can do a similar "trick" with ice, and without deceiving the audience. The weighted copper wire (Fig. 23.17) travels through the block, but the ice still remains in one piece. The pressure of the wire (like the pressure of the hands in the making of snowballs) lowers the freezing point of water. Thus when pressure is exerted, the ice melts because its temperature is now above its new freezing point. When the pressure is released the freezing point returns to normal, thus the ice refreezes. This is called *regelation*.

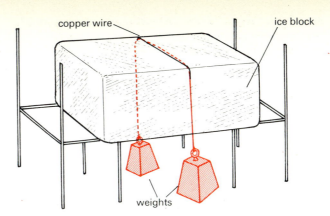

**Fig. 23.17**  *Experiment to show the regelation of ice*

The pressure exerted by the wire melts the ice; the wire sinks through the water; the water released from the pressure of the wire refreezes, and the block remains in one piece. In fact when the water above the wire refreezes, its latent heat of fusion is given out and absorbed by the ice below the wire to give it the necessary heat for melting.

QUESTION 10: Can you explain why the experiment is not a success when string is used instead of the copper wire?

On freezing, water expands and exerts a tremendous force on any container.

### 23.14  Why do car owners use anti-freeze?
Try this experiment.

A glass bottle with a screw cap is filled completely with water and the cap is screwed on very tightly. It is then placed in the centre of a tin and packed all round with salt and ice. After many hours the glass bottle is examined.

The water in the glass bottle has frozen and the glass has shattered.

This can happen to any container filled with water. If damage to the container is to be avoided, we must prevent the water from freezing. In the case of pipes, lagging is used; and in the case of car radiators, anti-freeze is added to lower the freezing point (this is often ethylene glycol.)

*QUESTION 11:* If you were building a concrete fish pond in your garden would you construct it like a box with sides straight and parallel, or would you slope the sides gently outwards? Give reasons.

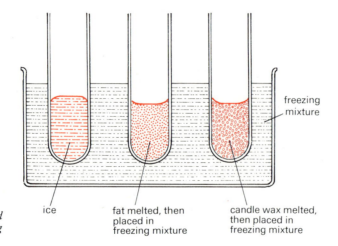

**Fig. 23.18** *Not all liquids expand on solidifying*

ice

fat melted, then placed in freezing mixture

candle wax melted, then placed in freezing mixture

freezing mixture

### 23.15 Do all liquids expand on solidifying?

The experiment shown in Fig. 23.18 investigates the effects of solidifying water, melted fat and melted candle wax by means of a freezing mixture. It is clear from measurements taken that whereas water expands on solidifying, fat and candle wax contract. This applies to a number of other liquids also and has a useful application in the making of moulded articles: a contracted solid can easily be removed from its mould.

254

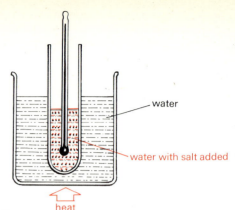

water

water with salt added

heat

**Fig. 23.19** *The presence of impurities in water raises the boiling point*

### 23.16 Do dissolved impurities have any affect on the boiling point of water?

Just as the presence of an impurity alters the freezing point, the presence of an impurity changes the boiling point. The presence of any "foreign" molecule upsets the normal molecular pattern and molecular forces, therefore changing the material's normal physical reactions.

By using the apparatus illustrated in Fig. 23.19 it can be discovered that the presence of an impurity such as salt raises the boiling point. A small quantity of salt is added to the water in the test tube. The temperature at which it boils is noted. This procedure is repeated as more salt is added.

### 23.17 Water can be boiled by cooling it!

Figure 23.20 shows the apparatus by which we can do this. The water is heated to boiling point and steam is allowed to escape from the flask. (The flask must be round and of thick glass to make the experiment safe.)

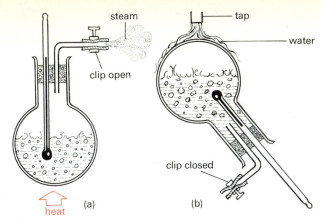

**Fig. 23.20** *Experiment to determine the effect of air pressure on boiling point*

After several minutes the clip is closed and the heating stopped. The flask is left for some minutes before being placed beneath the cold water tap. As cold water runs over the flask the water is seen to boil. It is possible to get the water to boil in this way when the flask is only just warm to the touch.

How can we explain this?

The escaping steam drives air out of the flask. When the flask is sealed and cooled the remaining water vapour in the flask is condensed. Thus the air pressure above the water in the flask is greatly reduced.

When the air pressure above water is reduced, the boiling point is lowered.

This effect can easily be explained if you think back to the explanation of vaporization in terms of the kinetic theory (page 246). In order to escape from the surface of the liquid during vaporization, the molecules have to overcome the force resulting from the downward pressure of the air. Obviously if the air pressure is less, the molecules need less kinetic energy to enable them to escape, and the water boils at a lower temperature.

*QUESTION 12:* People living at high altitudes above sea-level find it very difficult to cook foods thoroughly by boiling them in water. Explain why.

### 23.18 Water boils at a higher temperature if air pressure is increased

This fact follows from the discussion above. It is made use of in the pressure cooker. This cooks food quickly because the pressure inside can be twice that of normal atmospheric pressure and the temperature of the water is about 120°C.

In food manufacturing processes it is necessary to cook at very high temperatures in order to kill off bacteria. This combined cooking and sterilization process is carried out by passing *superheated* steam under pressure into large containers (autoclaves).

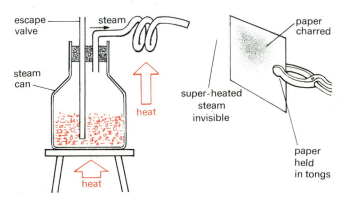

**Fig. 23.21** *Superheated steam*

Superheated steam can be produced quite easily by using the apparatus illustrated in Fig. 23.21. The steam will set fire to the paper because the molecules have so much energy and its temperature is very high. Such steam can be used to drive turbines, but we will discuss this in the chapter on heat engines.

## THINGS TO DO

**A.** Make a wet and dry bulb hygrometer.

Select two identical thermometers and clamp them upright. Under one thermometer, at a distance of approximately 10cm, position a small container that will hold water. Cut a rectangular strip of meat-cloth (ask your butcher if he can supply this) long enough to stretch from the bottom of the container to the top of the thermometer bulb. Wind the cloth round a pencil so that a long tube of muslin is formed. Slip one end of the tube round the bulb of one of the thermometers. The other end should dip below the surface of the water in the container. Keep the muslin in position by tying a piece of cotton around it near the top of the thermometer bulb. This "wet" bulb must be continuously supplied with water, therefore the level in the container must be maintained. Keep a record of the temperatures indicated on each of the thermometers. The difference between the two temperatures indicates the humidity of the air.

**B.** Make a solar still (Fig. 23.22).

Select a wooden or cardboard box, approximately 30 × 30 × 30cm (without a lid). Draw the diagonals on two opposite sides of the box. Saw down one set of opposite diagonals. Then you are left with the base of the still as illustrated in Fig. 23.22. Paint the inner surfaces of the box with white paint so that light will be reflected.

Choose a pane of glass that fits exactly across the sloping top of the box. Seal the glass to the top edge of the box using adhesive tape as a hinge. Obtain a piece of plastic rain guttering (or tube cut in half). Position this along the bottom edge of the glass so that it acts as a trough, catching the condensed vapour. Place a trough of salt water in the still. Stand the still out of doors so that it traps the sun's rays. (The word "still" is used to describe a form of apparatus where a process of vaporization and condensation takes place. This may be done to purify a liquid or to concentrate a solution.)

**C.** A refrigerator working in reverse is termed a "heat pump". Find out all you can about the use of a heat pump in modern forms of central heating, and its advantages.

**D.** Make a chart showing the melting points of different materials. Illustrate how their melting points may influence the everyday uses of the materials. (For example, steel is a suitable material for fireproof doors.) Include those materials that do not have a definite melting point and which become semi-molten. (Iron may be beaten into shape when it is semi-molten.)

**Fig. 23.22** *Construction of a solar still*

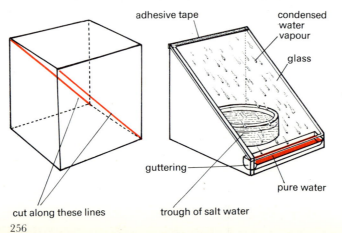

adhesive tape

condensed water vapour

glass

guttering

pure water

cut along these lines

trough of salt water

1. Explain the following statements:
   a. If butter is kept in a glass container which stands in a few centimetres of water inside a porous pot, it will not become soft in hot weather.
   b. The cooling of the butter (above) is more effective if the porous pot is placed in a draught.

2. Explain the following statements:
   a. The sea round our coastline very rarely freezes.
   b. The surface of fat that has solidified curves downwards in the centre, whereas the top of an ice cube curves upwards.

3. Compare and contrast the processes of boiling and evaporation.

4. Describe, using a diagram, how you would show that an evaporating liquid removes heat from its surroundings. Give an example of one application of this process.

5. Why does a wet and dry bulb hygrometer indicate how much moisture there is in the air?

6. Answer the following briefly:
   a. What is meant by the term "latent heat"?
   b. You observe the temperature as you heat a block of ice until it becomes steam. At what stages would you expect to see the temperature remain steady?
   c. What are the two different types of specific latent heat and to which changes of state do they apply?
   d. Describe the changes that take place in the molecular state of a liquid when it boils.

7. Explain the following statements:
   a. Different substances have different specific latent heats.
   b. A motorist possesses a car with a water-cooled engine, therefore he buys ethylene glycol in the winter.

   c. Plants and animals whose tissues contain a large percentage of water do not "freeze up" even when the water in their surroundings is frozen.
   d. A housewife sometimes cooks puddings by a process which she calls "steaming".

8. Describe an experiment that can be done in the laboratory to show that people at high altitudes can boil water at a lower temperature than 100°C (373K).

9. Explain the term "regelation". Why are people able to skate on ice but not on glass?

10. What is the substance that you would mix with ice in order to form a freezing mixture? Why does the ice melt fairly quickly after the substance is added?

11. Who cooks more economically, Mrs A, who cooks all her vegetables in a pressure cooker, or Mrs B who cooks them in saucepans? Explain your answer.

12. "Sweating is nature's way of cooling the body." Explain.

13. What is the scientific reasoning behind the statement that people often make on a hot sultry day: "What we need is a shower of rain to clear the air."

14. When my grandmother made ice cream she packed ice round the container and then poured salt on the ice. Why did she pour the salt on the ice?

15. On a hot summer's day I decided to cool my kitchen by leaving the refrigerator door open. Was this a good way to cool the kitchen? Give a reason for your answer.

**CHAPTER 24** | **24.1 James Prescott Joule**

# THE MEASUREMENT OF HEAT

In Chapter 12 we were concerned with one of the most important concepts in physics, namely, the conservation of energy.

The law of conservation of energy was first explicitly stated by an Englishman, James Prescott Joule, in 1843.

Joule devoted many years to finding the relationship between heat and mechanical energy. Even when he was on his honeymoon he spent many hours measuring the temperature difference between water at the top of a waterfall and that at the bottom! His many experiments led him to believe that energy is indestructible and that whenever energy is converted into heat an exact equivalence of heat energy is always obtained. This statement has been tested for all forms of energy, and all experiments have so far verified this principle. Indeed the law of conservation of energy is one of the foundation stones of science. One important aspect of Joule's work was that he showed that heat was a form of energy.

## 24.2 The unit of heat energy

Before taking a holiday abroad we change some of our money into the currency of the country we are to visit. The foreign money looks different, but it has the same value as the English money we converted.

Just as money can be converted from one currency to another so can energy be converted from one form to another (Fig. 24.1). However, in the case of currency it would be simpler for banks and travellers abroad if all the world dealt in the same unit of money! If you look again at (b) in Fig. 24.1 you will see that in the case of energy, scientists throughout the world do in fact deal in the same unit. This international unit is the joule (see page 114).

The joule (J) is the unit of all forms of energy including heat energy.

Heat energy may be taken up by a material, in which

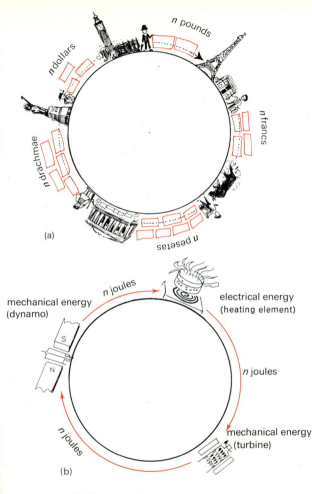

**Fig. 24.1**   *There is no universal unit of currency but there is a universal unit of energy: the joule*

case it becomes hotter; or it may be emitted, whereupon the material becomes colder. We observe these changes when we consider the temperature of the material.

Temperature is a measurement of the degree of hotness of a material.

If the same quantity of heat energy is given to different materials of the same mass, will their temperatures rise by the same amount? We can find out the answer to this question by means of an experiment which uses electrical energy to provide the heat. (Electricity is used because it is easily controlled: a constant supply can be ensured and we know exactly how much energy we are providing.) An ammeter is connected in series with the heater and a voltmeter across it; the current may be kept constant by using a variable resistance. (The circuit is the same as that in Fig. 18.17, page 182, with the heater in position PQ.) The energy liberated in a set time ($t$) can be calculated according to the formula:

$$\text{electrical energy} = \text{V}It \text{ joules (see page 184)}$$

Alternatively the heater may be connected to a joule-meter as is done in the following experiment.

### 24.3   Experiment to find out what happens when an equal quantity of heat energy is given to unit mass of different materials

a. *Water*

A metal container is weighed empty on a lever balance. Water is poured into the container until 1kg is present (Fig. 24.2). The initial temperature of the water is taken. An electric immersion heater in the form of a rod is placed in the water, together with a stirrer. A joulemeter is placed in the circuit of the immersion heater which is connected to a 12V a.c. supply. The initial joulemeter reading is noted.

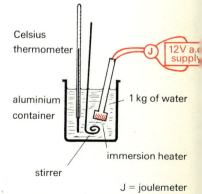

**Fig. 24.2**   *Experiment to determine the temperature rise of water (liquid) when it is given a known quantity of heat*

259

The heater is switched on and the water is stirred continuously. When a reasonable rise in temperature is achieved the heater is switched off and the highest temperature that is reached is recorded. The final joule-meter reading is noted so that the amount of energy supplied to the heater may be calculated.

The experiment may be repeated with equal masses of different liquids so that a comparison may be obtained.

b. *Aluminium*

The aluminium is provided in the form of a cylinder weighing 1kg. The cylinder has two holes drilled into it (Fig. 24.3). The immersion heater is placed in the central hole, and a thermometer in the second hole. The procedure already described above is repeated. The joule-meter is observed continually so that when it registers the same amount of electrical energy that was given to the heating of the water, the heater is switched off. The highest temperature that is reached is again recorded. The experiment may be repeated with 1kg masses of different solids.

The temperature rises for different materials are found to be quite different. For example the temperature rise for water is about a thirtieth of that for lead, and about a fifth of that for aluminium.

*QUESTION 1:* Lagging around the containers in the above experiments may produce better results. Why is this?

## 24.4   The "appetite" of materials for heat

In the above experiments we ensured that the heater in each case delivered the same quantity of heat energy to the same mass of material. Therefore, the difference in temperature rise can only be caused by the nature of the material itself.

As an illustration of this let us take a class of children,

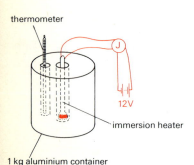

**Fig. 24.3** *Experiment to determine the temperature rise of aluminium (solid) when it is given a known quantity of heat*

thermometer

1 kg aluminium container

immersion heater

12V

**Fig. 24.4**   *Different children have different appetites for ice cream. In much the same way different materials have different appetites for heat.*

none of whom have eaten since breakfast. At noon we provide each of them with as many ice creams as they can eat (Fig. 24.4). Do you think that they will all eat the same number? Their appetite for ice creams varies enormously. Each child's stomach has a different capacity for ice cream!

In a similar fashion, individual materials have different "appetites" for heat energy. We call this *heat capacity.*

## 24.5   Specific heat capacity

In order to compare the heat capacity of different materials we define the *specific heat capacity* as follows.

The specific heat capacity of a material is the amount of heat energy (measured in joules) that is required to raise the temperature of 1kg of the substance by 1 kelvin (K).

The unit of specific heat capacity is therefore joules per kilogramme kelvin (J/kg K).

If the temperature rise is measured according to the Celsius scale, then (because 1°C = 1K) the specific heat capacity unit is J/kg °C.

The table on this page gives the values of the specific heat capacities of some common substances. If you look at these values you can see that the specific heat capacity for water is thirty times that of lead and five times that of aluminium. This bears out our results in the experiment on page 260.

## SPECIFIC HEAT CAPACITY VALUES

| Substance | J/kg K (or J/kg °C) | Substance | J/kg K (J/kg °C) |
|-----------|---------------------|-----------|------------------|
| Lead | 140 | Wood | 1680 |
| Mercury | 140 | Turpentine | 1800 |
| Brass | 370 | Paraffin | 2100 |
| Copper | 380 | Methylated | |
| Iron | 460 | spirits | 2400 |
| Glass | 670 | Brine | 3000 |
| Aluminium | 840 | Water | 4200 |
| Zinc | 380 | Ice | 2100 |

QUESTION 2: If paraffin and mercury are used in the experiment described in section 24.3, what rise in temperature do you expect for these materials compared with the temperature rise for water? (Refer to the table for the specific heat capacities of these materials.)

## 24.6   The measurement of heat energy

The specific heat capacity for water as given in the table

is 4200J/kg K. Therefore we can state that the temperature of 1kg of water rises by 1K (1°C) when supplied with 4200J.

From this statement we can calculate that

a. the temperature of 2kg of water rises by 1K (1°C) when supplied with (2 × 4200) = 8400J
b. the temperature of 3kg of water rises by 1K (1°C) when supplied with ? J
c. the temperature of 4kg of water rises by 1K (1°C) when supplied with ? J

OR

d. the temperature of 1kg of water rises by 2K (2°C) when supplied with (2 × 4200) = 8400J
e. the temperature of 1kg of water rises by 3K (3°C) when supplied with ? J
f. the temperature of 1kg of water rises by 4K (4°C) when supplied with ? J

OR

g. the temperature of 2kg of water rises by 2K (2°C) when supplied with 16 800J
h. the temperature of 3kg of water rises by 4K (4°C) when supplied with ? J
i. the temperature of 5kg of water rises by 10K (10°C) when supplied with ? J

QUESTION 3: What is the value of the heat energy in joules for parts b, c, e, f, h and i given above?

From this we are able to see that in order to calculate the number of joules required to heat a given quantity of water through a required temperature rise, we must multiply the specific heat capacity for water by the mass of water and its change in temperature.

This can be stated in the following form:

Quantity of heat = mass of water × specific heat capacity for water × temperature change

The above argument clearly applies to other substances. Thus we have a general formula by which we can calculate the quantity of heat needed to cause any given temperature change in any mass of substance, so long as we know the specific heat capacity.

**Fig. 24.5** *How to calculate heat energy*

Quantity of heat energy = mass × specific heat capacity × change in temperature

This may be abbreviated to
$$Q = mcT$$ (see Fig. 24.5)

where $c$ is used to denote specific heat capacity (J/kg K), $m$, mass (kg) and $T$, temperature change (K). The heat energy, $Q$, is in joules.

### 24.7 Calculations using the heat formula
(Refer to the table on page 261 for all specific heat capacities to be used in the calculations. In these calculations heat loss to the surroundings or to containers is ignored.)
1. A mass of water of 100g is to be heated from 285K (12°C) to 305K (32°C). What heat energy in joules is required?

Specific heat capacity for water = 4200J/kg K
Mass of water = 100g = 0.1kg
Temperature rise = 305K − 285K = 20K (20°C)
$$Q = mcT$$
$$= 0.1 \times 4200 \times 20J$$
$$= 8400J$$
The quantity of heat energy required is 8400J.

262

2. A 10kg mass of aluminium is given $3.36 \times 10^5$J of heat energy. What will be the temperature rise for the aluminium, and what will be its final temperature if its initial temperature is 300K (27°C)?

Specific heat capacity for aluminium = 840J/kg K
Mass of aluminium = 10kg
Quantity of heat supplied = $3.36 \times 10^5$J
$$Q = mcT$$
$$3.36 \times 10^5 = 10 \times 840 \times T$$
$$\text{Therefore } T = \frac{3.36 \times 10^5}{10 \times 840} \text{ K}$$
$$= 40K$$

The temperature rise is 40K.
Initial temperature of aluminium = 300K (27°C)
Therefore final temperature = initial temperature + temperature rise
$$= 300 + 40K$$
$$= 340K (67°C)$$
The final temperature is 340K.

*QUESTION 4*: (a) How many joules of heat energy are required to raise the temperature of 2kg of copper from 283K (10°C) to 308K (35°C)? (b) A heater supplies 2100 joules of heat per second. 1kg of water starts off at a temperature of 283K (10°C). What will be its final temperature if it is heated by this heater for one minute?

Where two substances at different temperatures are mixed together, there is a flow of heat energy from the one at a higher temperature to the one at a lower temperature. This continues until they are both at the same temperature, the final temperature of the mixture. In this case: heat given out by one material = heat absorbed by the other, providing there is no heat loss to the surroundings.

### 24.8 Experimental method to determine the specific heat capacity for a liquid
The method generally used is identical with that used in the experiment described in section 24.3a. Any mass of

liquid may be used as long as it is measured accurately on the lever balance. This is done by first weighing the container empty, then reweighing it when it contains the required mass of liquid.

The readings of the joulemeter and the temperatures are recorded before the current is switched on, and again just after it is turned off. The liquid must be stirred during the experiment.

In order to obtain accurate results we should consider the heat exchange more closely. Is there anything else that is accepting heat from the immersion heater? The container of the liquid (a calorimeter) absorbs some heat and we must take this into account. The formula for the heat exchange is thus:

Heat liberated by the heater = heat absorbed by the liquid + heat absorbed by the container
Heat liberated by the heater = final reading of joulemeter − initial reading of joulemeter

Readings:

Initial reading of joulemeter = 11 000J
Final     „          „        = 15 830J
Specific heat capacity of container = 380J/kg K
Mass of container = 150g = 0.15kg
Mass of liquid = 100g = 0.1kg
Temperature rise = 10K (10°C)

Calculations using readings:

Heat liberated by heater = heat absorbed by liquid + heat absorbed by container

$$= (mcT)_{\text{liquid}} + (mcT)_{\text{container}}$$

Thus $15\,830 - 11\,000 = (0.1 \times c \times 10) + (0.15 \times 380 \times 10)$
$$4830 = c + 570$$
Therefore $c = 4830 - 570$J/kg K
$$= 4260\text{J/kg K}$$

The specific heat capacity of the liquid is 4260J/kg K.

## 24.9 Experimental method to determine the specific heat capacity for a solid

An identical electrical method to that described in the experiment on page 260 in section 24.3b is used here.

Similar calculations to those above may be carried out to determine the specific heat capacity of the solid, using the following formula.

Heat liberated by heater = heat absorbed by the cylinder
$$= (mcT)_{\text{cylinder}}$$

*QUESTION 5:* (a) What is the specific heat capacity of turpentine in J/kg K when an experiment shows that $1.35 \times 10^{4}$J are required to raise the temperature of a mass of 0.5kg by 15K (15°C)? (b) In an experiment the temperature of a 1kg cylinder rises through 40K (40°C) when heated by a 60W heater for 9 minutes 20 seconds. What metal is the cylinder made of?

## 24.10 Uses and effects of materials having different specific heat capacities

The most important aspect of the high specific heat capacity of water is, of course, its influence on weather, and in particular the effect of the oceans not only on coastal climates but on the general world atmospheric circulation. You can find out more about this from your geography teacher.

Because of its high specific heat water has many industrial and domestic uses — in central heating for instance (it gives out more heat when cooled 1K than an equal mass of any other substance) and as a cooling agent for all forms of machinery (in comparison with other liquids its temperature rise is small for any given quantity of heat).

Other materials are useful because of their low specific heat. Aluminium and copper make good saucepans because they reach the required temperature quickly and absorb little heat in doing so. The low specific heat of mercury makes it a good liquid to use in a thermometer: it quickly reaches the temperature it is measuring and absorbs very little heat from its surroundings.

QUESTION 6: Before the days of hot water bottles and electric blankets hot cinders were used in copper bed-warming pans. If similar pans were constructed to hold a mass of water equal to the mass of cinders and at the same temperature as the cinders, would they heat the bed more efficiently?

### 24.11   What is heat capacity?

In the case of appliances such as kettles, saucepans and boilers, it is convenient for engineers to think in terms of the "appetite" of the complete mass of the object for heat.

The heat capacity of a body is the quantity of heat required to raise its temperature by 1 degree.

The unit for heat capacity is joules per kelvin.

If a body has a mass of $m$kg, and a specific heat capacity of $c$ J/kg K, then $mc$ joules will be required to raise its temperature through 1K.

$$\text{Heat capacity} = mc \ (\text{J/K})$$

### 24.12   Experiment to determine how much heat is needed to change ice to water

The specific latent heat of fusion of ice (page 246) may be determined by the method shown in Fig. 24.6.

The immersion heater is placed in the centre of a metal container which is packed full of well-crushed ice. The heater is switched on, the current is kept constant by means of a variable resistance and the time taken for the ice to melt completely is observed. The time taken for the temperature of melted ice to rise through 40K (40°C) is also noted. This final observation allows the rate at which the heater is supplying heat to be calculated.

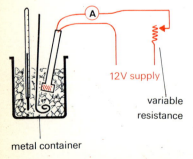

metal container

**Fig. 24.6** *Experiment to determine the specific latent heat of fusion of ice*

A
12V supply
variable resistance

Calculation:

If $m$ kilogrammes of melted ice take 600s to heat up through 40K

Then heat supplied by heater in 600s = heat taken up by water
$$= (mcT)_{\text{water}}$$
$$= m \times 4200 \times 40 \text{ joules}$$

Therefore heat supplied by heater in 1s $= \dfrac{m \times 4200 \times 40}{600}$ joules

If the same mass of ice takes 1200s to melt

Then heat supplied to melt $m$kg of ice $= \dfrac{m \times 4200 \times 40}{600} \times 1200\text{J}$

Therefore heat supplied to melt 1kg of ice $= \dfrac{m \times 4200 \times 40}{600 \times m} \times 1200\text{J}$
$$= 336\,000\text{J}$$

The specific latent heat for the fusion of ice is 336 000 J/kg.

QUESTION 7: At what rate would you expect the thaw of a heavy fall of snow to take place on a dull day?

### 24.13   Experiment to determine how much heat is required to change water to steam

The specific latent heat of vaporization for water may also be determined with the same apparatus as in the previous experiment. The metal container is weighed empty on a lever balance, and water is added until 0.5kg is present. The immersion heater is placed in the water and the rate at which it supplies heat is found by recording the time taken for the water to go through a definite temperature rise. The water is allowed to boil away until there is just enough left to cover the immersion heater element. The time taken for this to occur is recorded. The container and remaining water are reweighed in order to calculate the mass of water which has been boiled away.

If heater raises temperature of 0.5kg of water by 80K in 220s
Then heat supplied by heater in 220s = heat taken up by water

$$= (mcT)_{\text{water}}$$
$$= 0.5 \times 4200 \times 80\text{J}$$

Therefore heat supplied by heater in 1s $= \dfrac{0.5 \times 4200 \times 80}{220}\text{J}$

If the heater takes 600s to boil away 200g (0.2kg) of water
Then heat supplied to boil away 0.2kg of water

$$= \dfrac{0.5 \times 4200 \times 80 \times 600}{220}\text{J}$$

Therefore heat supplied to boil away 1kg of water

$$= \dfrac{0.5 \times 4200 \times 80 \times 600}{220 \times 0.2}\text{J}$$
$$= 2\,290\,000\text{J/kg}$$

The specific latent heat of vaporization for water is 2 290 000J/kg. (Normally accepted value: 2 260 000J/kg)

You can now see what a tremendous amount of heat energy is stored within 1kg of steam and why it is such a valuable "working fluid" in engines.

*QUESTION 8:* Explain why 1g of steam will probably produce a more serious burn when it comes into contact with human flesh than 1g of boiling water.

### 24.14 Experiment to compare mechanical energy and electrical energy by using the heating effect that each produces

This comparison may be carried out using the apparatus illustrated in Fig. 24.7. The pulley is attached to a spindle which is housed in a split block. The wings on the housing are used for adjustment. For example, if they are tightened, the housing grips the spindle more tightly and greater friction and heating are produced. The temperature rise may be noted on a thermometer positioned within a central hollow in the spindle. Within a spiral groove in the spindle there is a coil of resistance wire. This wire is connected to terminals on the pulley.

**Fig. 24.7** *Apparatus used to compare the heating effect of mechanical and electrical energy*

A measured length $d$, 8 to 10m, of cord is wound around the pulley. A spring balance, calibrated in newtons, is attached to the free end of the cord. The cord is pulled and the wing nuts are adjusted so that a convenient tension $F$, required to rotate the pulley, is recorded on the spring balance. A suitable reading for $F$ would be about 70 newtons. This procedure of fixing the tension will have caused the spindle to heat up and it must be allowed to cool down before beginning the experiment. Its initial temperature is then recorded.

The length of cord $d$ is rewound onto the pulley. The cord is drawn out under constant tension $F$. The highest temperature reached by the spindle is recorded.

Thus the rise in temperature brought about by the heating effect of the mechanical work can be calculated.

The mechanical work expended $= Fd$ joules (see page 114).

The apparatus is allowed to cool. The initial temperature is recorded, then the terminals are connected to a joulemeter and a 12 volt electrical supply. The current is switched on. It is switched off when the same number of joules (recorded on the joulemeter) of electrical energy have been fed into the apparatus. The highest temperature is recorded.

The rise in temperature brought about by the electrical energy will compare very closely with that brought about by the mechanical work. As the mass and specific heat capacity of the spindle is constant in each case, we conclude that the heating effect of an equal quantity of mechanical and electrical energy is the same. This is yet another instance where we have verified the law of conservation of energy.

A. Find out the heat capacities and specific heat capacities of the teapots and kettles in your home, using the following method.

Weigh the teapot empty, then pour boiling water into it. Stir the water and record the temperature when it is steady. Reweigh the teapot to find the mass of the water. The heat energy absorbed by the teapot = heat energy given out by the boiling water (assuming that there is no heat loss to the surroundings).

B. Find the specific latent heat of fusion of ice by using a funnel (there is sure to be one in your kitchen or garage). Pack it with ice at 0°C. Obtain a large metal bolt which should be weighed and then suspended by a thread in boiling water for several minutes, until its temperature is 373K (100°C). Quickly transfer the bolt to the ice in the funnel and collect the melted ice that drips through. The mass of the melted ice should be measured. The specific latent heat of fusion may be calculated by assuming that the heat given out by the bolt on cooling to 273K (0°C) = heat taken in by the mass of ice in order to melt.

C. Measure the power provided by an electric kettle.

To do this find how long it takes a known mass of water at a measured temperature to come to the boil. Use the fact that 1kg of water needs 4200J to raise its temperature 1°C to calculate the heat supplied to the water. The number of joules supplied to the water every second is the power. Is your answer smaller or greater than the power of the element? Can you explain why?

D. Make up a *Who's Who* file, with illustrations, on the life and work of James Prescott Joule.

## THINGS TO WORK OUT

(For all calculations in this section use the constants given in this chapter and on page 247.)

1. What quantity of heat energy is required to raise the temperature of (a) 10g of copper through 50K (50°C), (b) 1kg of water through 1K (1°C), (c) 10kg of water through 1K (1°C), (d) 100g of lead through 10K (10°C), (e) 150g of aluminium through 0.5K (0.5C)?

2. Explain why the following statement is true: A bath full of water at body temperature contains more heat energy than a tablespoonful of boiling water.

3. A and B are two masses of water at 273K (0°C) which receive the same quantity of heat energy. When the 200g (A) mass of water reaches boiling point, the temperature of the second mass (B) has only reached 323K (50°C). What is the mass of B?

4. Place the following substances in order, according to which gives out the greatest quantity of heat energy on cooling through 55K (55°C): (a) 1kg of lead, (b) 0.5kg of wood, (c) 1kg of glass, (d) 0.2kg of water.

5. What quantity of heat energy is given out by the following?
   a. 100g of lead cooling from 373K (100°C) to 293K (20°C)
   b. 20g of copper cooling from 300K (27°C) to 280K (7°C)
   c. 1kg of mercury cooling through 10K (10°C)
   d. 10kg of paraffin cooling from 373K (100°C) to 263K (−10°C)

6. Define specific heat capacity. Describe an experiment to find the specific heat capacity of water.

7. 1kg of coke provides 33 600 000 joules of heat energy when it burns. What mass of coke in grammes would be required to heat 120kg of water from 293K (20°C) to 353K (80°C)?

8. Define heat capacity. 100g of boiling water is poured into 150g of water contained in a calorimeter. The calorimeter and water are intially at a temperature of 293K (20°C). The final temperature of the mixture is 323K (50°C). Assuming that there is no heat loss to the surroundings what is the heat capacity of the calorimeter?

9. Explain the following statements:
   a. During the day at the seaside the sand is much hotter than the water although each is receiving the same amount of heat energy per unit area.
   b. Sweating is nature's way of keeping the body cool.

10. Why does it take longer to boil away a quantity of water than to bring the same quantity of water to its boiling point?

11. Crushed ice is placed in a funnel. Cylinders of different masses and different materials are heated in boiling water for a period of time so that it can be assumed that they are at a temperature of 373K (100°C).
   a. Certain cylinders of lead, copper, aluminium, wood, iron and brass, all having the same mass, are placed in turn in the funnel. List them in descending order, according to the quantity of ice melted by each.
   b. Which would melt the greater quantity of ice, a 50g copper cylinder or a 100g copper cylinder? Explain your answer.
   c. What mass of ice would the 100g copper cylinder melt?
   d. What mass of aluminium would melt the same quantity of ice as a 100g lead cylinder?

12. Define the specific latent heat of fusion and the specific latent heat of vaporization. Describe a method to determine the specific latent heat of vaporization of water.

13. What quantity of heat will be required (a) to melt 10g of ice, (b) to turn 250g of water at its boiling point into steam at the same temperature?

14. Describe a method to determine the specific latent heat of fusion of ice. A lump of metal of material X is placed on a block of ice. The metal weighs 1g and is initially at the temperature of boiling water. 0.125g of melted ice is collected. What is the specific heat capacity of the metal?

15. How much heat energy is required to change 0.1kg of ice at 273K (0°C) to steam at 373K (100°C)?

16. What quantity of heat energy is given off when 1.5kg of steam at 373K (100°C) condenses and then cools to a temperature of 303K (30°C)?

17. You are in a hurry to drink your cup of coffee, but it is far too hot. (a) If milk has a lower specific heat capacity than water, will milk or water (using equal quantities) cool the coffee more effectively? (b) Explain why the coffee is cooled down by water. In your explanation use the theory that all molecules possess kinetic energy and that temperature measures the average kinetic energy of the molecules. (c) What is the final temperature when 20g of water at 293K (20°C) is mixed with 100g of tea at 353K (80°C)? (Assume that tea has the same specific heat capacity as water.)

18. 100g of mercury at 343K (70°C) is poured into water at 293K (20°C). The final temperature of the mixture is 313K (40°C). (a) Which of the liquids absorbs heat energy? (b) What is the mass of the water?

## CHAPTER 25

**Fig. 25.1** *Transport in the late eighteenth century*

# HEAT ENGINES

**Fig. 25.2** *Transport 1970s*

## 25.1 Prime movers

Look at the contrasting drawings in Figs. 25.1 and 25.2. They give us some idea of how methods of transport have developed during the past 180 years. Good engines are essential for modern transport, and in this chapter we shall see how a number of modern engines work.

Among the many different types of machine there are what are called prime movers. *A prime mover is a machine that converts the energy available in nature into mechanical energy.*

Such a machine is used to drive other secondary machines. In earlier times the prime movers were animals, wind and water. Then in 1712 Newcomen

**Fig. 25.3** *Stirling Cycle hot air engine, designed to show the direct conversion of heat into work. The transfer cylinder has sides of Pyrex glass, thus the heating (electric) element is visible. The piston in this cylinder operates the power cylinder and thus, in turn, the crankshaft.*

"harnessed" heat energy and used it to produce mechanical energy. This was the first time a heat engine had been used to drive machinery, but it was not the first heat engine. Hero of Alexandria had built a simple steam turbine in the first century AD. It was 2000 years before the conversion of heat energy into mechanical energy was used to drive machinery. (See Fig. 25.3.)

## 25.2 The efficiency of heat engines

We saw in Chapter 14, that machines can never be 100 per cent efficient. This of course applies to heat engines, in fact in their case rather more so. Some heat engines, the old express steam locomotive for example, are only 8 per cent efficient. Why is the efficiency of heat engines so low? Consider the following problem.

*QUESTION 1:* A and B are two groups of 20 children each. Group B is in lines of 5: Group A is not in any sort of order. When the whistle blows group A has to form 4 lines of 5 children and group B has to scatter. Which group will be first to fulfil instructions? (See Fig. 25.4.)   **Fig. 25.4**

group A trying to line up                    group B scattering

It is well known that it is easier to fall into disorder than to create order. The heat engine used to drive machinery has, so to speak, to line up its molecules to apply a regular mechanical force. It is hardly surprising that a lot of energy is wasted in the process. The result is often an extremely low efficiency.

Generally we can say that in the most efficient engines about one-third of the heat energy is converted into mechanical energy. The remaining two-thirds of that heat energy is lost to the exhaust gases and in the cooling of the engine.

Gradually the flame gets smaller and smaller as the gas is burnt, and air enters the can through the hole in the bottom. The flames appear to go out, then a loud explosion occurs, blowing the lid completely off with considerable force.

A mixture of gas and air in certain proportions is explosive. The hot expanding gases in the tin blew off the lid.

Hot expanding gases produced when a fuel mixture is ignited provide the necessary force to drive the moving parts of a petrol or diesel engine.

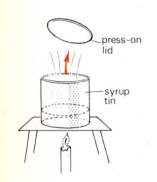

**Fig. 25.5** *Steam provides a driving force*

### 25.3   The working of heat engines
The following two experiments will give you some idea of how mechanical forces are developed from heat energy. Both should be conducted with care and the second should be done in your school laboratory with the assistance of your teacher.

1. *Experiment to study the force of steam (Fig. 25.5)*
Place a little water in the bottom of a tin that has a press-on lid. Replace the lid, but not too tightly. If the water in the tin is boiled the lid will be blown off by the force of the steam. Stand well away from the tin while it is being heated.

2. *Experiment to study the explosive force of gas and air (Fig. 25.6)*
A small hole is made in the centre of the lid of a syrup tin. The lid is placed, not too tightly, on the tin. A hole about 1.5cm in diameter is made in the centre of the base and through this a piece of rubber tubing is inserted to pass coal gas into the tin. When it is full of gas (excess gas can easily be detected coming from the can) the tubing is quickly removed and the gas is lit as it comes from the lid of the can.

### 25.4   Classes of heat engine
The classification of engines is determined by the method by which the "working fluid" is given its heat. "Working fluid" is the term used to describe the gases that cause the movement within the engine.

*In an internal combustion engine* the fuel is burnt inside the working cylinder.

*In an external combustion engine* the fuel is burnt in a furnace which is external to the working parts of the engine.

Within each of these classes of engine two methods are used for converting the kinetic energy of the hot expanding gases into mechanical energy. These are: (a) a *piston-cylinder* arrangement, which is used in *reciprocating* engines (backwards and forwards motion of the piston); (b) a *paddle-wheel* arrangement which is used in *turbines* (rotary motion).

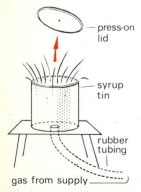

**Fig. 25.6** *An explosive mixture of gases provides a driving force*

# INTERNAL COMBUSTION ENGINES: RECIPROCATING

## 25.5  Basic requirements

When an engine is designed certain basic requirements have to be taken into consideration. Four of these requirements are as follows:

a. There must be a means of introducing the correct mixture of fuel and air into the engine.

b. The mixture must be compressed.

c. There must be a means of igniting the compressed mixture.

d. The spent gases must be removed from the engine.

The distance travelled by the piston from one end of the cylinder to the other is called the *stroke* of the engine. (Newcomen's first engine did 6 strokes in 60 seconds whereas a modern sports car engine will do 100 revolutions per second). Internal combustion engines are built to carry out the above basic requirements in a *4-stroke* or *2-stroke* cycle.

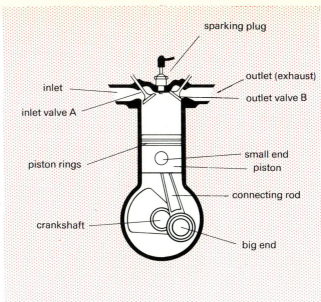

**Fig. 25.7**  *Cross-sectional diagram of a cylinder of a 4-stroke engine*

## 25.6  4-stroke petrol engine

Figure 25.7 shows in section a single cylinder of a 4-stroke petrol engine. The *inlet valve* A is located at the top of the cylinder and in between it and the *outlet valve* B is the sparking plug. You may have heard a garage mechanic or your father talking about adjusting the *tappets*. This is the adjustment on the mechanism that controls the amount by which the valves open and shut. The *piston* which moves within the cylinder is joined by a connecting rod to a *crankshaft*.

*QUESTION 2:* What is the purpose of the crankshaft? (Hint: carry out the simple procedure shown in Fig. 25.8.)

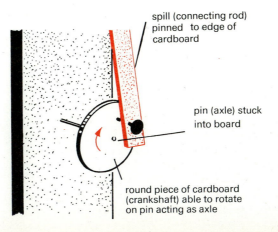

**Fig. 25.8**  *Move the spill (piston rod) up and down. What happens to the crankshaft?*

**Fig. 25.9** *The working of a 4-stroke cycle*

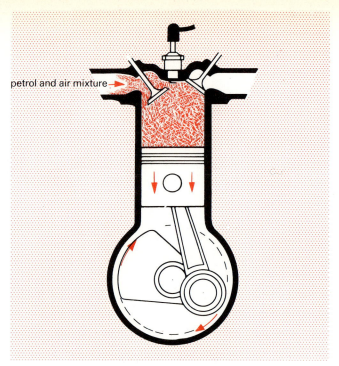

petrol and air mixture →

The 4-stroke cycle is illustrated and described in Fig. 25.9 (a) to (d).

(a) *Induction* (suction) *stroke*. When starting the engine, the starter motor in the car (or the kickstart in a motor cycle) turns the crankshaft which rotates and causes the piston to move downwards. At the same time mechanisms connected to the engine are brought into action to open the inlet valve and close the exhaust valve. Obviously there is a greater pressure outside the cylinder than inside, thus a petrol and air mixture is drawn in from the carburettor. The inlet valve closes when the pressure in the cylinder is equal to air pressure.

(b) *Compression stroke*. As the crankshaft completes its first revolution, it drives the piston upwards. The two valves are closed at this point and hence the gases are compressed. When the piston is almost at the top of the cylinder the *induction coil* (see page 323) and the *distributor* function to cause a spark between the points of the sparking plug. A spark ignites the gases, and if the engine is running efficiently a temperature between 450°C and 850°C is obtained.

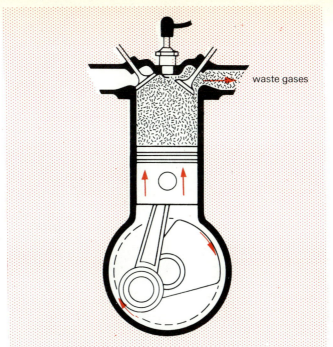

waste gases

(c) *Expansion or power stroke*. The valves still remain closed. The force of the hot expanding gases drives the piston downwards and the crankshaft moves round on its second revolution.

(d) *Exhaust stroke*. As the crankshaft completes its second revolution, it drives the piston upwards. The exhaust valve is opened, so that the waste gases of combustion are driven out.

Once started, the cycle repeats itself as long as fuel, air and electricity are supplied. Although there is only one power stroke, this provides enough energy to drive the crankshaft which in turn drives the moving parts of the machinery. A flywheel on the crankshaft stores some of this energy in order to maintain the other three strokes of the cycle.

You will have observed that the crankshaft revolves twice for just one power stroke in a single cylinder engine. This is suitable for concrete mixers, but in a car there is usually a minimum of four cylinders. These are arranged to fire (spark) in the order 1–3–4–2 as shown in Fig. 25.10. Thus there is a power stroke for every half revolution of the crankshaft.

**Fig. 25.10** *Arrangement of cylinders within a 4-cylinder engine*

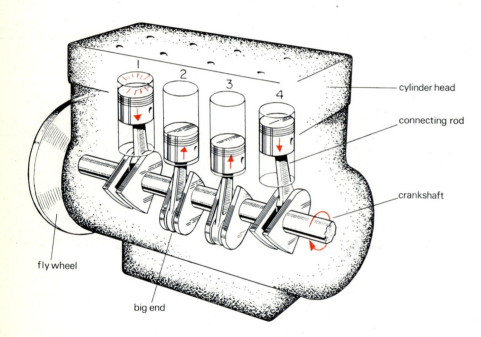

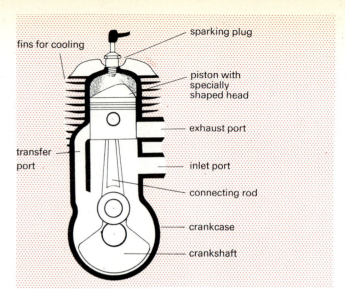

**Fig. 25.11** *Cross-sectional diagram of a cylinder of a 2-stroke engine*

### 25.7 2-stroke petrol engine

Figure 25.11 shows in section a cylinder of a 2-stroke petrol engine.

The 2 stroke cycle is illustrated and described in Fig. 25.12 (a) to (c).

*QUESTION 3*: Can you give two reasons why car engines have four cylinders instead of a single cylinder?

A single-cylinder engine can usually be sufficiently cooled by a simple circulation of air around it, but a water-cooling system is usually required for a greater number of cylinders (see Fig. 21.18).

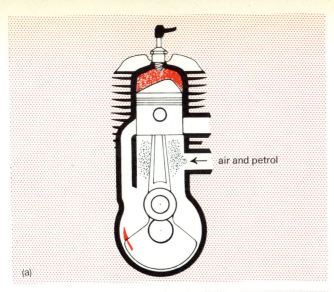

(a)

air and petrol

*Stroke 1.* The piston is at the top of the cylinder (top dead centre) after rising and compressing the gases above it. The exhaust and transfer ports are closed but the inlet port is open so that an air and petrol mixture enters the crankcase. The spark ignites the gases above the piston, driving the piston down (*power stroke*). As the piston descends it closes all the ports, and the fuel mixture below it is compressed in the crankcase. When the piston reaches the furthest point in its decent (bottom dead centre) the transfer port is opened so that fresh fuel is forced into the cylinder above the piston and the burnt gases pass out through the exhaust valve.

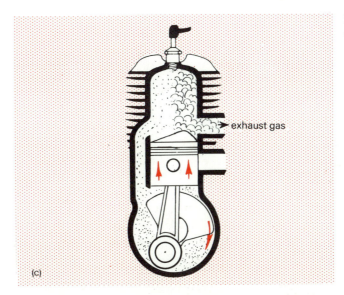

exhaust gas

(c)

**Fig. 25.12** *The working of a 2-stroke cycle*

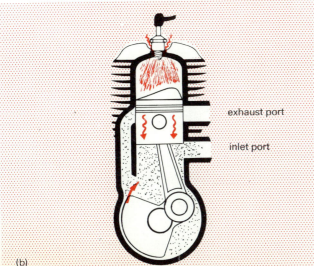

exhaust port

inlet port

(b)

*Stroke 2.* As the piston moves upwards the removal of the waste gases is completed, the ports are closed and the gases above the piston are compressed.
The cycle is repeated.

## 25.8 Differences between 2-stroke and 4-stroke petrol engines

The differences between a 2-stroke engine and a 4-stroke engine can be seen by comparing Figs. 25.7 and 25.11.

1. Note that a 2-stroke engine has openings called *ports* on the sides of the cylinder, whereas a 4-stroke engine has valves. The inlet port opens into the crankcase (the section below the cylinder). A third port called the *transfer port* acts as a connecting passage between the crankcase and the cylinder.

2. In a 2-stroke engine the top part of the piston is curved to create the correct circulation of gases in the cylinder and to ensure that the fresh fuel and the exhaust gases do not mix.

*QUESTION 4:* Can you see two other differences?

A 2-stroke engine gives twice as many power strokes as a 4-stroke engine running at the same speed. We might therefore expect it to have twice as much power. But in practice such a high revving engine does not produce this expected high power. To help you to understand this look at Fig. 25.13. If the horses were changed round the trotter would probably not make much impact on the laden cart! Designers of motor cars have to consider similar problems of power-to-weight ratios. For a 2-stroke engine to work at maximum efficiency the bodywork has to be kept as light as possible, hence its use in powering mopeds, motor scooters, small boats and lawnmowers.

The thermal efficiency of a petrol engine is in the region of 28 per cent.

## 25.9 Oil/diesel engine

An oil engine was first built in 1892 by Rudolph Diesel but it was not until 1897 that such an engine ran efficiently.

A diesel engine can be either 2-stroke or 4-stroke. Figure 25.14 shows diagrammatically a diesel cylinder in cross-section. Compare this with Fig. 25.7.

**Fig. 25.13** *The fast-moving trotter is not producing more power than the slow-moving cart horse. In the same way the faster-revving 2-stroke engine does not produce more power than the 4-stroke engine*

(a) trotter  (b) cart-horse

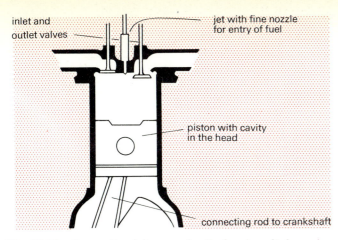

inlet and outlet valves

jet with fine nozzle for entry of fuel

piston with cavity in the head

connecting rod to crankshaft

**Fig. 25.14**  *Cross-sectional diagram of a diesel engine cylinder*

increases. Similarly the temperature of air in a diesel engine increases as it is compressed.

The rise in temperature of the air is such that it is above the ignition temperature of the oil. Thus when the oil is pumped into the cylinder just at the end of the compression stroke, the necessary ignition of the gases takes place. The resulting expanding hot gases create the required power stroke. The cycle of the diesel engine is therefore the same as that of a petrol engine except for the method used to introduce and ignite the fuel.

High-compression engines have to be made of strong materials and are therefore heavy, tending to have poor acceleration. However, their thermal efficiency of approximately 36 per cent is higher than that of any other type of heat engine, and they are usually very reliable.

*QUESTION 5:* What is it that is vital to the ignition of the gases in a petrol engine but is not found in a diesel engine?

Note two other points about the diesel engine:
1. The oil is *pumped* into the cylinder by an external pump, through a fine nozzle, thus forming a spray inside the cylinder.
2. The piston is capable of compressing the air drawn in on the suction stroke to one-sixteenth of its original volume. (It is called a high-compression engine because the compression is much greater than that in a petrol engine.)

What effect does this high compression have on the air? The following simple experiment will give you the answer.

Put your finger over the end of a bicycle pump and then pump hard. You will find that the air forced out is hot, and that the temperature of the barrel of the pump

### 25.10  Wankel rotary piston engine

This is a type of intermediate engine; it has a piston, but instead of reciprocating it rotates. (The piston may be termed a rotor.)

The engine is shown in Fig. 25.15. This engine is relatively simple in construction and in a twin rotor, 4-litre-capacity engine there are only approximately 154 moving parts compared to approximately 388 in a 8-cylinder reciprocating engine of the same power.

**Fig. 25.15**  *A twin-rotor Wankel engine*

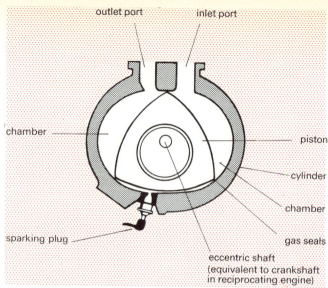

**Fig. 25.16**  *Cross-sectional diagram of a Wankel rotary engine*

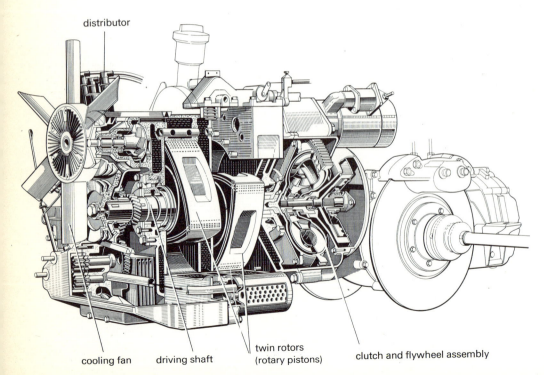

Figure 25.16 shows in diagrammatic form the engine in cross-section. You will see immediately that the cylinder is like an oval with two slight indentations at its middle. This is called an *epitrochoidal* shape. The rotary piston is triangular in shape, with the sides curved. It is connected by concentric gearing to the *driving shaft*. The apexes of the piston have gas seals so that it creates three separate gas-tight chambers within the cylinder. When the piston rotates the volume of each chamber is changed. The inlet and outlet ports are on the opposite side of the cylinder to the sparking plug.

The engine works on a 4-stroke cycle with three stages of this cycle taking place at the same time but in different chambers of the cylinder. The cycle for the chamber enclosed by the piston side AB is illustrated and described in Fig. 25.17(a) to (d).

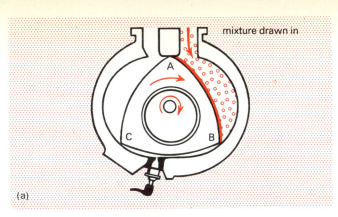

mixture drawn in

(a)

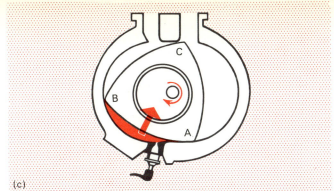

(c)

(a) *Induction stroke*. The inlet port is open, thus the mixture of gas and air is drawn into the chamber which is at its greatest volume.

(c) *Power stroke*. The expanding gases drive the rotary piston round.

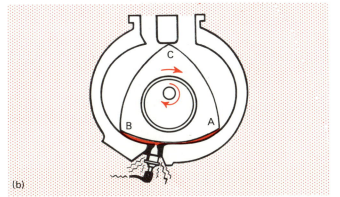

(b)

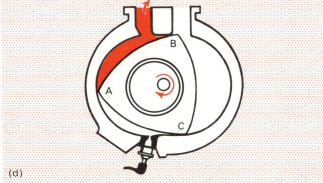

(d)

(b) *Compression stroke*. The piston rotates clockwise with the result that the volume of this chamber gets less. The gases are compressed. Eventually the rotary piston reaches the point where AB is above the sparking plug. The volume of this chamber is then at its minimum and the sparking plug operates to ignite the gases.

·(d) *Exhaust stroke*. The piston rotates and the side AB completes one revolution, the volume of this chamber is decreased so that the exhaust gases are driven out of the open exhaust port.

**Fig. 25.17**  *The working of a rotary piston cycle*

*QUESTION 6:* What advantage does this engine have over reciprocating engine with respect to (a) moving parts, (b) assembly, (c) wear on the engine, (d) size of engine?

### 25.11   Internal combustion turbines

The paddle-wheel arrangement used in turbines creates an efficient and smoothly running engine. This is because the energy of the "working fluid" is applied directly to the rotational moving parts.

A turbine has two main parts: (i) a *nozzle*, or nozzles, through which are forced the hot expanding gases. The gases, because of the shaping of the nozzles leave at very high speed and they are then directed onto (ii) the numerous shaped *blades* of the paddle wheel. The change in direction and speed of the high-speed gases after they have hit the blades, causes the blades to rotate. These in turn cause the rotor axle to rotate.

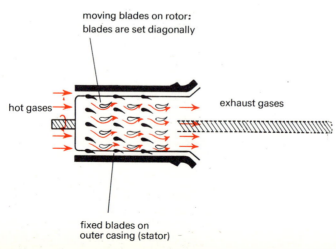

**Fig. 25.18**   *Cross-sectional diagram of a turbine engine*

moving blades on rotor: blades are set diagonally

hot gases

exhaust gases

fixed blades on outer casing (stator)

Figure 25.18 shows diagrammatically and in cross-section these two parts of a turbine. In fact the nozzles are formed by a series of fixed blades which are part of the outer casing. These are called the *stator*. The rotating paddle wheels are the *rotor*.

### 25.12   Gas turbine engines

It was the Second World War that stepped up investigation into and development of the gas turbine engine. It was first used for aircraft and later in industry and for road, rail and sea transport.

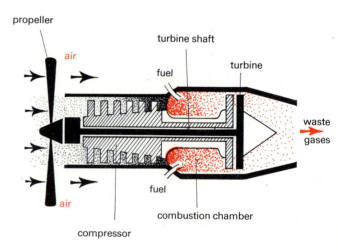

propeller

air

turbine shaft

fuel

turbine

waste gases

fuel

air

compressor

combustion chamber

**Fig. 25.19**   *Cross-sectional diagram of a turbo-propeller aircraft engine*

### 1.   *Turbo-propeller aircraft engine*

This engine is illustrated in Fig. 25.19. It is set in motion by an electric motor which rotates the compressor, thus drawing in air. The compressed air passes on to the combustion chamber. Here, fuel is injected into the chamber and the temperature of the air is sufficient to ignite it. The hot expanding gases leave the chamber and enter the turbine which is thus rotated. The turbine is connected by the rotor axle to the *propeller*, and also to the compressor which thus draws in more air. The cycle therefore is repeated without further need of the electric motor. The exhaust gases leave at the rear of the engine and help a little in the movement of the aircraft, although the main thrust forward is developed by the propeller.

How do the exhaust gases assist the movement of the plane? Consider the following problem.

boy jumps forward

gate opens inwards

**Fig. 25.20**   *As the boy jumps forwards the gate moves backwards*

QUESTION 7: Look at Fig. 25.20. The boy is jumping forwards off the top of the gate, which opens in towards the field. If the catch is open, what will happen to the gate when the boy jumps?

Gas turbine engines have now been incorporated into both cars and ships. The efficiency of such engines has been increased by preheating the intake of air. A lorry using such a gas turbine engine is seen in Fig. 25.21.

**Fig. 25.21**   *Gas turbine engine in a lorry*

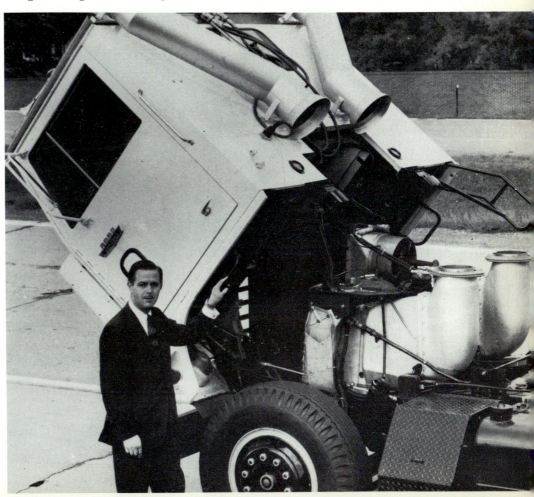

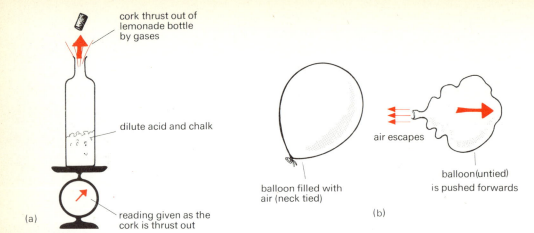

cork thrust out of
lemonade bottle
by gases

dilute acid and chalk

reading given as the
cork is thrust out

(a)

air escapes

balloon filled with
air (neck tied)

balloon(untied)
is pushed forwards

(b)

**Fig. 25.22** *Experiments to show that "to every action there is an equal and opposite reaction"*

**Fig. 25.23** *Cross-sectional diagram of a jet aircraft engine*

Jet engines are shown in Figs. 25.23 and 25.24. The intake and burning of the fuel is exactly the same as that for the turbo-prop engine. However, in jet engines and rocket engines the turbine is only used to drive the compressor since there is no propeller. The movement of the aircraft is entirely the result of the thrust made upon it from the gases which escape through the *jet pipe*. A narrowing nozzle is used to get the gases to travel at high speeds and therefore it delivers a larger thrust on the engine.

A great amount of thrust is needed to raise a rocket carrying a satellite away from the earth's surface (see Fig. 12.12 on page 119). In the case of the American *Saturn V* rocket, 48 262kg of fuel are burnt in the first-stage engines to produce enough thrust to get it away from the earth.

*QUESTION 8:* A jet aircraft uses oxygen from the air in order to burn its fuel. A rocket however may travel in outer space where there is no atmosphere. From where does it get its oxygen?

Nowadays the use of jet engines in other forms of transport is being developed. The French are already testing a monorail system using jet-powered vehicles.

### 3. *Ramjet engine*

A ramjet engine has no compressor unit. When working at full efficiency the compression of the air is achieved by the engine moving at such a high speed that shock waves are created at the point of the *ram dome*. The shock wave passes close enough to the air intakes of the engine to "ram" the air into the engine. This happens again as the air meets a second shock wave actually inside the air intake. Both of these "rammings" cause the temperature of the air to rise considerably. Thus it is at a sufficiently high temperature to ignite fuel that is sprayed into the combustion chamber. The gases leave the rear of the engine and create the forward thrust.

### 2. *Jet and rocket engines*

Look at Fig. 25.22 (a) and (b). The experiments illustrate quite convincingly the following fact:

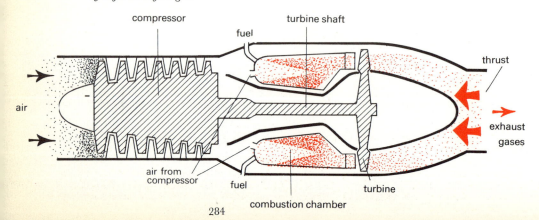

compressor

fuel

turbine shaft

thrust

air

exhaust
gases

air from
compressor

fuel

combustion chamber

turbine

**Fig. 25.24** *A jet aircraft, showing the central ram dome.*

# EXTERNAL COMBUSTION ENGINE: RECIPROCATING

### 25.13  Steam locomotive or reciprocating steam engine

Steam is an excellent working fluid in an external combustion engine because water is readily available almost everywhere. The steam can easily be used again after condensing. Vast quantities of steam are produced by a small quantity of water. One teaspoonful will give enough steam to fill a 10 litre petrol can.

Many of you will wonder why we should bother to consider steam locomotives when they are no longer in use on our railways. But the present age owes much to the steam locomotive. It was a splendid achievement and will always be a symbol of the great advances made by the scientists and engineers of the nineteenth century.

In Fig. 25.25 you see one of the most famous of steam locomotives pulling an equally famous train, "The Flying Scotsman". Figure 25.26 (a) and (b) shows the action of the steam engine.

**Fig. 25.25** *A4-class Pacific locomotive* Mallard, *holder of the world record for steam traction (126mph or 203km/h in 1938)*

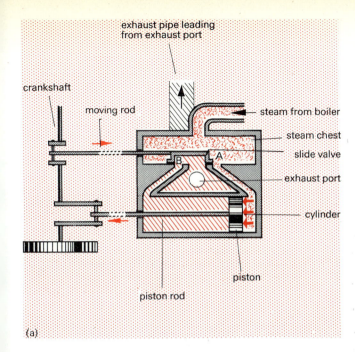

(a)

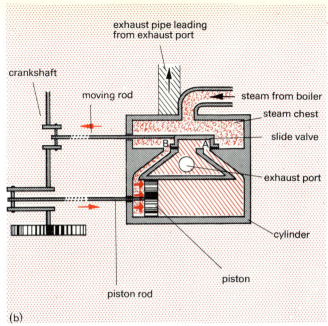

(b)

**Fig. 25.26** *Cross-sectional diagrams of a steam engine*

Steam from the boiler is passed into the *steam chest* and from there enters one side of the engine cylinder by means of the inlet port A which is open. The expanding steam thus forces the piston to move from right to left within the cylinder. This delivers the power stroke to the piston rod, which is connected directly or indirectly to a crankshaft. The crankshaft is rotated and in turn rotates the wheels. An *eccentric rod* is also connected to the crankshaft, its other end is connected to the *slide valve* mechanism which opens and closes one valve at a time. When the piston moves from right to left so the eccentric rod moves the slide valve mechanism across from left to right. Thus inlet port A is closed and inlet port B is opened. The steam can now enter the other half of the cylinder

via inlet port B and forces the piston along the cylinder from left to right. This also serves to push the exhaust gases out through the exhaust pipe. The pipe leads either to the air or to a condenser. Thus the cycle is repeated and continuous motion is achieved.

Although the steam locomotive has disappeared in this country the steam engine as such will always have its place in industry, particularly where a controlled slow-running form of power is required, for example in steam rollers and winding gear. On the continent, steam engines which burn oil as fuel are used extensively. It is also possible, in view of air pollution by petrol engines, that there will be a redevelopment of an efficient steam car.

**Fig. 25.27** *Low pressure turbine shaft for a generator*

# EXTERNAL COMBUSTION ENGINE: TURBINE

### 25.14  Steam turbine

The first practical steam turbine was built by Charles Parsons in 1885. In 1889 Parsons used larger turbines in a power station which was built to supply electricity to Newcastle. In 1897 he used a similar turbine in a boat (the *Turbinia*) and demonstrated its abilities when the British fleet was assembled at Spithead for Queen Victoria's Jubilee. The performance of the boat was so impressive that by 1914 nearly the whole of the British fleet was powered by steam turbines. The efficiency of a steam turbine is in the region of 30 per cent. The largest steam turbines are usually found in power stations providing electricity for the National Grid.

Figure 25.27 shows a turbine under construction. What do you notice about the size of the rotor blades? They increase in size because as the steam passes through the stator its pressure is converted into an increase in speed and at the same time there is a corresponding increase in volume. Thus as a series of stators and rotors are used, so the series of blades must become larger to take full advantage of this expansion.

# FUTURE ENGINES FOR SPACE TRAVEL

### 25.15  Nuclear rocket engines

Chemical rockets are likely to be superseded by nuclear rockets which are more efficient and powerful. Figure 25.28 shows a nuclear rocket engine on its test site in the Nevada desert, USA. Liquid hydrogen is pumped around the nuclear reactor, becomes heated, and reaches a very high temperature. This hot gas is then released from the nozzle of the engine, thrusting it forwards.

**Fig. 25.28**  *Nuclear rocket development station, Nevada USA*

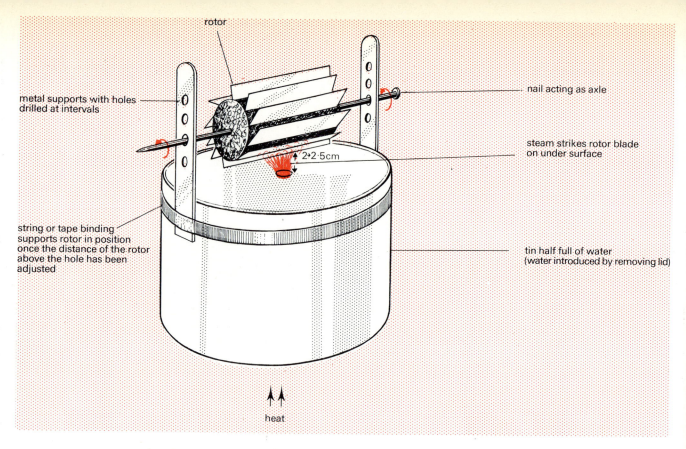

rotor

metal supports with holes
drilled at intervals

nail acting as axle

steam strikes rotor blade
on under surface

2-2·5cm

string or tape binding
supports rotor in position
once the distance of the rotor
above the hole has been
adjusted

tin half full of water
(water introduced by removing lid)

heat

**Fig. 25.29** *Construction of a model steam turbine*

A.  Make a model steam turbine (Fig. 25.29).

Select a medium-sized tin with a press-on lid. Punch a hole in the lid as shown in the diagram. Take another tin and, using tin snips, remove the top and bottom. Slice it vertically downwards along its join and flatten the resulting sheet of metal. From this metal cut 10 pieces of tin, each the size of a razor blade. Stick these at regular intervals into the circumference of a cork which is the same length as the pieces of tin, and 2 to 3cm in diameter. Select a nail that is 5cm longer than the diameter of the tin. Push the nail through the centre of the cork to act as an axle. You have now constructed the turbine rotor.

Support the rotor by two pieces of metal as indicated in the diagram. It should be in such a position that when water is boiled in the tin the steam from the hole in the top strikes the edge of the lower blade. Thus the rotor is driven round by the steam.

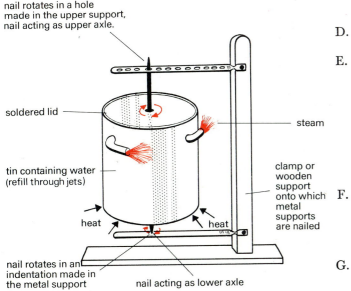

nail rotates in a hole made in the upper support, nail acting as upper axle.

soldered lid

steam

tin containing water (refill through jets)

clamp or wooden support onto which metal supports are nailed

heat

heat

nail rotates in an indentation made in the metal support

nail acting as lower axle

**Fig. 25.30** *Construction of a model jet*

B.    Make a model jet (Fig. 25.30).

Obtain a medium-sized tin with a press-on lid. In the sides near the top, make two holes opposite to one another. Either pierce these holes at an angle so that the steam will leave the tin along its sides, or solder on two pieces of curved metal tubing as shown in the diagram. Support the tin as shown, and apply heat.

C.    Make a form of jet transport.

Use an empty truck from a large plastic model train set, or construct a similar container on four wheels. Blow up a sausage-shaped balloon and clip the end with a paper clip to seal it. Anchor the balloon onto the truck by means of two rubber bands and place it on the floor. Release the clip and watch the progress on your "jet-propelled" truck.

D.    Visit a local transport museum and study the construction of the engines on display.

E.    Make a chart showing the history of steam as a form of power. Bring it up to date by looking out for references in newspapers and magazines to the future use of steam cars. For instance, the state of California has announced already that it is aware of pollution dangers and that accordingly the use of steam and electric cars will be encouraged in preference to petrol-driven vehicles. Why are such changes taking place?

F.    Construct a wall newspaper on the work of the people responsible for the development of any form of heat engine, e.g. Newcomen, the Stephensons, Charles Parsons, Charles Benz, Frank Whittle, the Wright brothers, James Watt.

G.    Make an illustrated file on the use of heat engines in air travel. Bring it up to date with details of engines that are to be used in man's exploration of space.

1.  Compare the workings of a steam engine and a steam turbine. Point out their advantages and disadvantages.

2.  What are the major differences in the workings of a diesel engine and a petrol engine? (a) Why is a diesel engine more efficient than a petrol engine? (b) Why do motor cars have at least four cylinders?

3.  Explain, with the aid of diagrams, how a 2-stroke petrol engine differs from a 4-stroke petrol engine in respect of (a) the structure of the cylinder, (b) the shaping of the piston, (c) the cycle that takes place within the cylinder.

4.  With the help of a diagram, explain the action of a jet engine. (a) In what ways does the jet engine differ from a turbo-prop engine? (b) What is the major difference between a rocket engine and a jet engine? Which is able to fly beyond the earth's atmosphere?

5.  Explain the function of each of the following in a reciprocating steam engine: (a) boiler, (b) slide valve, (c) cylinder and piston, (d) eccentric rod.

6.  What are the advantages of a rotary engine over a 4-stroke internal combustion engine?

7.  State and explain the 4 terms that apply to the 4 stages in the 4-stroke cycle of an internal combustion engine.

8.  Write a few sentences about the action of each of the following in an internal combustion engine: (a) crankshaft, (b) sparking plug, (c) inlet valve, (d) exhaust valve, (e) piston.

9.  To what do the terms "stator" and "rotor" apply in a turbine? Illustrate your answer by very simple diagrams.

10. Explain why steam is a most suitable "working fluid" within an engine.

# METERS MOTORS AND GENERATORS

**Part Five**

*An Electricar milk float* (Courtesy of Brush
Electrical Engineering Co. Ltd)

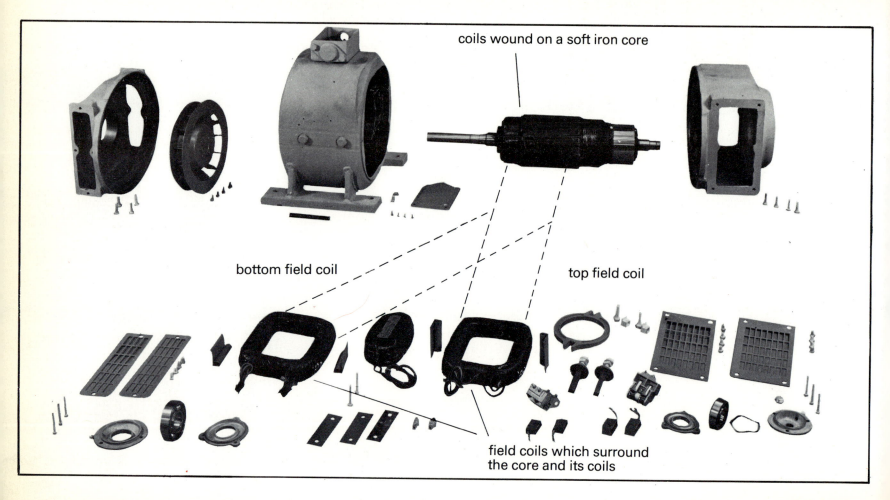

coils wound on a soft iron core

bottom field coil

top field coil

field coils which surround
the core and its coils

**Fig. 26.1** *Exploded view of a d.c. electric motor*

294

## CHAPTER 26

# ELECTRIC METERS AND MOTORS

## 26.1 The motor effect

The milk float in the frontispiece puts to good use the mechanical energy that can be obtained from electrical energy. This conversion of energy is carried out by means of an electric motor.

Look at the exploded view of an electric motor in Fig. 26.1. Notice the circular arrangement of the field coils. These carry electricity and are, in fact, electromagnets (see page 194) which set up a magnetic flux. In the centre of this magnetic flux is the soft iron core with several coils of wire wound round it. An electric current flows through these coils. In an electric motor we have, therefore, a current flowing in a conductor which is in a magnetic flux. Let us see what happens under these circumstances by using the apparatus illustrated in Fig. 26.2.

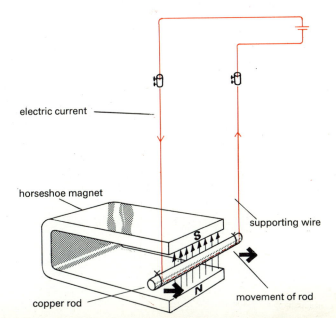

electric current

horseshoe magnet

supporting wire

copper rod

movement of rod

**Fig. 26.2** *Experiment to demonstrate the motor effect*

**Fig. 26.3** *Fleming's left-hand rule*

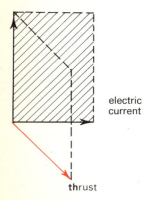

direction of magnetic flux

electric current

thrust

out of the page, towards you

(a)

In this experiment a copper rod, acting as the conductor, is suspended between the two poles of a horseshoe magnet so that it can move freely in a horizontal plane (Fig. 26.2). It is essential that there is good electrical contact between the rod and its supports (wires) which are connected to the terminals of a low voltage d.c. supply.

Direct current (d.c.) is an electric current in which the electrons flow in one direction only.

In such a position the rod (and thus the electric current) is at right angles to the magnetic flux that exists between the two poles of the horseshoe magnet.

The rod is observed while a current is passed through it. Then the battery is reversed so that the current flows in the opposite direction through the rod. The experiment is then repeated with the magnet turned over, so that the direction of the magnetic flux is reversed.

When the current is flowing through the rod, the rod is seen to swing either backwards or forwards between the poles of the magnet. On the first occasion it moves in the direction indicated by the arrows in Fig. 26.2. When *either* the direction of the current *or* the magnetic flux is reversed, the movement takes place in the opposite direction. This illustrates what is known as the *motor effect*.

In Chapter 19 we saw that there is a magnetic flux associated with an electric current. Any magnet placed near a conductor carrying an electric current experiences a force resulting from this magnetic flux. There is an equal and opposite force on the conductor, which is the motor effect.

Motor effect: when an electric current flowing within a conductor cuts a magnetic flux there is a force on the conductor.

The experiment also shows that the force or thrust is in a direction that is at right angles to both the direction of the electric current and that of the magnetic flux (Fig. 26.3 (a)). We may use a simple rule to help us determine the direction of the movement of the conductor; it is known as Fleming's left-hand rule and is illustrated in Fig. 26.3.

Fleming's left-hand rule: put forefinger, second finger and thumb of left hand as shown in Fig. 26.3(c); **f**orefinger points in direction of **f**lux (two **f**s), se**c**ond finger points in direction of **c**urrent (two **c**s), **th**umb indicates direction of **th**rust (two **th**s).

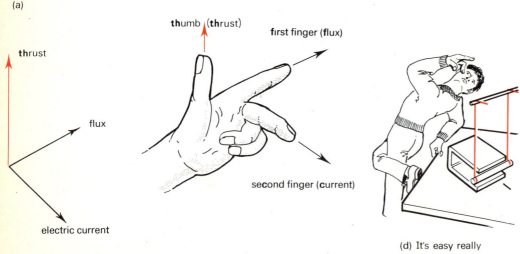

thrust

flux

electric current

(b)

thumb (thrust)

first finger (flux)

second finger (current)

(c)

(d) It's easy really

## 26.2 Why is there a force on a conductor that is cut by a magnetic flux?

We know from our previous work that a magnetic flux is set up round a conductor when an electric current flows through it. The reaction between this magnetic flux and the magnetic flux of the permanent magnet produces the force that moves the conductor. In fact it is the magnetic forces of attraction and repulsion that set the conductor in motion.

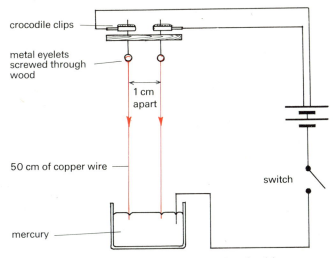

**Fig. 26.4**  *A force exists between wires carrying electricity*

## 26.3  The use of the motor effect in defining the unit of current, the ampere

A similar force to that seen in the above experiment is observed when an electric current is passed through two 50cm copper wires suspended parallel to one another and about 1cm apart (Fig. 26.4). When the current is passed in the same direction through both wires there is a force of attraction between the wires. When the current in one wire is reversed, then a force of repulsion is observed between the two wires.

The forces set up between the magnetic fluxes created around the conductors in this sort of situation have given the scientist an accurate method of determining and defining a measurement of current flow.

1 ampere is the current which, when flowing in each of two infinitely long parallel straight wires, positioned 1 metre apart in a vacuum, produces between the two wires a force of repulsion or attraction of $2 \times 10^{-7}$ newtons per metre of wire.

## 26.4  Experiment to show the force that exists between two coils, each carrying a current

As illustrated in Fig. 26.5 two coils are suspended side by side. An electric current is passed around each coil in the same direction. Then the current is reversed in one of the coils.

When current flows round the coils in the same direction the coils attract each other. When current flows round the coils in opposite directions they repel each other.

QUESTION 1: Fig. 26.6 is a diagram showing the two coils used in the experiment. The direction of the

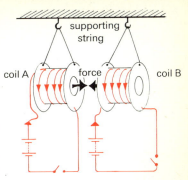

**Fig. 26.5**  *A force exists between two coils each carrying an electric current*

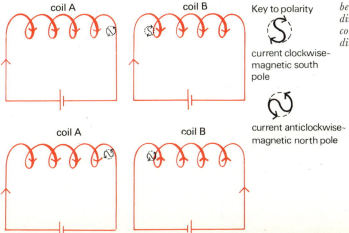

Key to polarity

current clockwise~ magnetic south pole

current anticlockwise~ magnetic north pole

**Fig. 26.6**  (*a*) *Force of attraction between coils when current flows in same direction* (*b*) *Force of repulsion between coils when current flows in opposite directions*

current is shown in each case. The direction of the current determines the magnetic polarity of the ends of the coils. Explain in these terms why there is a force of attraction in one case and a force of repulsion in the other.

In everyday life the motor effect has many important applications. We will consider some of these in respect of (a) current-detecting and current-measuring instruments, and (b) electric motors.

## 26.5 The galvanometer

The magnetic flux produced by a current flowing in a conductor is proportional to the strength of the current.

Some current-detecting instruments are based upon this principle. A current-detecting instrument is called a galvanometer.

Figure 26.7 illustrates a simple galvanometer that you can construct for yourselves by fixing a pocket compass to the centre of the base of a shallow cardboard box. Thirty turns of insulated wire are then wound around the box to form a coil around the compass. The whole apparatus is set up so that the wire of the coil and the direction of the compass needle are parallel.

By connecting the galvanometer in series with a variable resistance and a low voltage d.c. supply, it can be shown that a small current flowing through the coil will set up a magnetic flux sufficient to cause a deflection of the magnet.

*QUESTION 2:* In this simple galvanometer we have used the results of the important experiment using a single wire and magnet (Chapter 19). (a) What was the name of the scientist who carried out this experiment? (b) What important fact, unknown in his day, did his experiment reveal?

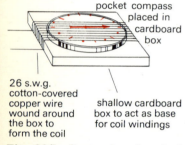

pocket compass placed in cardboard box

26 s.w.g. cotton-covered copper wire wound around the box to form the coil

shallow cardboard box to act as base for coil windings

**Fig. 26.7** *Construction of a simple galvanometer*

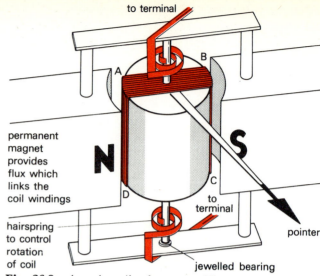

to terminal

permanent magnet provides flux which links the coil windings

hairspring to control rotation of coil

jewelled bearing

to terminal

pointer

**Fig. 26.8** *A moving-coil galvanometer. The path of the current is shown in red*

## 26.6 A moving-coil galvanometer

A coil of wire is pivoted in a vertical position between the curved poles of a permanent magnet, as in Fig. 26.8.

At the centre of the coil there is a soft iron core. The core is fixed but the coil can rotate around it. The curved magnetic poles and soft iron core produce a magnetic flux which is concentrated throughout the coil.

The current enters and leaves the coil by the means of hair springs positioned at the top and bottom of the coil.

When the current flows around the coil windings it cuts the magnetic flux in the space between the magnetic poles. The current flows down one side of the coil and up the other side (e.g. down AD and up CB). This means that the forces acting on each side of the coil are in opposite directions to one another and produce a couple (a turning effect) twisting the coil on its axis. The rotation of the coil is controlled by the size of the current and the couple in the hair springs which opposes the motion. The pointer is attached to the axis of the coil. Its movement over the scale indicates the value of the current.

If when the pointer is at rest it is over the centre of the scale (a *centre zero galvanometer*) then the direction of the deflection will indicate in which direction the current is flowing around the coil.

If the galvanometer deflection is large for a very small current, then the galvanometer is said to be highly sensitive.

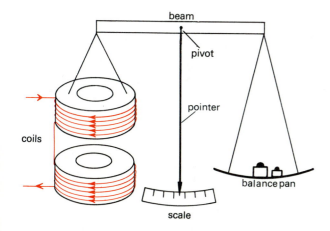

**Fig. 26.9**  *A simple current balance*

## 26.7  Calibration of the galvanometer

The galvanometer may be calibrated by connecting it in series with an instrument called a *current balance*, one type of which is illustrated in Fig. 26.9. The same current is passed through both coils and the attractive force between them is balanced by putting weights on the balance pan. The greater the current that flows, the greater is the force existing between the coils. Once the value of the

force between the coils is known, the value of the current flowing in them may be calculated. This calculation is achieved by using the definition of the ampere (page 297), a knowledge of the length of the wire in the coils, and the distance between the coils.

A galvanometer with a scale calibrated in amperes and used to measure the size of the current, is called an *ammeter*. It is connected in series in a circuit.

A galvanometer with the scale calibrated to measure potential difference is called a *voltmeter*. It is connected in parallel in the circuit.

*QUESTION 3:* A galvanometer has a scale showing 10 divisions per milliampere. (a) What is this instrument called? (b) What is the value of an unknown current which gives a deflection of two divisions? (c) What happens when an alternating current (a.c.) is passed through a moving-coil meter?

## 26.8  Moving-iron meters (two main types)

*1. Repulsion-type meter*

The circular coil that carries the current has two small soft iron rods arranged within it (Fig. 26.10). One is

**Fig. 26.10**  *A repulsion-type meter*

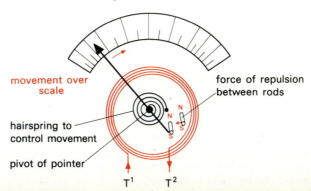

fixed, the other is attached to the end of a pointer which is pivoted so that it can move in a horizontal plane across a scale. The rods become induced magnets when the current flows around the coil. Thus like poles of the induced magnets are together and the two iron rods repel each other. The strength of the current determines the strength of the magnetic flux and thus the force of the repulsion. A large current produces a strong force of repulsion. A hair spring fixed at the pivot of the pointer controls the movement of the pointer.

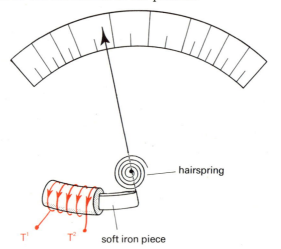

**Fig. 26.11**  *An attraction-type meter*

### 2. *Attraction-type meter*
A piece of soft iron (Fig. 26.11) is attached to the end of a pointer which is pivoted so that it can move in a horizontal plane across a scale. A hair spring is fixed at the pivot and controls the movement of the pointer over the scale. The current is passed through a long coil which is positioned next to the piece of soft iron. The magnetic flux set up around the long coil attracts the iron into the coil. The greater the current, the greater is the attraction and the further the pointer moves over the scale.

300

*QUESTION 4:* A current from a battery is passed in one direction through both types of moving-iron meters. The movement of the pointer is observed. The battery is turned round in the circuit so that the current flow through the meter is reversed. (a) Explain why the pointers continue to move in the same direction when the battery is turned round. (b) Could you use moving-iron ammeters to measure alternating currents?

### 26.9  Can a milliammeter be adjusted to read a higher value of current?
Some galvanometers only take a maximum current of 15mA, but we may want to use such instruments to measure much greater currents. This can be done by allowing only a small fraction of the main current to flow through the meter. The "alternative route" is provided for the larger part of the main current through a thick wire of low resistance placed in parallel with the meter. This resistance is termed a *shunt*. Different resistances are used to take different known fractions of the main current. For example, if a shunt takes nine-tenths of the current, then the meter will take one-tenth, and in order to find the value for the total current we must multiply the meter reading by 10.

### 26.10  How is a galvanometer converted to a voltmeter?
A voltmeter measures the potential difference in volts. To act as a voltmeter a galvanometer must be connected parallel to the particular section of the circuit for which the potential difference is required. In order to prevent a large current flowing through a sensitive galvanometer a known high resistance is placed in series with it. This series resistor is termed a *multiplier*. The small

current flowing through the galvanometer is directly proportional to the potential difference across it. The scale may be calibrated directly in volts. A voltmeter must have a high resistance otherwise the current and the p.d. are changed by a significant amount when the voltmeter is put into the circuit.

## 26.11 Electric motor (d.c.)

We have seen how a force is produced on a conductor which is cut by a magnetic flux, but so far we have not obtained any form of continuous motion. Can we obtain continuous motion by using such a force? We can if we construct a motor.

At the beginning of this chapter we looked briefly at the structure of a motor. We discovered that basically it consists of a coil carrying an electric current which is placed in a magnetic flux.

The continuous motion of such a coil is brought about by connecting its ends to a *commutator*. The simplest form of commutator is two halves of a brass (or copper) ring (a split ring). These two halves are kept apart on a non-conducting core. Each end of the coil is connected to one half of the split ring.

The split rings of the commutator make good electrical contact with the leads of the external electrical circuit by means of two carbon brushes, which are held in position by springs.

On the next page is a diagram of a motor. Study it together with the table on the action of a commutator and you will soon see how it works.

We can see that the commutator's job is to reverse the direction of the current flowing in the coil when it passes the vertical position. By this means the forces that act on the coil are always in the same direction and cause it to rotate continuously in that direction.

When the coil is in the vertical position the couple on it is at a minimum. The couple increases to a maximum when the coil is in a horizontal plane. This means that a motor consisting of one coil would not provide very smooth motion. In order to overcome this, commercial motors have many coils wound at angles to one another on a soft iron core which is attached to the axle. This structure is called the *armature*. The commutator is made up of a number of equal segments (Fig. 26.12). At any point in the rotation of the armature there is always one coil that is under a maximum turning force. Thus a

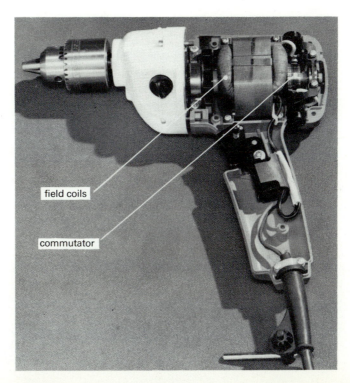

field coils

commutator

**Fig. 26.12** *Commutator in the electric motor of an electric drill*

| Position of coil | Split ring and carbon brush in contact | Direction of current | Force on AB | Force on CD | Movement of coil |
|---|---|---|---|---|---|
| Horizontal Fig. 26.13(a) | F with X<br>E with Y | ABCD | downwards | upwards | starts to rotate anti-clockwise |
| Vertical Fig. 26.13(b) | no contact between split rings and brushes | no current flowing | zero | zero | coil's momentum keeps it rotating |
| Just past the vertical Fig. 26.13(c) | F with Y<br>E with X | DCBA | upwards | downwards | continues to rotate anticlockwise |
| Horizontal Fig. 26.13(d) | F with Y<br>E with X | DCBA | upwards | downwards | anticlockwise |
| Vertical | No contact between split rings and brushes | no current flowing | zero | zero | coil's momentum keeps it rotating |
| Just past the vertical and so on . . . . . . | F with X<br>E with Y | ABCD | downwards | upwards | anticlockwise |

**Fig. 26.13**

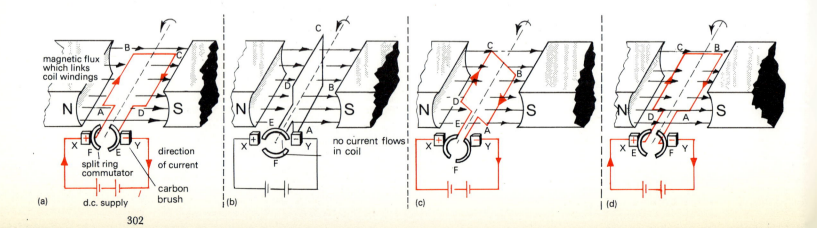

smooth rotary movement is obtained. The curved magnetic pole faces and the soft iron core ensure that there is a concentrated magnetic flux.

Figure 26.14 shows a very simple electric motor, one that we could construct at home or in the laboratory. In this model the commutator consists of two pins which are in contact with bare wires from the electrical circuit.

### 26.12 Can a d.c. electric motor work if supplied with a.c. electricity?

If the permanent magnet of the d.c. electric motor is replaced by electromagnets called field coils, then the motor will work on a.c. as well as d.c. supply. Most household appliances, such as vacuum cleaners and washing machines use a.c./d.c. motors. (Induction motors will be discussed in the next chapter.)

The terms *rotor* and *stator* may be met in connection with commercial motors. The armature forms the rotor (rotating section) and the field poles the stator (stationary part).

You can even get a motor like the one in Fig. 26.14 to work from an a.c. supply as long as the coil is spun at a speed which synchronizes (keeps in step) with the alternations of the current. A synchronous a.c. motor constructed on this basis is used for electric clocks. In this the rotor is the permanent magnet and the alternating current flows through the field coils.

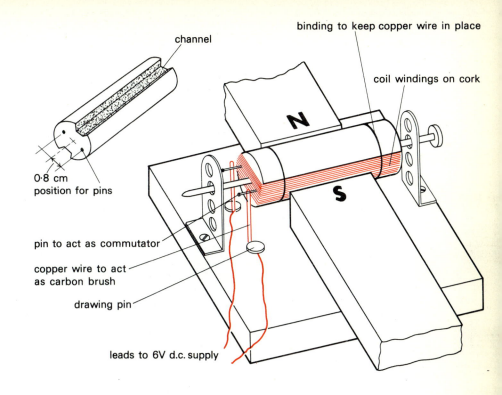

**Fig. 26.14** *Construction of a simple d.c. electric motor*

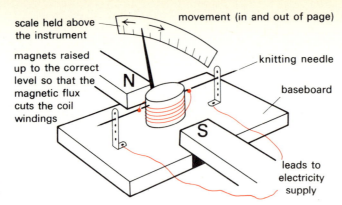

scale held above
the instrument

movement (in and out of page)

magnets raised
up to the correct
level so that the
magnetic flux
cuts the coil
windings

knitting needle

baseboard

leads to
electricity
supply

**N**

**S**

**Fig. 26.15** *Construction of a moving-coil meter*

A.     Make a moving-coil meter (Fig. 26.15).

Construct the meter support from a rectangular piece of wood and two metal uprights. Drill holes (just large enough to allow a thin knitting needle to rotate freely within them) in the supports. Screw the supports to the base to keep them in position. Take a cork about 2.5cm diameter to act as core for the coil.

Construct the coil by winding regularly 10 to 15 turns of insulated copper wire (approximately 28 s.w.g.) round the base of the cork. Break a long thin knitting needle (pointed at both ends) in half. Push one half into one side of the cork so that it is just above, and parallel to, the plane of the coil. It should be pushed in far enough to hold firmly, but make sure that the needle does not reach the centre. Push the second half of the needle into the opposite side of the cork, in the same way, so that the two halves make a straight line. (The two halves must not meet in the centre of the cork. Why?)

Place the ends of the knitting needle through the appropriate holes in the uprights to make the horizontal axis of the coil. As the knitting needle and the metal supports are part of the circuit for the flow of electricity to the coil, the insulation from a small section of the two ends of the copper wire must be removed and the bare wire wound tightly round the knitting needle to ensure good electrical contact.

Cut a pointer from a piece of card and place this between the coil and the cork, so that it can rotate freely over a scale when the coil rotates. Support two magnets on either side of the coil as shown in Fig. 26.15, with opposite poles facing one another to provide the magnetic flux threading the coil windings. Calibrate the meter by passing a known current through it. You will find that such a meter will read up to 2amps, and you can adjust the sensitivity by removing sections of the cork from below the coil.

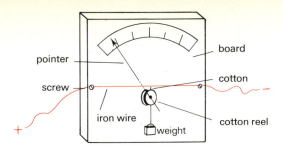

**Fig. 26.16** *Construction of a hot-wire ammeter*

B.  Make a hot-wire ammeter (Fig. 26.16).

This is an application of the fact that substances expand when heated, and that electricity flowing through a conductor causes a heating effect.

To the centre of a wooden board (30cm × 25cm) nail a cotton reel in such a way that it is able to rotate freely. Mark out a horizontal line approximately 6cm above the circumference of the cotton reel. Insert two screws near to the edge of the board at either end of this line. Next tightly stretch a thin piece of iron wire between these two screws and secure each end by winding it around the screws. Secure one end of a length of thin cotton to the centre of the wire above the cotton reel. Bring the cotton down vertically and pass it around the reel once. Suspend a sufficiently heavy weight to the free end of the cotton to keep the cotton and the wire above it taut.

Connect wires from the circuit to each screw, as shown. Fix a pointer on the cotton reel so that it can move over a scale when the reel rotates. This will happen when electricity passing through the iron wire causes it to heat and expand; the slack is taken up by the weight moving downwards. Calibrate the scale by passing a known current through the meter.

C.  Make an electric motor (see Fig. 26.14).

Take a cork of 2.5cm diameter and using a sharp knife (or razor blade) cut two fairly shallow channels along the length of the cork and on opposite sides. Push a knitting needle through to act as the axis of the cork. Wind 30 turns of insulated copper wire (approximately 26 s.w.g.) onto the cork. Start winding at one end of the cork. Move across to the other end then back again, winding on top of the first layer. Finish at the end of the cork at which you first started. Keep the windings in position by two strips of sticky-tape or masking-tape around the cork. On one end of the cork position two pins, each at a distance of approximately 0.8cm from the axis and opposite each other. Push the pins into the cork for about half their length. Connect the ends of the coil to the pins. Use baseboard and uprights as for the moving-coil meter. Support the coil horizontally by the uprights, and turn it to check that it can rotate smoothly. Construct two brushes from copper wire. To do this measure the distance from the baseboard to a point slightly above the pin commutator. Having obtained this measurement select the same length from two copper wires, stripping the insulation from the sections likely to contact the pin commutator. At the baseboard end, wind the wire once round the shaft of a drawing pin and place each copper wire in such a position that when it is pulled straight and vertical it will just come in contact with one of the pins. Push home the drawing pins into the baseboard. Position a strong horseshoe magnet, or two magnets as illustrated in the moving-coil meter, across the coil. Connect a 4V or 6V battery to the two lengths of copper wire and give the coil a flick to start it rotating.

D.  Construct the simple synchronous electric motor (in the form of a roundabout) illustrated in Fig.

**Fig. 26.17** *Construction of a simple attraction-type motor*

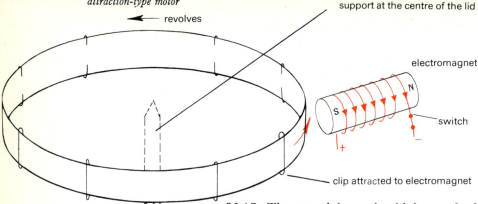

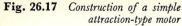

← revolves

support at the centre of the lid

electromagnet

N

S

switch

+
−

clip attracted to electromagnet

26.17. The roundabout should be made from a circular cardboard lid with a deep enough side for the paper clips to be slipped onto it. Construct the electromagnet in the manner discussed in Chapter 19. Position the electromagnet a little to one side of one of the paper clips and switch on the current for a fraction of a second to attract the clip, thereby rotating the roundabout. Repeat this procedure every time a paper clip is to one side of the electromagnet. You will find a similar synchronous motor in an electric clock.

E.  Make an illustrated catalogue entitled "The use of meters in our everyday lives". See that each type of meter is correctly labelled.

F.  Compile an illustrated *Who's Who* file on the life and work of Michael Faraday, "the father of electricity".

G.  Make an illustrated catalogue on "The use of electric motors in our everyday lives". Find out what type of motor is used in the appliances you list. Put in a section on the possible use of electric and induction motors in transport of the future.

straw

magnet

spring

coil

magnet

**Fig. 26.18** *A simple ammeter*

H.  Construct an ammeter like the one in Fig. 26.18. Magnadur magnets have their poles on the large faces. Calibrate the scale.

Use similar components to construct an electric motor. Fig. 26.19 should help you do this. Make sure you bare the ends of the wires from the ends of the coil which form the commutator.

If a friend has constructed a meter and a motor like the ones shown in 26.15 and 26.14, compare the construction and how well the different designs operate.

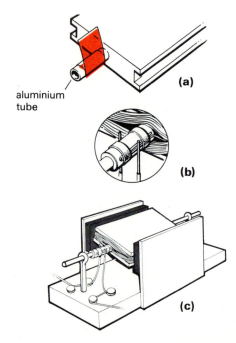

(a)

aluminium
tube

(b)

(c)

**Fig. 26.19** *An electric motor. The commutator end of the aluminium tube must be insulated with sellotape otherwise the two ends of the coil will be shorted together.*

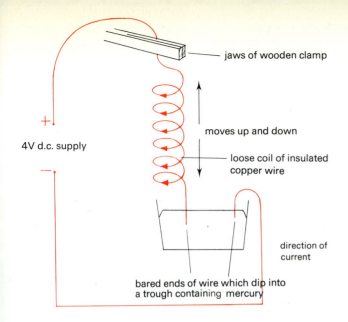

**Fig. 26.20**  *Roget's Jumping Spiral*

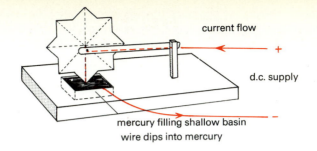

**Fig. 26.21**  *Barlow's wheel*

1. Figure 26.20 illustrates Roget's Jumping Spiral. When an electric current is passed through the coil it is seen to move up and down, making and breaking the circuit at the point of contact with the mercury surface. Explain why there is this form of motion in the spiral.

2. How would you use Barlow's wheel, the apparatus illustrated in Fig. 26.21, to demonstrate that a force is exerted on a conductor carrying an electric current and threaded by a magnetic flux. State the rule for determining the direction of this force.

3. Figure 26.22 illustrates a piece of aluminium foil wrapped on a flexible piece of card. A current is passed through the foil, flowing in the direction indicated in the diagram. The north pole of a magnet is brought towards the foil as shown in the diagram. Which way does the foil move? Can an identical movement of the foil be produced by changing the position of the magnet and the direction of flow of the current?

4. A coil of wire is positioned horizontally between two magnetic poles with the north pole situated on the left. A current flows in a clockwise direction around the coil when viewed from above. Explain, with a diagram, the movement of the coil. How may a commutator be used to make the coil function as an electric motor producing continuous movement?

5. Fig. 26.18 shows a model ammeter. When a current flows through the coil the pointer deflects. Which of the following would make the pointer move further: (a) putting more turns on the coil, (b) making the springs at the ends stiffer, (c) increasing the strength of the magnets?

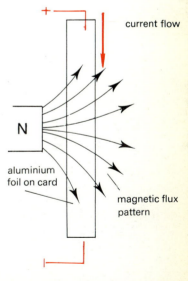

**Fig. 26.22**  *Which way does the foil move as the magnet is brought up to it?*

307

6.　A form of electrical "cannon", which fires steel missiles, may be constructed using a series of coils wound on the barrel of the cannon (Fig. 26.23). The flow of electricity is so arranged that each one of the coils is switched on and off in turn.

a. If the missile is placed at the point A at the bottom of the cannon and coil B is switched on then what will happen to the missile?

b. If the electricity is not switched off in coil B what will happen to the missile?

c. If the electricity is switched off in coil B just as the missile enters it, what will happen to the missile?

d. If the missile is to be kept moving, at what position within B must the missile be when the current is switched off in B and on in C? If this process is repeated for all the coils along the length of the cannon what effect will this have on the speed of the missile?

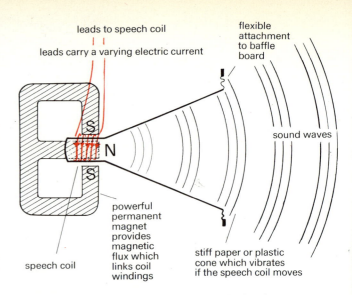

Fig. 26.24　*The structure of a moving-coil loudspeaker*

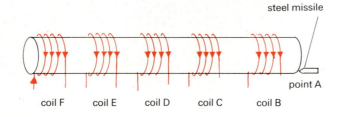

**Fig. 26.23**　*A form of electric cannon*

7.　Figure 26.24 illustrates a moving-coil loudspeaker. The current fed into the speech coil varies in strength in a way that corresponds to the original sound waves in the air. How does the loudspeaker convert this electrical energy into sound energy?

8.　(a) An electric motor is never 100 per cent efficient. In what form is most of the energy wasted? (b) An electric drill operates on 250V a.c. mains. It takes a current of 0.5A. If the power output of the drill is 75W, what is its efficiency? (For a definition of efficiency see page 136.)

9.　A milliammeter has a resistance of $15\Omega$. A current of 0.05A produces a full scale deflection of the pointer. A resistor (shunt) of resistance $R$ is connected in parallel with the meter (Fig. 26.25). A current of 2A arrives at the junction A and leaves at the junction B. If 0.05A flows through the meter:

a. What current flows through the shunt?

b. What is the p.d. across the shunt? (Give your answer in terms of the resistance of the shunt, $R$.)

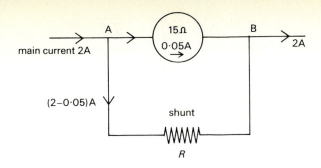

**Fig. 26.25**

c. What is the p.d. across the meter? (Give your answer in terms of the resistance of the meter, that is 15Ω.)

d. The meter and the shunt are in parallel. Therefore the p.d. across each is the same. Equate your answers to b and c in order to determine the value $R$.

e. How would you convert this meter of resistance 15Ω which reads 0 to 0.05A, to a meter which reads 0 to 2A?

10. A milliammeter has a resistance of 15Ω. A current of 0.05A produces a full-scale deflection. If this instrument were to be calibrated as a voltmeter, what would be the maximum reading of the instrument?

11. What multiplier (series resistor) would you connect to the milliammeter in question 10 so that it will give readings of p.d. up to 50V? (Hint: remember that the same current passes through the multiplier and meter, Fig. 26.26).

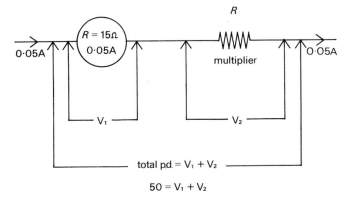

**Fig. 26.26**

12. How would you convert a 5Ω meter which reads 0–5mA to a meter which reads (a) 0–5A, (b) 0–5V?

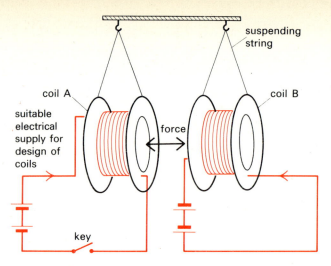

## CHAPTER 27

**ELECTRO-MAGNETIC INDUCTION**

**Fig. 27.1** *A force exists between two coils each carrying an electric current*

### 27.1 Experiment with two coils

The apparatus illustrated in Fig. 27.1 will be familiar to you because we used it in the previous chapter to show the forces of attraction and repulsion that exist between two coils when they are carrying an electric current.

What happens when the leads from coil B are disconnected from the battery and the electric current is switched on in coil A?

As we would expect from our previous work, there is no movement of the coils. An electric current is required to flow in each coil before there is a force between the two coils.

But if the two free leads from coil B are joined together to make good electrical contact (Fig. 27.2) and the electric current is switched on in coil A, then something happens that we do not expect! A force is seen to exist between the two coils. It is not a continuous force, however. It exists for a short time as the current is switched on in coil A, and again when the current is switched off

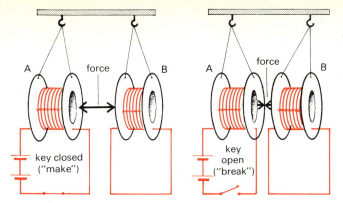

**Fig. 27.2**  *A force exists when the current is switched on and off in coil A*

(the force is small and a large current is needed). The two forces are in opposite directions on the "make" and "break" of the circuit of coil A. (Switching electricity on and off is often referred to as the make and break of the circuit.)

Our previous experiments have shown that for such forces to exist there must be an electric current flowing in both coils. So we must conclude that the switching on and off of an electric current in coil A, causes (induces) an electric current to flow in coil B. This can easily be checked by placing an instrument to detect such a current in the circuit of coil B.

*QUESTION 1:* What type of instrument would you place in the circuit of coil B?

## 27.2  What produces the induced electric current?

The induced electric current in coil B is dependent in some way upon the magnetic flux formed by the electric current flowing in coil A.

*Electromagnetic induction* is the term used to describe the production of an electric current or potential difference by a changing magnetic flux.

For many years after Oersted had shown that an electric current produces a magnetic flux (see page 192) scientists sought for evidence of the opposite effect: i.e. that a magnetic flux could produce (induce) an electric current. The first scientist to demonstrate this effect was Michael Faraday in the year 1831, when he was Director of the Royal Institution in London. The type of apparatus he used is illustrated in Fig. 27.3. In his experiment Faraday wound the two coils on a common circular iron core. When the current was switched on and off in the *primary coil* (A in the diagram) then a current was induced in the *secondary coil* (B) and detected on the galvanometer.

Faraday realized that the induced electric current only flowed when there was a changing electric current in the primary coil. It is the changing magnetic flux produced by the changing current in the primary coil which induces the current in the secondary coil.

**Fig. 27.3**  *Apparatus similar to that used by Faraday to demonstrate electromagnetic induction*

## 27.3  Two experiments to find out more about electromagnetic induction

*Experiment A*
A current flowing in a known direction is passed through a centre zero galvanometer in order to find out the direction of movement of the pointer. We discover that a deflection occurs to the right when there is a current flowing from the left to the right. Insulated wire is then wound around a cardboard tube to give a regular long coil. The coil is connected to the centre zero galvanometer (Fig. 27.4). A permanent bar with its north pole pointing in the direction of the coil, is moved into the coil, then held still. Finally the permanent magnet is removed from the coil. When the north pole is moved

**Fig. 27.4** *Experiment to study electro-magnetic induction*

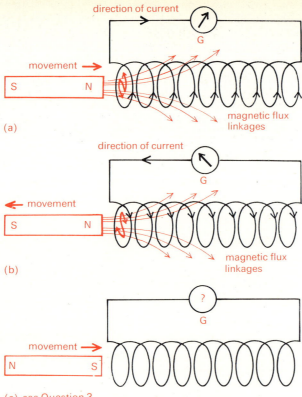

direction of current

G

movement →

S          N

magnetic flux
linkages

(a)

direction of current

G

← movement

S          N

magnetic flux
linkages

(b)

?

G

movement →

N          S

(c)  see Question 3

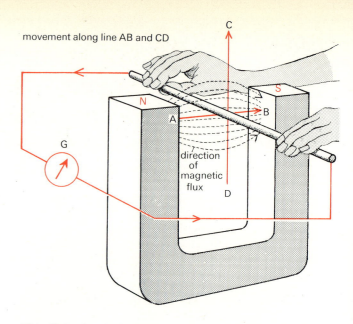

movement along line AB and CD

C

N          S

A          B

G

direction
of
magnetic
flux

D

**Fig. 27.5** *In which direction is the wire moving to produce this deflection in the galvanometer?*

into the coil the needle of the galvanometer moves to one side of the centre zero. It moves to the opposite side when the magnet is removed. This indicates that an electric current is induced in the coil while the magnet is moving. The induced current flows first in one direction, then in the opposite direction, but flows only when the magnet is moved. When the magnet is stationary there is no movement of the needle and we know therefore that no current is flowing. The same results are produced when the coil is moved instead of the magnet.

*Experiment B*
A straight piece of wire is connected to the terminals of a centre zero galvanometer. The wire is first held stationary between the poles of a strong horseshoe magnet (Fig. 27.5). Then the wire is moved up and down along the line CD, and sideways along the line AB.

When the wire is stationary there is no movement of the galvanometer needle. There is also no movement, and therefore no induced current, when the wire is moved sideways along AB. But the needle moves to one side of the centre zero when the wire is moved upwards, and it moves in the opposite direction when the wire is moved downwards.

We may summarize the results as follows:

When (a) the magnetic flux through a coil is changed or (b) a conductor moves across a magnetic flux, then there is an e.m.f. (electromotive force) induced within the conductor. If the circuit containing the conductor is closed then the induced e.m.f. produces an electric current.

*QUESTION 2:* Why is there no movement of the galvanometer needle when the wire moves sideways along the line AB?

## 27.4 Do experiments A and B tell us anything about the direction of the induced current?

Both experiments showed that when the movement was reversed, the direction of the induced current was also reversed.

In 1834, Henry Lenz made similar observations concerning the direction of induced currents. He stated what is now known as Lenz's law (yet another example of the principle of conservation of energy).

Lenz's law: the direction of the induced current is such that it opposes the change producing it.

We can apply Lenz's law to the experiments illustrated in Fig. 27.4. When the north pole of the magnet is moved towards the coil, as in Fig. 27.4(a), the induced current flows in such a direction as to produce a temporary north pole at the end of the coil facing the magnet. When the magnet is moved away, as in Fig. 27.4(b), the induced current flows in such a direction as to produce a temporary south pole at this end. In the first instance the temporary north pole is a repelling force, and opposes the movement of the magnet towards the coil. Similarly the temporary south pole exercises an attracting force, and opposes the movement of the north pole away from the coil.

*QUESTION 3:* When the south pole of the bar magnet is moved towards the coil, as in Fig. 27.4(c), will the deflection of the galvanometer needle be to the left or right of the zero mark? Explain your answer.

The results of the experiment illustrated in Fig. 27.5 indicate that the direction of the induced current in a straight wire is in a direction at right angles to both the direction of the magnetic flux and the movement. This gives us a simple rule known as Fleming's right-hand rule.

Fleming's right-hand rule (Fig. 27.6): position the first finger, second finger and thumb of your right hand at right angles to one another. In this position, when the **f**irst finger points in the direction of the magnetic **f**lux, the thu**m**b shows the direction of the **m**otion, and the se**c**ond finger points in the direction of the induced **c**urrent.

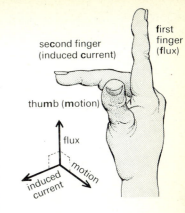

**Fig. 27.6** *Fleming's right-hand rule*

## 27.5 What factors affect the strength of the induced electric current?

1. *Experiment to study the effect of the strength of the magnetic flux*

Using the apparatus illustrated in Fig. 27.4, bar magnets of different strengths are moved at the same speed into the coil. The size of the induced current is observed by the deflection of the needle of the galvanometer.

A stronger magnetic flux is seen to produce a stronger induced electric current.

Most of us would expect such a result when we remember that the induced current is dependent upon the changing of the magnetic flux. What other factors might affect this process and therefore alter the strength of the induced electric current? Does the number of turns or the speed of cutting affect the size of the induced electric current?

2. *Experiment to study the effect of the number of turns of the coil*

A bar magnet is moved into a number of coils in turn.

Each coil is made up of a different number of turns, for example 300, 600, 900, 1200. The size of the induced current is observed by the deflection of the needle of the galvanometer.

The induced electric current is found to be greatest when the magnet is moved into the coil with the largest number of turns.

*3. Experiment to study the effect of the speed of the change*
A bar magnet is moved into and out of a coil a number of times. In each case the speed of the magnet is varied and the deflection of the needle of the galvanometer is observed.

When there is a rapid movement of the magnet, a greater induced current is seen to be produced.

In fact, experiments in which detailed measurements have been taken have shown that the magnitude of the induced e.m.f. is proportional to the rate of change of the magnetic flux.

*4. Experiment to study the effect of setting up a magnetic circuit*
Two similar circuits are set up as shown in Fig. 27.7. The throw on the galvanometer is observed when the switches are closed. The deflection is much greater when the coils are wound on a soft iron core. The circuit of magnetic flux in the iron core increases the induced e.m.f.

The results of these experiments have important applications in the construction of appliances such as dynamos.

A large induced current is produced when (a) there is a high rate of change of magnetic flux, (b) coils with a large number of turns are used, (c) there is a complete magnetic circuit.

The statement above and the one on the previous page are sometimes combined and stated as *Faraday's law of electromagnetic induction*:

314

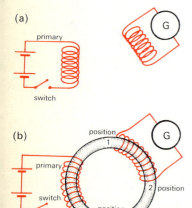

**Fig. 27.7** *The setting up of a magnetic circuit increases the induced current*

When the magnetic flux cutting a conductor changes, then the induced e.m.f. is proportional to the rate of change of flux.

## 27.6 Methods of obtaining a continuous induced current

In all the above experiments the induced current is not a continuous one. In order to achieve this we must produce a continuously changing magnetic flux. A very easy way of doing this, in the laboratory and at home, is to attach one end of the magnet to a spring clamped vertically above the coil (Fig. 27.8). When the spring is set vibrating the magnet moves up and down within the coil inducing a continuous alternating current.

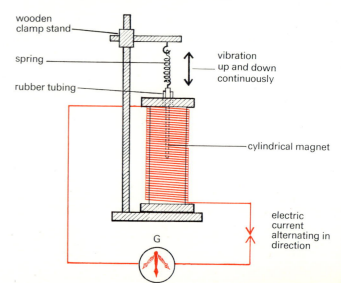

**Fig. 27.8** *A way of producing a continuous induced current*

In commercial appliances such as generators and dynamos (sometimes called alternators), which are used to convert kinetic energy to electrical energy, a continuous rotational movement is produced. This can be demonstrated with the apparatus illustrated in Fig. 27.9.

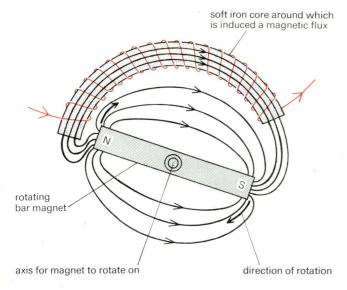

**Fig. 27.9** *Rotational movement of the magnet causes a changing flux in the coil. The current reverses (i.e. it is an alternating current).*

The rotating magnet produces a continuously changing flux through the coil and hence an alternating current flows in the coil.

## 27.7 A bicycle dynamo
This small generator of electricity is familiar to us all. When a cyclist needs light he moves the dynamo so that its driving wheel comes into contact with the bicycle wheel rim. Thus the dynamo's driving wheel rotates and in turn drives a permanent magnet. The magnet rotates inside a coil of insulated wire wound on a U-shaped soft-iron core. One lead from the coil is earthed through the frame of the bicycle. The other lead goes to a terminal, which has connections to the front and rear lights.

A bicycle dynamo can be used to light a 6V bulb. You can try many other methods of driving it, by water wheel, for instance, or by gears or a driving belt as shown in Figs. 27.10(a) and (b). In each case when the dynamo

**Fig. 27.10**(*a*) *A bicycle dynamo for use in the laboratory*

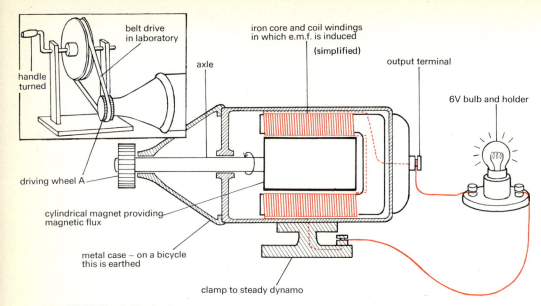

**Fig. 27.10**(b) *A bicycle dynamo used to light a lamp*

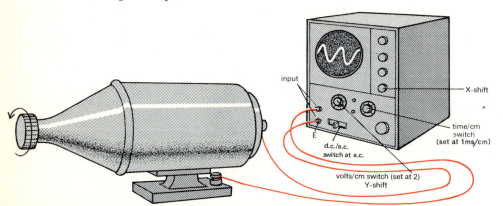

**Fig. 27.11** *Wave form of an alternating current generated by a bicycle dynamo and displayed on an oscilloscope*

316

driving wheel is turned slowly the lamp is seen to flicker, but as the rotation is speeded up the flickering effect disappears. If a centre zero galvanometer is placed in the circuit instead of the lamp, a deflection of the needle to the right and left of the centre zero is observed as the dynamo is rotated slowly. These results and our previous experiments show us that the induced current is an alternating current.

If the bicycle dynamo is connected to the input (Y-plates) of a cathode ray oscilloscope (see page 334), the change in direction of the current can be seen on the screen (Fig. 27.11). The oscilloscope in fact draws a graph of e.m.f. against time.

## 27.8 The construction and action of an a.c. generator (dynamo)

Figure 27.12 illustrates a dynamo in its simplest form. The coil is positioned between the curved pole pieces of a permanent magnet. When the coil rotates it cuts across the magnetic flux between the poles. Thus the magnetic flux through the coil is changed and an e.m.f. is induced in the coil. When the coil is connected to a closed external circuit, then an induced current flows.

The direction of the induced current can be determined by applying Fleming's right-hand rule to either side of the coil. In the table opposite it is applied to the side AB. A summary of what occurs during one complete revolution of the coil is given in the table and in Fig. 27.12 (a) to (e).

*QUESTION 4:* Why is the induced e.m.f. at a maximum when the coil is horizontal?

The summary in the table showing how an a.c. generator works shows that when the coil has completed one revolution the induced e.m.f. and current have gone

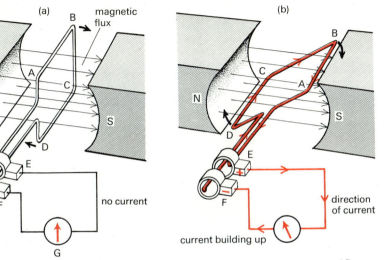

(a)

magnetic flux

slip rings

E and F carbon brushes

no current

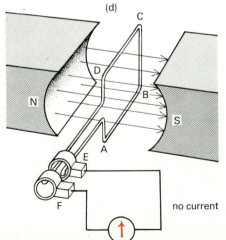

(b)

direction of current

current building up

| Coil position | Movement of side AB | e.m.f. (letters apply to Fig. 27.13) | Direction of current |
|---|---|---|---|
| Vertical Fig. 27.12(a) | — | zero (V) | none |
| Vertical–horizontal Fig. 27.12(b) | downwards | building up (V–W) | DCBA |
| Horizontal Fig. 27.12(c) | downwards | at a maximum (W) | DCBA |
| Horizontal–vertical | downwards | decreasing (W–X) | DCBA |
| Vertical Fig. 27.12(d) | — | zero (X) | none |
| Vertical–horizontal Fig. 27.12(e) | upwards | building up (X–Y) | ABCD |
| Horizontal | upwards | at a maximum (Y) | ABCD |
| Horizontal–vertical | upwards | decreasing (Y–Z) | ABCD |
| Vertical | — | zero (Z) | none |

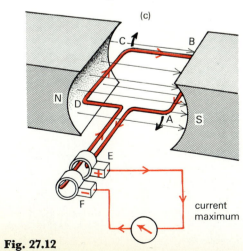

(c)

current maximum

(d)

no current

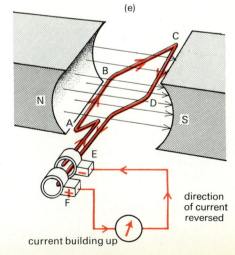

(e)

direction of current reversed

current building up

**Fig. 27.12**

317

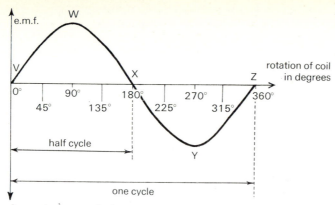

**Fig. 27.13** *Variation in e.m.f. of an a.c. generator as the coil is rotated*

through one *cycle* (represented by V–Z on the graph in Fig. 27.13).

Every time the coil passes through the vertical position, the side AB reverses the direction in which it is cutting the flux, and the direction of the current is reversed. Thus an alternating current which flows first in one direction and then the other, is obtained from the coil.

The frequency of an electric current is the number of complete cycles every second, measured in hertz (Hz).

The frequency of the a.c. mains in England is 50 hertz.

*QUESTION 5:* What will happen to the frequency of the a.c. if the rotation of the dynamo is speeded up?

A commercial a.c. generator consists of many turns of wire wound on a soft iron core. The coil and core are called the *armature*. The armature is rotated by means of a driving wheel connected to its axle. The ends of the coil are connected to *slip rings*, which are rings of copper or brass attached to the axle of the armature. Carbon (graphite) brushes are kept in contact with the slip rings by means of springs. In this way the current is fed from the coil into the external circuit.

318

**Fig. 27.14** *In the foreground, an industrial generator and exciter unit*

## 27.9 Industrial generators or alternators

The alternator (Fig. 27.14) generates an alternating current by means of rotating magnets and stationary coils. The magnets providing the magnetic flux are electromagnets. There are many pairs of these and they form what is termed the *rotor* (rotating part). The d.c. current to the electromagnets is supplied by a small generator called the *exciter*. The reason for having a rotating magnetic flux is that difficulties arise if you try (a) to collect very high voltages through carbon brushes, (b) to rotate a large armature at a high speed. The field poles rotate close to stationary coils which form the *stator* (stationary part). The output current is collected from terminals on the outer framework of the stator.

### 27.10  Direct current (d.c.) generator

A d.c. generator is structurally the same as a d.c. motor. The electromagnets form the stator, and the coils the rotor. The most important difference between a d.c.

generator and an a.c. generator is the use of a split ring commutator instead of slip rings for feeding the electric current into the external circuit.

Figure 27.15 shows a d.c. generator in its simplest form. When the coil rotates an e.m.f. is induced and the current flows around the closed circuit as in (b). After the coil has completed a half-revolution (from vertical to vertical) and the e.m.f. is at zero just before it is reversed, each split ring comes into contact with the other carbon brush as in (c). The action of this commutator is very similar to that of the commutator of the electric motor (page 302).

**Fig. 27.15**  *The action of a commutator on a d.c. generator*

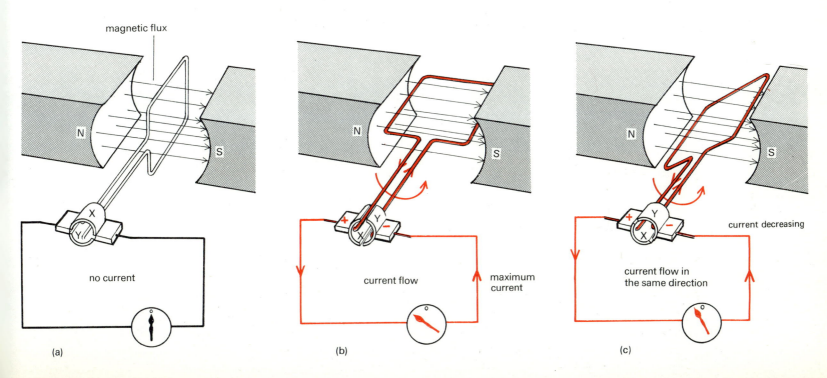

magnetic flux

N    S

no current

(a)

N    S

current flow    maximum current

(b)

N    S

current decreasing

current flow in the same direction

(c)

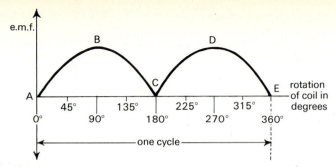

**Fig. 27.16** *Variation in e.m.f. of a d.c. generator as the coil is rotated*

The wave-form of a direct current is illustrated in Fig. 27.16.

### 27.11 How can a smoother direct current be obtained?

This is achieved by having a large number of coils wound onto the armature at angles to one another. The commutator is made up of many segments. The ends of each individual coil are connected to two segments on opposite sides of the commutator. The connections between the coils and commutators are such that a steady d.c. with only a small ripple (Fig. 27.17) is obtained in the external circuit.

**Fig. 27.17** *A smoothed direct current*

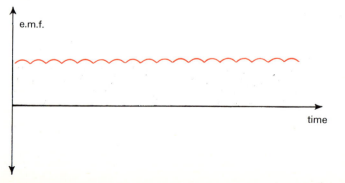

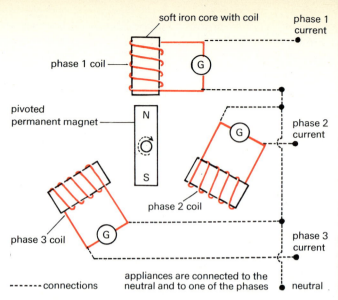

**Fig. 27.18** *Experiment using three coils to produce a 3-phase a.c. supply*

### 27.12 What is a 3-phase a.c. supply?

A 3-phase a.c. supply is obtained when a magnet is rotated in the centre of a system of three coils separated by angles of 120° (Fig. 27.18). The electric currents that are obtained are "out of step" with one another: i.e. their wave forms are at different phases (Fig. 27.19).

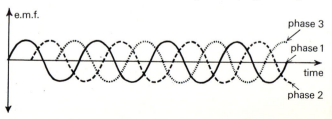

**Fig. 27.19** *A 3-phase a.c. supply*

Alternating current is highly adaptable and has an important place in the everyday supply of electricity all over the country. Let us look at some of its applications by starting again with the coils in Fig. 27.1.

### 27.13 Mutual induction

When coil A in Fig. 27.1 is connected to a low voltage a.c. supply and coil B is short circuited, a very large force is seen to exist between the two coils. This indicates that an e.m.f. can be induced in the secondary coil by a constantly alternating (changing) e.m.f. in the primary coil.

Mutual induction is the production of an e.m.f. in the secondary coil as a result of a changing current in the primary coil.

*QUESTION 6:* Will the galvanometer needle move to the right or the left of the zero mark when the switch in the primary circuit is closed (Fig. 27.20)? (Hint: remember Lenz's law.) The needle moves to the left if the current flows from right to left.

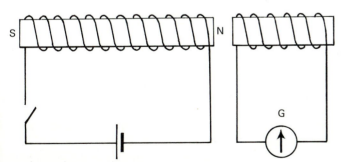

**Fig. 27.20**  *Experiment to demonstrate mutual induction*

Mutual induction is used in the design of appliances known as *transformers*.

### 27.14 Transformers

These are constructed of a primary coil and a secondary coil, both of which are wound on a core of soft iron. Often the core is laminated, consisting of numerous strips or stampings of iron.

We saw on page 314 that the size of the induced e.m.f. depends upon the number of turns on the coil. Your teacher can help you to study this fact in relation to a transformer by using the assembly transformer kit in the following experiment (Fig. 27.21). ANY EXPERIMENT INVOLVING MAINS SUPPLY, AS THIS DOES, CAN BE DANGEROUS AND NEEDS EXPERT HANDLING.

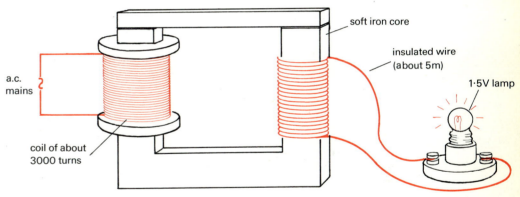

**Fig. 27.21**  *A transformer core and coil, to show the action of a step-down transformer*

An insulated coil of many turns (e.g. 3000) which can be connected to the a.c. mains is positioned around one arm of the U-core of the transformer. A 1.5V lamp is connected in series with a long length (e.g. 5m) of insulated wire, one turn of which is wound around the other arm of the transformer core. The lamp is observed as more and more turns of wire are wound onto the arm.

When there are only a small number of turns the lamp does not light. However, when more turns are added (in this case when there are about 18), then the lamp lights. But the primary coil is connected to the mains of 250V,

321

whereas the voltage induced in the secondary coil, with a small number of turns is only 1.5V. Thus our transformer has reduced the voltage and it is called a *step-down transformer*.

### 27.15   What is a step-up transformer?

It is clear from our previous experiment that provided the e.m.f. and the number of turns on the primary coil remain constant, the induced e.m.f. will depend upon the number of turns on the secondary coil. Why is this? Consider the experiment shown in Fig. 27.22.

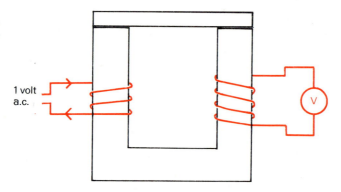

**Fig. 27.22**  *Experiment to show the action of a step-up transformer*

Two turns of ordinary connecting wire are wound around one arm of the transformer core in order to form the primary coil. The primary is connected to a 1V a.c. supply. Two turns of another piece of wire are wound around the other arm of the transformer core in order to form the secondary coil. A voltmeter is also connected across the secondary coil and its reading recorded. The voltmeter reading is recorded when the number of turns on the secondary is 2, 4, 6, 8 and 10.

The recorded readings of the voltmeter are 1, 2, 3, 4 and 5 volts respectively. From this we draw the following conclusion:

322

The induced e.m.f. in the secondary coil is directly proportional to the number of turns on the secondary coil provided that the e.m.f. and number of turns on the primary coil remain constant.

Since each turn on the secondary coil cuts the magnetic flux it is to be expected that the same e.m.f. is induced in each turn. Thus if an e.m.f. of 0.05V is induced across one turn, then an e.m.f. of $0.05 \times 20V = 1.0V$ is induced across 20 turns.

A further experiment using coils with a different number of turns as both the primary and secondary coils will demonstrate that (neglecting losses):

$$\frac{\text{e.m.f. across secondary coil (output)}}{\text{e.m.f. across primary coil (input)}}$$

$$= \frac{\text{number of turns on secondary coil}}{\text{number of turns on primary coil}}$$

Thus if the number of turns on the secondary coil is greater than that of the primary, the output e.m.f. will be higher than the input e.m.f. This is the principle for the construction of a *step-up transformer*.

Note that in all transformer constructions it is essential that the coils are well insulated from the core.

### 27.16   Does the conservation of energy apply to transformers?

The experiment illustrated in Fig. 27.23 helps us to answer this question.

A simple step-up transformer is made from a primary coil of 20 turns and a secondary coil of 50 turns wound on a common soft iron core. The primary is connected to a low voltage a.c. supply of 2 volts. An ammeter and voltmeter are connected in the primary circuit and the

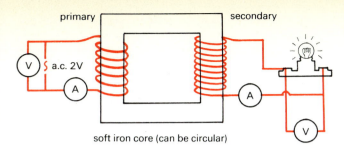

primary      secondary

a.c. 2V

soft iron core (can be circular)

**Fig. 27.23** *Transformers and the conservation of energy: the energy used in the secondary cannot be more than the energy fed into the primary*

secondary circuit. Readings are taken from the two sets of instruments.

As was stated on page 184, electrical power is measured by the product of volts and amperes (energy = V$It$ joules). Thus if we assume the transformer to be 100 per cent efficient, i.e. if all the electrical energy supplied to the primary is converted into electrical energy in the secondary.

Then:

Power supplied to primary (input) = power obtained from secondary (output).

Therefore:

Primary e.m.f. × primary current = secondary e.m.f. × secondary current.

The above experiment shows us that this relationship is very nearly true, and that only a small quantity of energy is lost (it is converted into heat energy).

## 27.17    The use of transformers within the grid system

If a large current is transmitted through a wire there is a large conversion of electrical energy to heat energy. This is because the heat developed is proportional to the square of the current flowing (see page 184). To reduce this effect it is simple to lower the current and transmit the electricity at high voltage (high tension). A high tension of approximately 132 000V is found in the grid system. Thus there is an extensive use of both step-up and step-down transformers at substations.

## 27.18    Why do transformers get hot when used?

The heating effect created within the core of the transformer is caused by *eddy currents*. This is the name given to the induced currents set up within a mass of a conducting material (solid, liquid or gas). The currents are visualized as being like the swirling eddy currents seen in water.

If a transformer core is laminated this reduces the eddy currents and hence the heating effect within the core. However, the heating effect of eddy currents can be put to good use — for the melting of steel within an induction furnace, for instance. The furnace acts as a solid conductor and it is surrounded by a coil through which high frequency alternating currents are passed.

The rotational movement of the metal disc in some car speedometers and domestic electric meters is caused by the setting up of eddy currents within the disc. There is then a reaction between the magnetic fields resulting in motion.

## 27.19    Induction coil

A transformer will not work when a d.c. supply is passed through it, but an induction coil, an appliance which is similar in principle to a step-up transformer, can use d.c. supply.

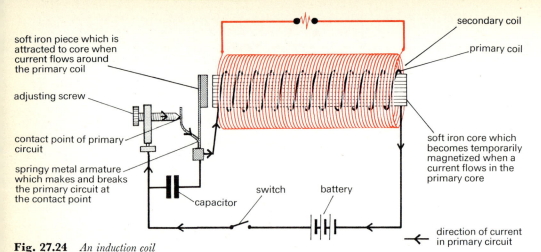

**Fig. 27.24** *An induction coil*

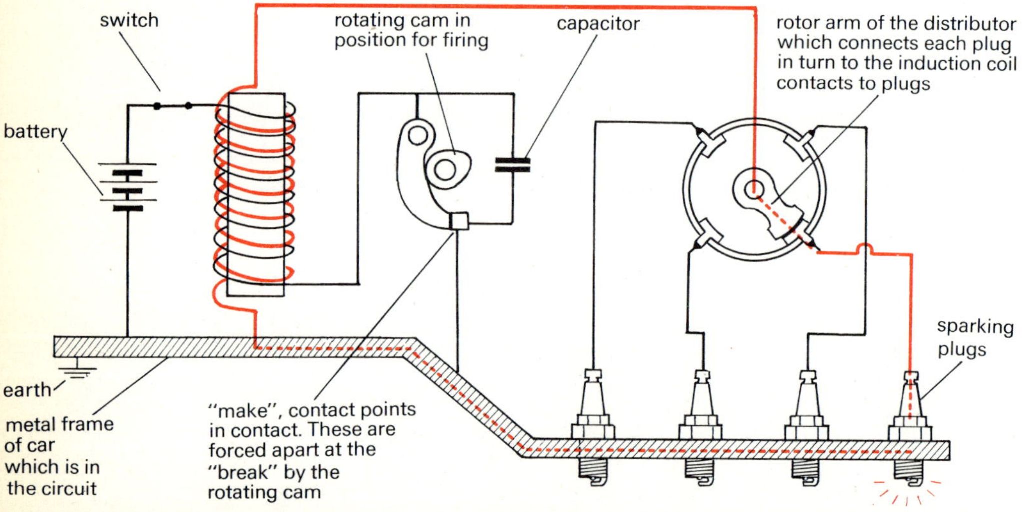

soft iron piece which is attracted to core when current flows around the primary coil

adjusting screw

contact point of primary circuit

springy metal armature which makes and breaks the primary circuit at the contact point

capacitor

switch

battery

secondary coil

primary coil

soft iron core which becomes temporarily magnetized when a current flows in the primary core

direction of current in primary circuit

switch

battery

earth

metal frame of car which is in the circuit

rotating cam in position for firing

capacitor

"make", contact points in contact. These are forced apart at the "break" by the rotating cam

rotor arm of the distributor which connects each plug in turn to the induction coil contacts to plugs

sparking plugs

**Fig. 27. 25** *Ignition system of a motor car*

A continuously changing current is produced by a mechanism in the primary coil which makes and breaks the circuit (Fig. 27.24). The action of this make and break mechanism is identical with that already described in connection with the electric bell (page 195). The make and break occurs in the region of 50 times per second and on average a 6 to 12V input will produce an output in the region of 100 000V.

Such an induction coil is used within the ignition system of a motor car (Fig. 27.25).

### 27.20   Induction motor

Induction motors are generally used where a large power output is required. Such motors are probably responsible for approximately 90 per cent of the motive power obtained through electric motors in everyday use.

An induction motor has a stator which produces the necessary changing magnetic field. The stator is made of three sets of coils wound on soft iron cores and arranged symmetrically in a circle (Fig. 27.26). Each set of coils is connected to a single phase of the 3-phase a.c. supply. The rotor generally consists of a number of copper bars held together by end rings forming a cage (thus the term "squirrel cage rotor"). The copper cage is placed around an iron core (Fig. 27.26).

When the current flows in the stator coils an induced current is set up in the rotor. The "force of induction" which sets the rotor revolving is the force between the magnetic fluxes of the two electric currents, one of which has induced the other. No commutators or carbon brushes are involved. Thus the number of moving parts in contact with one another is less than in d.c. motors and the wear and tear is much less than that in a d.c. motor.

A *linear induction motor* can be thought of simply as a rotary induction motor with the stator cut and rolled out flat. These are likely to be much used in future rail travel.

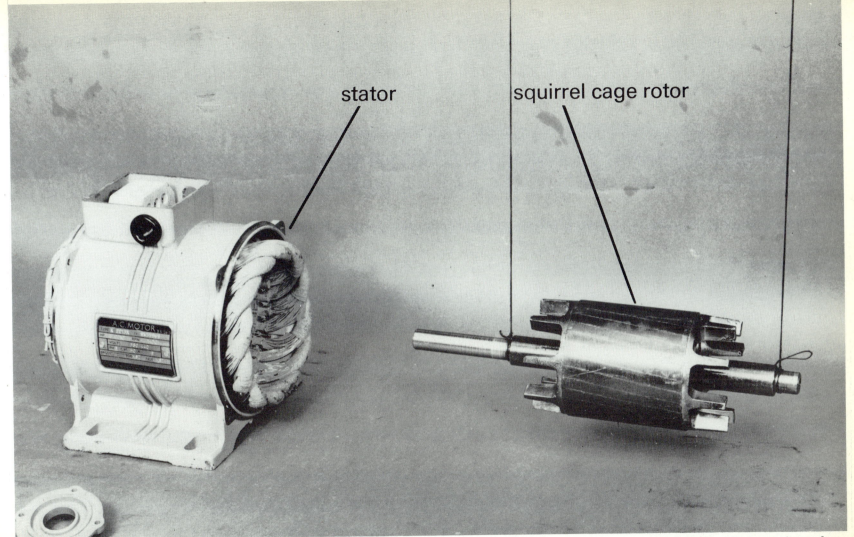

stator

squirrel cage rotor

**Fig. 27.26** *The stator and rotor of an induction motor*

**THINGS TO DO**

**A.** Make a d.c. dynamo.

The d.c. motor made according to the instructions given in "Things to do" Chapter 26, page 305 or the one on page 306, may be adapted to form a simple d.c. dynamo. Instead of connecting the leads to a battery, connect them to a galvanometer, or light bulb. Rotate the axis of the armature by hand, or more quickly by winding thread around it once and pulling as shown in Fig. 27.27.

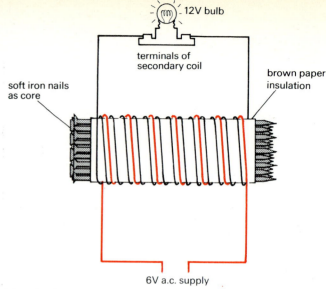

**Fig. 27.28** *Construction of a simple transformer*

**Fig. 27.27** *How to rotate the axle of a model dynamo*

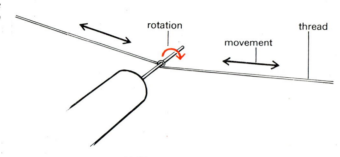

**B.** Make a transformer.

As illustrated in Fig. 27.28 tie together 8 to 10 soft iron nails (approximately 15cm in length) in a bundle. Around the bundle of nails wrap strips of brown paper. Form the primary coil by winding 100 turns of insulated copper wire (approximately 24 s.w.g.) around the core as indicated in the diagram. Wrap more layers of brown paper around this coil before starting to wind on the secondary coil. A similar wire is used for the secondary coil, which should consist of 400 turns.

Connect the primary coil to a 6V a.c. supply and place in series with the secondary coil a 12V car bulb. Such a transformer will give a secondary voltage of approximately 10V. Some energy will also appear in the form of heat.

**C.** Make d.c. and a.c. supply visible.

Electricians use this method for detecting the polarity of terminals because it reveals the positive terminal. Make up your detection paper by soaking thick blotting paper in a thinly running paste of starch (or flour) mixed with potassium iodide (a chemist's shop or your chemistry department at school will provide the latter). Allow excess solution to run off the paper before placing it on top of an upturned metal pastry case (Fig. 27.29). Connect the case by a crocodile clip to the negative terminal of a 12V battery (car battery). From the positive terminal take a length of connecting wire and by means of a crocodile clip connect it to the metal shaft of a screwdriver. Hold the screwdriver by its insulated handle and draw its metal tip across

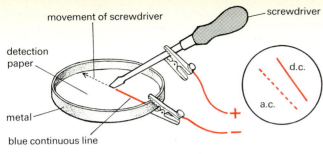

movement of screwdriver

screwdriver

detection paper

d.c.

metal

a.c.

blue continuous line

**Fig. 27.29** *Making a.c. and d.c. visible*

the detection paper. Repeat the procedure but this time remove the battery and connect the circuit up to a 12 a.c. supply (low-voltage power-pack).

By a chemical action the electricity causes iodine to be released at the positive pole where it reacts with the starch giving a dark blue compound. This change in colour can be detected. A d.c. supply produces one continuous line, whereas an a.c. supply produces a series of dashes.

D. Use a small dynamo to set up a working model of a power plant. Use different forms of energy to drive the dynamo.

E. Find out the positions of power plants in England and how they are linked into the grid system. Draw a general map of this system and enlarge your own location to show the positioning and use of sub-stations, etc.

F. List and illustrate possible uses of transformers in the home.

G. How is electricity supplied from your local power station? If possible go on a tour of the power station and then construct a file dealing with processes seen on your visit.

H. The electricity generated at our power stations is carried across the countryside by an electric grid. All of you will have seen the pylons which support the cables around the countryside (Fig. 18.5, page 176). Find out all you can about the grid system. In particular try and answer the following questions. (a) Why is the electricity transmitted at 132 000V or even higher voltages? (b) What would you expect to find at a grid supply point? (c) Why does not each town have its own generating station?

I. Find out how a speedometer works.

# THINGS TO WORK OUT

1.  a. What is a cycle?
    b. Draw a graph to distinguish between a d.c. supply and an a.c. supply of electricity.
    c. State three ways in which you could increase the current produced by a dynamo.

2.  Describe an experiment that you would carry out, using a galvanometer, coil of wire and a permanent magnet, to demonstrate electromagnetic induction. What factors affect the strength of the induced current?

3.  Describe, with a diagram, the structure and action of a step-up transformer. If a transformer has a secondary coil of 2000 turns and a primary of 500 turns, what will be its output voltage when the input voltage is (a) 5V, (b) 250V?

4.  Lawnmowers that operate on petrol have a magneto working in unison with the engine. What is the purpose of the magneto?

5.  Why are the cores of transformers and induction coils laminated?

6.  Look at "Things to do" number C, pages 326–7.
    Explain why a continuous line was produced on the detection paper by the d.c. supply, and a line of dashes by the a.c. supply.

7.  A transformer has a primary of 100 turns and a secondary of 600 turns. What input voltage will be necessary in order to produce output voltages of (a) 2400V, (b) 9V?

8.  Draw a diagram and describe the action of a simple a.c. generator. How would you convert this to produce d.c.? What is meant by a smoothed d.c. supply of electricity?

9.  If you are given two voltmeters, a dismountable transformer core and several coils consisting of different numbers of turns, how would you demonstrate the differences between a step-up and step-down transformer? State the formula on which the design of such transformers is based.

10. A transformer has a power output of 25W and an output voltage of 10V. When its primary is connected to the 250V mains, what is the value of the current in (a) the secondary coil, (b) the primary coil. Assume that there is no electrical energy loss?

11. Write a short description of an induction coil and how it operates. A bell transformer may be made into a shocking-coil by connecting one of the coils to a 2V d.c. supply and the other coil to two metal handles. When the two people at the ends of a line of people all holding hands grasp the metal handles they will not feel any shocks. What must be inserted into the circuit of the primary coil in order to get the transformer to function as a shocking-coil? State when the shocks will be felt.

12. What is the main advantage of alternating current over direct current for supplying electrical power? Why is it more economical to transmit electricity over the grid system at a high voltage?

13. Why can a hum usually be heard when a transformer is working? If the two C-cores forming the soft iron core of a transformer are slowly pulled apart a loud chatter can be heard. Explain this.

14. Figure 27.30 shows an aluminium disc which can rotate in a vertical plane. Without the magnet present it rotates freely, but when the magnet is put in position the rotation is heavily damped and the disc quickly comes to rest.
    a. What is meant by the words "heavily damped"?
    b. What causes the damping?
    c. Suggest how a disc of aluminium could be constructed so that the damping due to the magnet would be much less. (Hint. What is done to reduce the heat losses in the core of a transformer?)

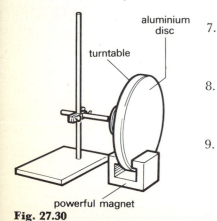

aluminium disc

turntable

powerful magnet

**Fig. 27.30**

# SOME OTHER TOPICS

**Part Six**

*Television camera and monitors*

# ELECTRONICS

## 28.1 Introductory experiment

So far in this course we have assumed the existence of electrons. We have thought of an electric current as a flow of electrons. We shall now consider some of the evidence that leads us to believe in the existence of these tiny invisible particles.

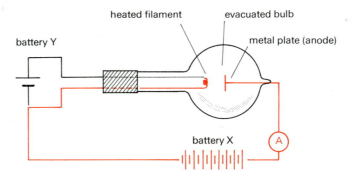

**Fig. 28.1** *A current flows only when the metal plate is positive and the filament is heated*

The experiment illustrated in Fig. 28.1 consists of an evacuated bulb containing a coil of wire (the *filament*) and a metal plate called the *anode* (Greek = upward way) because it is connected to the positive terminal of the battery X. The filament is heated by connecting it to the battery Y. When the batteries are connected as shown in the diagram the ammeter gives a reading indicating that a current is flowing in the red part of the circuit. If either (a) the battery X is reversed, or (b) the filament is not heated by the battery Y, then no current flows. Look carefully at Fig. 28.1 and work out the polarity (positive or negative) of the filament and metal plate.

*QUESTION 1:* Which of the following statements would explain the observed results: (a) positive charge is given off by the heated filament, (b) negative charge is given off by the heated filament?

The fact that a current only flows in the red circuit when the filament is heated and the plate is positive, is strong evidence for the fact that something negative (electrons?) is "boiled off" when the filament is heated. Let us see by further experiments whether we are correct in assuming that it is negative charge that is being "boiled off".

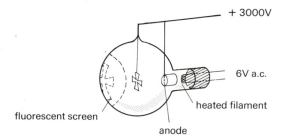

**Fig. 28.2** *Maltese cross experiment. What does it tell us about the properties of cathode radiation?*

### 28.2 Maltese cross experiment

Figure 28.2 shows an evacuated bulb with a heated filament and anode. The emission from the heated filament passes through the hole in the anode. A piece of metal, in the shape of a Maltese cross, is fixed in the middle of the bulb and connected to the positive terminal of a high voltage supply (e.g. 3000V). The front of the tube is covered with a fluorescent material which glows green when electrons strike it.

When the 3000V supply is disconnected from the Maltese cross and the filament heated, a shadow of the cross is cast on the screen by the light rays emitted from the filament. When the 3000V supply is connected to the cross, fluorescence occurs on the screen and a shadow is cast by the emission from the filament. If a magnet is brought up to the tube, the fluorescent shadow is seen to move, but the optical shadow produced by the light remains stationary.

332

*QUESTION 2:* What can you deduce about the properties of the emission from the heated filament by (a) the formation of a shadow and (b) the movement caused by the magnet? (Hint see page 296).

If the north pole of the magnet is brought up to the tube (Fig. 28.2) so that the lines of flux travel into the paper, the beam moves upwards. By using the rule set out on page 296 we can deduce that the emission from the filament is negatively charged. The filament is called the *cathode* (Greek = downward way) because it is connected to the negative terminal of the battery.

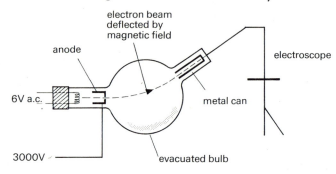

**Fig. 28.3** *The Perrin tube. What is the sign of the charge that collects on the electroscope?*

### 28.3 Perrin tube

In the Perrin tube (Fig. 28.3) the emission from the cathode (called *cathode radiation*) passes through an evacuated bulb and is deflected by a magnetic field into a metal can. The can is connected to an electroscope and the electroscope becomes charged. If a negatively charged rod is brought up to the electroscope the leaves diverge further showing that the electroscope is negatively charged. Whatever is emitted from the heated filament has a negative charge.

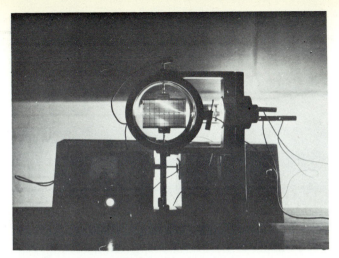

**Fig. 28.4** *Teltron deflection tube*

## 28.4 Deflection of cathode radiation by electric and magnetic fields

Figure 28.4 is a photograph of a Teltron deflection tube and Fig. 28.5 is a diagram of the tube. The large coils (not shown on Fig. 28.5) outside the tube enable a magnetic field to be applied across the evacuated bulb. The cathode radiation passes obliquely over the fluorescent screen. When a p.d. is applied across the horizontal metal plates EE′ so that E is negative, the beam is deflected downwards. If E′ is negative the beam is deflected upwards. From this we may conclude that the beam is negatively charged.

*QUESTION 3:* If a current flows in the coils producing a magnetic field into the paper, which way will the beam move? (Use the rule on page 296.)

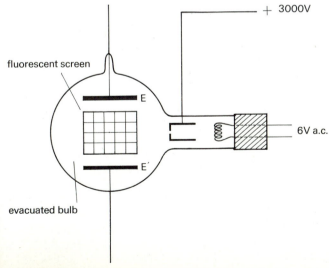

**Fig. 28.5**

## 28.5 What is cathode radiation?

So far our experiments have shown that cathode radiation is "something" negatively charged that flows from the heated cathode. Direct evidence concerning the nature of cathode radiation came first from experiments conducted by J. J. Thomson. By deflecting the radiation in electric and magnetic fields he showed that it behaved as if it were a stream of negatively charged particles. We call these charged particles electrons.

In 1909 R. A. Millikan demonstrated that there was a fundamental unit of charge, and that this charge was the charge on an electron (see page 163). He did this by suspending oil drops between two charged plates and measuring the charge on each drop. No drop was ever found to have a charge of less than $1.6 \times 10^{-19}$C. He calculated the charge on a very large number of drops, and always found that it was a multiple of $1.6 \times 10^{-19}$C. Whenever a drop gained or lost charge, the gain or loss was always a multiple of $1.6 \times 10^{-19}$C.

We conclude, therefore, that cathode radiation is a stream of electrons each of which has a charge of $1.6 \times 10^{-19}$C.

**Fig. 28.6** *The action of a diode*

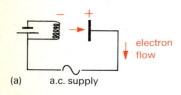

(a)    a.c. supply

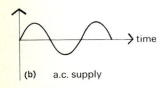

(b)    a.c. supply

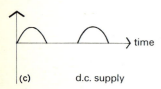

(c)    d.c. supply

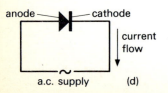

(d)    a.c. supply

## 28.6   The diode

### 1. *The diode valve*

The experiments described so far have led us to the conclusion that negative electrons are emitted by a heated filament. This emission of electrons by a heated metal surface is known as the *thermionic effect*.

Look again at Fig. 28.1 which illustrates the principle of the *diode valve*. Electrons flow from the filament called the cathode, which is connected to the negative terminal of the battery, to the metal plate called the anode, which is connected to the positive pole of the battery. In such a valve electricity can only flow in one direction.

If an alternating electrical supply is put in place of the battery X, electrons will only flow when the plate is positive and hence the valve works as a *rectifier*, converting the alternating current to a direct current (see Fig. 28.6(a)). Figure 28.6(b) shows the a.c. input, and Fig. 28.6(c) shows the current flowing in the circuit containing the valve. The current only flows during the part of the cycle when the anode is positive.

### 2. *The solid state diode*

Substances with properties somewhere between those of good insulators and good conductors are known as semiconductors (they do not conduct well enough to be called true conductors). A diode which is cheap to construct may be made by joining together two different kinds of semiconductor. The symbol for such a diode is shown in Fig. 28.6(d). Such a device has a very high resistance in one direction and a very low resistance in the other. This "one-way only" property means that it behaves in a very similar way to a diode valve. The current flowing in Fig. 28.6(d) is illustrated in Fig. 28.6(c).

334

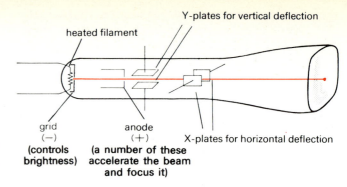

**Fig. 28.7** *Cathode ray tube*

## 28.7   Cathode ray oscilloscope (CRO)

When you look at television you are looking at the front of a cathode ray tube. The principle of a cathode ray tube is illustrated in Fig. 28.7.

A beam of electrons is produced by a heated filament and anode (the arrangement used to produce the beam is called an *electron gun*). The beam hits a fluorescent screen at the opposite end of the tube and produces a spot of light. On its way to the screen the beam passes through vertical and horizontal pairs of plates across each of which a p.d. may be applied. By applying a p.d. between the Y-plates the beam may be deflected up or down, and by applying a p.d. between the X-plates the beam may be deflected sideways. The deflection is proportional to the p.d. between the plates. The instrument may therefore be used as a voltmeter by applying the p.d. across one pair of plates and putting a scale calibrated in volts on the front of the screen.

*QUESTION 4:* In what way is an oscilloscope a better voltmeter than a moving coil instrument? (Hint: why does a good voltmeter have to have a high resistance?)

One important use of a CRO is the examination of various wave forms (see page 347). In order to do this a special circuit called a time-base circuit is connected to the X-plates. The time-base circuit produces a gradually increasing p.d. across the X-plates which causes the spot of light to move at a uniform speed across the screen from left to right. When it gets to the right-hand end of the screen it immediately starts from the left-hand end again. The spot is usually moving so fast from left to right that, because of the persistence of vision (see page 60), the observer sees a continuous line as shown in Fig. 28.8(a). If at the same time an alternating p.d. is applied across the Y-plates a curve like the one shown in Fig. 28.8(b) is traced out on the screen.

(a)                                    (b)

**Fig. 28.8** (a) *The screen of a CRO when the time base is switched on* (b) *The effect of connecting an alternating p.d. across the Y-plates*

There are usually three anodes in the CRO, and the knob that focuses the beam adjusts the potentials of these anodes. The brightness is controlled by adjusting the negative potential of the grid (see Fig. 28.7). This is a negatively charged cylinder. How do you think it controls the brightness?

## 28.8  Television
In a television tube the spot of light moves across the tube from left to right (as in a CRO), but it also moves

downwards as illustrated in Fig. 28.9. Most countries use a 625-line system, that is, the spot traces out 625 lines in its passage from the top to the bottom of the screen. The spot covers the screen 25 times a second. The incoming signal controls the intensity of the electron beam and hence a picture is built up of dark and bright spots. (Look at a picture in the daily newspaper with a magnifying glass and you will see that it also is a series of dark and bright spots.) The television picture is therefore traced out as a series of horizontal lines varying in brightness along their length. Because of the persistence of vision the viewer sees a continuous picture. We have already discussed the principle of colour television on page 49.

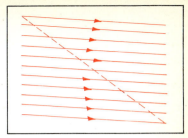

**Fig. 28.9** *Movement of a spot of light across and down a TV screen*

## 28.9  The photo-electric effect
We have seen that electrons are emitted from a metal surface when it is heated. Electrons are also emitted from some metals when ultra-violet light (see page 50) is shone onto them. In the experiment illustrated in Fig. 28.10 a zinc plate is attached to an electroscope.

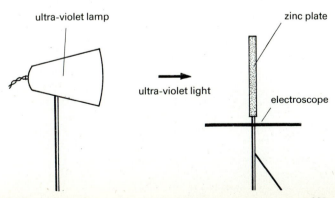

ultra-violet lamp

zinc plate

ultra-violet light

electroscope

**Fig. 28.10** *The photo-electric effect*

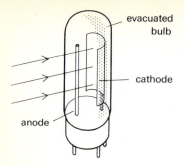

evacuated bulb

cathode

anode

**Fig. 28.11** *A photocell*

When the electroscope is charged negatively and the zinc plate is irradiated with ultra-violet light the electroscope quickly loses its charge. If the electroscope is charged positively and the zinc is irradiated with ultra-violet light the electroscope does not lose its charge.

*QUESTION 5:* Can you explain the above observations?

Some metals, such as caesium and antimony, emit electrons when visible light falls on them. This phenomenon of the emission of electrons by metals when either ultra-violet light or visible light falls on them is known as the *photo-electric effect*.

In a typical photocell (Fig. 28.11) light falls on a photosensitive cathode, close to which is an anode. When light falls on the cathode, electrons are emitted which are attracted to the anode. The magnitude of the photo-current depends on the intensity of the incident light. Such a cell is made use of in light-meters used by photographers and in many automatic devices.

Figure 28.12 shows a device that detects reflective tape on parcels. The parcels are loaded onto a single conveyor belt at the main store. The storekeeper puts reflective tape on the parcel, the position of the tape depending on the ultimate destination of the parcel. The diversion of the parcel to other conveyor belts takes place automatically at various points throughout its travel. The small heads (sensors) on the pillars are of the combined projector/receiver type using the principle that light projected is bounced back to the receiver unit from the reflective tape.

The solar cells shown in the photographs on pages 205 and 227 are another example of the photo-electric effect. The light energy incident on the solar cells is converted into electrical energy.

### 28.10 The transistor

A transistor consists of 3 pieces of semiconductor material and it has three wires coming from it. Its symbol is shown in Fig. 28.15. It is a current amplifier; a small change in the current through the base produces a large change through the collector. It is this property which makes it useful in an amplifier (Fig. 28.17). The best way to learn about transistors is to do experiments such as D on the next page.

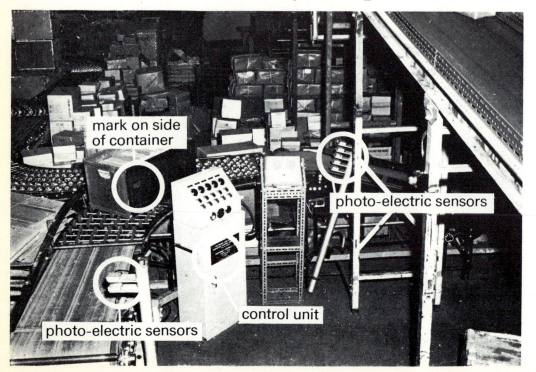

mark on side of container

photo-electric sensors

photo-electric sensors

control unit

**Fig. 28.12** *The photo-electric device automatically operates the diverting mechanism: the final destination of the parcel is determined by the position of the reflecting tape*

A. If you own or can borrow an exposure meter, investigate its action under different light intensities. Using a bright lamp, investigate what happens to the intensity of the light when the distance of the meter from the lamp is doubled.

B. If you have access to an electronics kit containing a photo-electric cell, build a miniature burglar-alarm system and show it to your teacher. (There are many such electronic kits on the market.)

C. a. Connect a torch bulb and a semi-conductor diode (a semi-conductor device with two terminals) in series with a battery. Reverse the battery connections. What happens?
b. Connect up the circuit shown in Fig. 28.13. Observe what happens when the battery connections are reversed.
c. Look at Fig. 28.14. Write down what you think will happen (i) when the battery is connected as shown and (ii) when the battery is reversed. Then wire up the circuit and see if you were right.

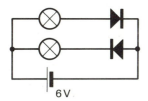

**Fig. 28.13**

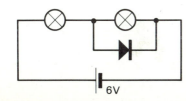

**Fig. 28.14**

D. Find out all you can about transistors and what they can do. A few ideas for experiments may help you to get started. Figure 28.15 is the symbol used for a transistor in electrical circuit diagrams. A transistor is basically a three electrode semi-conducting device, sealed in a very small case. The three wires coming from the case are connected to the base, the collector and the emitter respectively. You will have to refer to the manufacturer's catalogue to determine which wire is which in any transistor.

Set up the circuit of Fig. 28.16. $B_1$ and $B_2$ are bulbs. Adjust the values of $V_b$ and $V_c$ until the bulb $B_2$ lights. The current flowing in $B_1$ is too small to light the bulb. Unscrew the bulb $B_1$. What happens to the bulb $B_2$? Can a current flow in the collector circuit (shown in red) if there is no current flowing in the base circuit (the one containing $B_1$)? Investigate what happens to the bulbs as the value of $R$ is changed from $1k\Omega$ to $22k\Omega$. (Do not make $R$ less than $1k\Omega$ or you will damage the transistor.) Replace $B_1$ with a microammeter and $B_2$ with a milliammeter. For a fixed value of $V_c$ change the value of $V_b$ and record the readings of the meters. Does the transistor behave as a current amplifier? Does changing the value of $V_c$ make any difference? Make $V_c$ very small and gradually increase its value while $V_b$ is kept fixed.

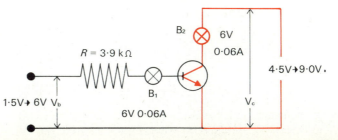

**Fig. 28.16**

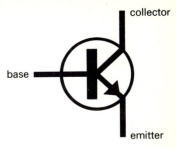

**Fig. 28.15**

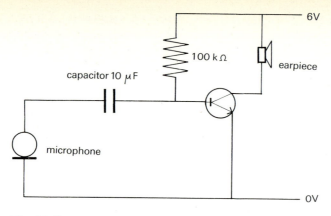

**Fig. 28.17**

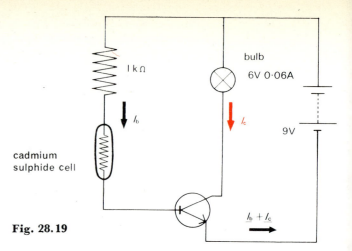

**Fig. 28.19**

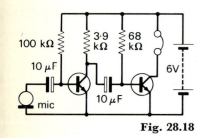

**Fig. 28.18**

Figure 28.17 is a circuit of an amplifier for you to build.

When your amplifier is working put a signal generator in front of the microphone and set the frequency to 100Hz. Connect the Y-input of a CRO across the earpiece. Observe the trace on the CRO and compare it with the trace when the CRO is connected across the microphone. Finally connect the CRO across the 100kΩ resistor.

See if your microphone will detect the beating of your heart. If not you can make the circuit more sensitive by adding another transistor as shown in Fig. 28.18.

Figure 28.19 is a circuit containing a cadmium sulphide cell. When light shines on the cell its resistance goes down. The bulb in the collector circuit only lights up when light is shining on the cell. This is because a current flows in the collector circuit (the current shown by the arrow $I_c$) only when the current $I_b$ in the base circuit exceeds a certain value. Only when light is shining on the cell is the current $I_b$ large enough for current to flow in the collector circuit.

There are many more exciting experiments you can do with transistors and there are many kits on the market which supply the parts, circuits and instructions. *The Wireless World* lists the names of many suppliers of electronic parts. A suitable cadmium sulphide cell is ORP 12 and any general purpose npn transistor (e.g. 2N3708 or BC 108) which will dissipate 1 watt will work satisfactorily and there are hundreds of different types available.

E. Find out about the work of Baird and how his first television set worked.

1. State two properties of cathode radiation and describe an experiment to demonstrate each of them.

2. Draw a diagram of a cathode ray tube and describe how it works.

3. What is meant by thermionic effect? Describe an experiment to demonstrate it.

4. What is the photo-electric effect? How can it be demonstrated? Describe one application of this effect.

5. Look back at Fig. 28.7. (a) Describe what happens to the electron beam when a battery is connected across the Y-plates. What happens when it is connected across the X-plates? In each case state clearly to which of the two plates the positive pole of the battery is connected. (b) In what direction would a magnetic field have to be applied to deflect the beam upwards?

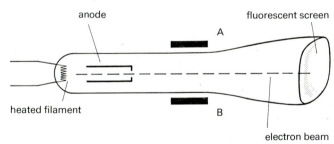

**Fig. 28.20**

6. Figure 28.20 shows a vacuum tube which produces a beam of electrons which impinge on a fluorescent screen, and cause a spot of light on the screen. If A and B are metal plates in a horizontal plane with A at a positive potential in respect of B, what change will this cause in the position of the spot on the screen? How will the spot move if A is a north pole of a magnet and B a south pole of a magnet? What will be observed if an alternating p.d. is connected across the two plates A and B?

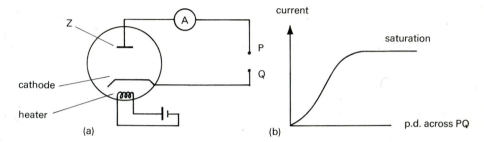

**Fig. 28.21**

7. In Fig. 28.21 (a) the cathode of the valve is heated by a small heating coil. (a) What will be emitted from the cathode? (b) What is the part Z called? (c) If a battery is connected across P and Q, what happens when P is positive? (d) What will the milliammeter A read when P is negative and Q is positive? (e) Fig. 28.21 (b) is a graph of the current through the milliammeter as the p.d. across P and Q is gradually increased. Why is there a limit to the maximum value of the current?

8. The time base of an oscilloscope is switched on. Describe what you would see on the screen when (referring to Fig. 28.22) the points AB, BC and CD are in turn connected across the Y-plates.

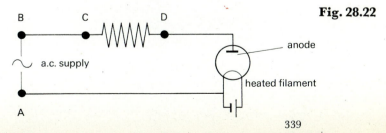

**Fig. 28.22**

## CHAPTER 29

# SOUND

**Fig. 29.1** *An effects staircase at Broadcasting House; at the top is an effects door with a variety of locks*

## 29.1   What is sound?

Figure 29.1 shows a staircase within a special room at Broadcasting House. Have you any idea what it is used for? Note that every tread on the staircase has three

340

parts. These are of concrete, wood and iron, and each of these materials has a different sound when stepped on. This is the effects staircase in the sound-effects room, where the everyday sounds that have to be put into broadcast programmes or plays to make them lifelike are produced.

We tend to take many sounds for granted, and we can even ignore some sounds and select others to concentrate on. For example, when rushing to get ready for school you may look at the clock and think that it must have stopped because it is not so late as you thought. In order to make sure you probably consciously listen for the tick of the clock to make sure that it has not stopped. Normally you would ignore this ticking noise! We usually become more conscious of sounds when we try, or are asked, "not to make a sound".

*QUESTION 1:* Without making a sound turn the page of this book.

Did you manage it?

Why is it so difficult to do?

## 29.2 Sound as energy

Sound is a form of energy, and it arises when there is movement, i.e. energy of motion.

Let us look more closely at these movements which produce sound.

*QUESTION 2:* Imagine that you are asked by a radio producer to create the sounds for his play. Using just the everyday things around you at home or in the classroom how would you make the sounds to represent (a) going out of a room, (b) having cold hands, (c) releasing an arrow from a bow, (d) the start of a football match? Can you think of one word in each case to describe the movement that you had to make to create those sounds?

## 29.3 Vibrations

Place some small particles of chalk or coal (1 to 2mm across) on a tin or box lid. Bang the lid gently with a pencil and you will see that the particles jump up and down, showing that the lid also moves upwards and downwards when it is struck.

This type of movement, where a particle or object moves to and fro about its rest or mean position, is called a vibration. Probably the simplest and commonest example of a vibration is a pendulum in motion. Sounds are produced by such vibrations.

The tips of our fingers are very sensitive to vibrations. If you put your fingertips on the sides of your mother's washing machine when the spinner is in action, or on the bonnet of your father's car when the engine is running, you will feel a tingling sensation. In the laboratory a light pith ball can detect the vibrations of a tuning fork, which are very hard to see (Fig. 29.2).

*QUESTION 3:* What happens to the vibration of a tuning fork (a kitchen fork held at the end and tapped on a solid surface can be used) when a hand is placed on it?

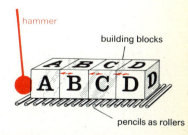

**Fig. 29.2** *The vibrations of a tuning fork set a pith ball vibrating*

## 29.4 Waves

A vibrating object is a source of sound, but how does the sound reach our ears? The following experiment will show you how sound travels.

Four blocks of wood (40mm³) — you could use babies' building bricks — are placed on a flat surface made of pencils laid side by side to act as rollers (Fig. 29.3). The end block, A, is hit sharply as shown in the diagram.

The vibration set up in the end block is passed on to the other blocks. Although it does not last long, there is enough time to see that the vibrations of the blocks cause unequal spaces to appear between them, some blocks are close together, the next group further apart, and so on

**Fig. 29.3** *A pulse is seen passing down a line of bricks*

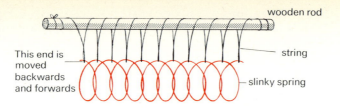

**Fig. 29.4** *A "slinky" shows us the passage of a longitudinal wave as a series of compressions and rarefactions*

wooden rod

This end is moved backwards and forwards

string

slinky spring

down the line. We can say that a pulse has passed down the line of bricks. A type of spring known as a slinky will also show this passage of a pulse if it is suspended as shown in Fig. 29.4.

A wave is a series of pulses at regular intervals.

If the end of the slinky is pulled and released so that it moves backwards and forwards it can be seen that a wave takes the form of a series of compressions and rarefactions. *Compressions* are areas where the vibrating particles are closer together, and *rarefactions* are where they are further apart. The particles in each case vibrate about a mean position and always return to this original position when the pulse has passed. Air molecules are set vibrating in this manner around a vibrating object.

A sound wave is a series of compressions and rarefactions moving through the air.

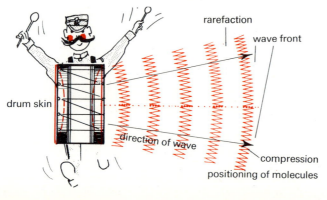

rarefaction

wave front

drum skin

direction of wave

compression

positioning of molecules

**Fig. 29.5** *A sound wave in air*

When a sound wave (Fig. 29.5) reaches the ear the compressions (regions of high pressure) and the rarefactions (regions of low pressure) set the ear drum vibrating in sympathy with the pulses of energy. A sound wave is one form of *longitudinal* wave.

In a longitudinal wave the particles vibrate backwards and forwards along the direction in which the wave is travelling.

Another type of wave can be seen when you drop a pebble into water, or when a rope is shaken at one end, as shown in Fig. 29.6. If you place a cork on the water and then drop in the pebble, or attach a piece of cotton to one spot on the rope, you can easily observe the vibration of the particles. The direction of the vibration of the particles is different from that seen in a longitudinal wave.

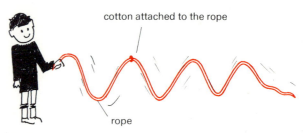

cotton attached to the rope

rope

**Fig. 29.6** *A rope shaken in this manner illustrates a transverse wave*

*QUESTION 4:* The cork placed on the water bobs up and down but does not move horizontally. What does this movement indicate about the direction of the vibration of the particles?

The waves seen on water and in the rope are both *transverse* waves.

In a transverse wave the particles vibrate backwards and forwards in a direction which is at right angles to the direction in which the wave is travelling.

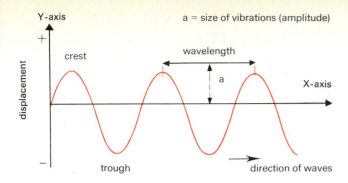

**Fig. 29.7** *A wave-form diagram*

Figure 29.7 is a wave-form diagram which may be drawn to represent any wave motion. The peaks of the wave form represent the positions where the particles are at maximum displacement. Points where the wave form crosses the X-axis represent the positions where the particles are at their mean position (zero displacement).

Wave-length is defined as the distance between two successive particles which are at the same point in their displacement. It is the distance between two crests or two troughs.

Wave forms may be viewed on an oscilloscope screen (see page 347).

## 29.5 Transmission of sound through other materials

Your doctor uses a stethoscope when he listens to your heart beat. Why is this effective? The following very simple experiment will suggest the answer.

Get a friend to scratch the edge of a table very gently with a fingernail. Very little sound can be heard. Now put your ear close to the table, and you will find that the sound is much louder.

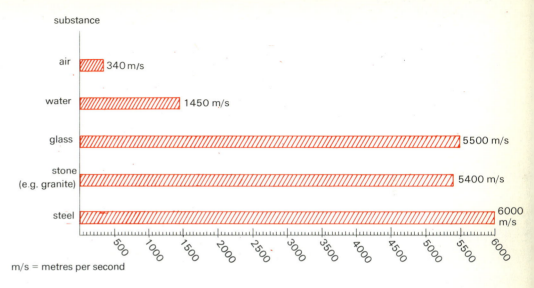

m/s = metres per second

**Fig. 29.8** *A chart comparing the speed of sound in materials at 20°C*

Sound travels well in hard materials (e.g. hardwood and iron) because the molecules transmit the vibrations more readily. The same is true for liquids.

The speed of sound in air is shown in Fig. 29.8 in relation to its speed in other materials.

Naturally a small amount of sound energy is used up when it is spread out in a large volume of air, but less energy is used up in a denser material. The approach of a train may be detected long before it is heard normally, by "listening" to the track — a method not recommended to intending passengers!

*QUESTION 5:* How does the stethoscope detect the heart beat so easily? (Hint: place the tips of your fingers on your chest.)

The experiment illustrated in Fig. 29.9 can be used to show that sound cannot travel across a vacuum.

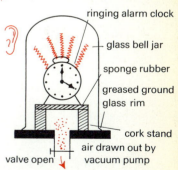

**Fig. 29.9** *Sound cannot travel through a vacuum: when air is completely withdrawn no sound from the clock can be heard*

*QUESTION 6:* Explain what is wrong with this statement: "The astronaut, who had just landed on the moon, turned suddenly as he heard something move behind him."

The fact that sound cannot travel through a vacuum may be demonstrated in the laboratory by hanging an electric bell inside a bell-jar and evacuating the air from the bell-jar (Fig. 29.9). When all the air has been evacuated the ringing of the bell can no longer be heard.

### 29.6   Refraction of sound

The temperature of air affects the speed at which sound travels. Sound travels faster if the air is hot. Refraction occurs when the speed of a wave motion is changed as it passes from one medium to another. Therefore refraction of sound occurs when there are air layers at different temperatures and therefore at different densities. (Fig. 29.10).

*QUESTION 7:* What effect does this have on our hearing of sounds at night?

**Fig. 29.10** *Temperature differences in layers of air cause refraction of the sound waves*

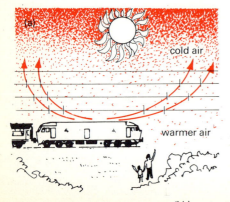

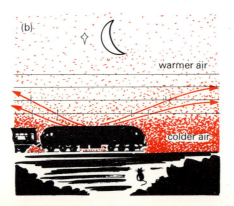

**Fig. 29.11** *The BBC Symphony Orchestra at the Royal Albert Hall*

### 29.7   Music

Figure 29.11 shows a group of people whose everyday job is to produce music.

*QUESTION 8:* Look at Fig. 29.11 and list four musical instruments that are being used. In each case state what it is in the instrument that vibrates in order to produce the sound.

There are three types of musical instrument, *wind*, *percussion* and *string*. Each instrument is able
  1. to produce high or low notes, i.e. to change *pitch*;
  2. to alter the strength (energy) of the vibration, i.e. to vary the degree of *loudness*;
The various instruments sound different even when they are producing the same note, i.e. there is a variation in *quality*.

How is this done? You can find out either by using standard instruments or by constructing some of your own at home, as illustrated in Fig. 29.12. Play each type of instrument and decide what it is that affects the pitch and loudness of that instrument. The following sections will help you to see whether you are correct in your observations.

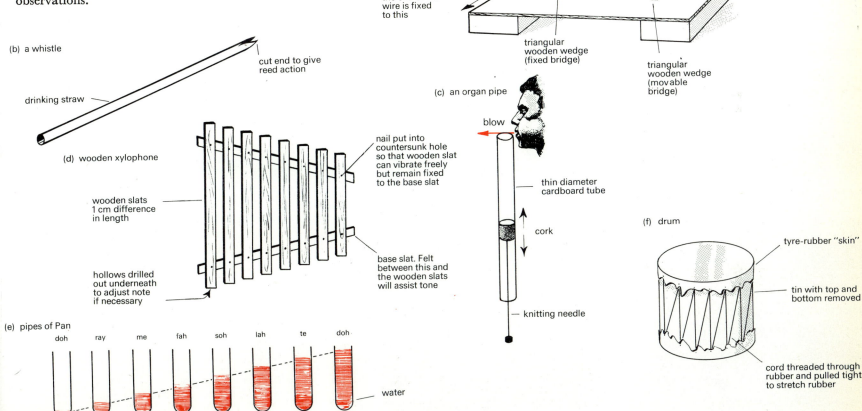

(a) a sonometer

wire wound around a second screw which can be turned to tighten or loosen the wire

screw in the baseboard – the end of the wire is fixed to this

100 cm

15 cm

wooden baseboard

triangular wooden wedge (fixed bridge)

triangular wooden wedge (movable bridge)

(b) a whistle

cut end to give reed action

drinking straw

(c) an organ pipe

blow

nail put into countersunk hole so that wooden slat can vibrate freely but remain fixed to the base slat

(d) wooden xylophone

thin diameter cardboard tube

cork

wooden slats 1 cm difference in length

hollows drilled out underneath to adjust note if necessary

base slat. Felt between this and the wooden slats will assist tone

knitting needle

(f) drum

tyre-rubber "skin"

tin with top and bottom removed

cord threaded through rubber and pulled tight to stretch rubber

(e) pipes of Pan

doh    ray    me    fah    soh    lah    te    doh

water

**Fig. 29.12** *Construction of some simple musical instruments*

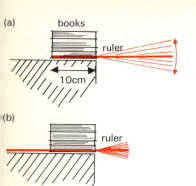

books
ruler
10cm

(b)
ruler

**Fig. 29.13** *Experiment to determine how the number of vibrations change the pitch of a note*

### 29.8 What determines pitch?

a. In string instruments
    i. tension: a tight wire produces a high note,
    ii. length: a short wire produces a high note,
    iii. thickness: a thin wire produces a high note.

b. In wind instruments
    length: a short column produces a high note.

c. In percussion instruments
    size: a small surface area produces a high note.

What is the common factor behind all these changes? It appears that in all of them the vibration is altered.

The following experiment will help you to understand how vibrations change the pitch of a note. Put the last 10cm of a long ruler under a pile of books on the edge of a table (see Fig. 29.13). Set the remaining length of the ruler vibrating and listen to the pitch of the note that is produced. Repeat, shortening the vibrating length by pushing more and more of the ruler under the books.

In this experiment we notice that we can detect the individual vibrations produced by a long section of the ruler but not those produced by a short section. This indicates that in the same period of time a short rod vibrates with a greater number of vibrations than a long rod. The short vibrating length also produces the higher note. This means that the greater the number of vibrations per second the higher the pitch of the note.

### 29.9 Frequency, wave-length and the speed of sound

The number of vibrations or cycles per second is called *frequency*.

The unit of frequency is the *hertz* (Hz).

    1 hertz = 1 vibration or cycle per second.

The wave forms of a note of high frequency and one of low frequency are shown in Fig. 29.14(a) and (b). Notice that the *wave-lengths* of these two frequencies are different.

A high-frequency (high-pitched) note has a short wave-length. A low-frequency (low-pitched) note has a long wave-length.

The time taken for one complete vibration is called the *periodic time* of the vibration.

The tables which follow show the frequencies of various musical notes and the frequency range of various sound-producing instruments.

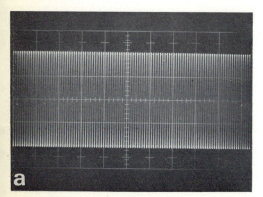

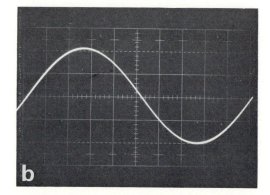

**Fig. 29.14** (a) *Wave form of a high frequency note, 10KHz* (b) *Wave form of a low frequency note, 100Hz*

**TABLE 1**

| Notes of the major diatonic scale of C (standard laboratory set of 8 tuning forks) | Frequency |
|---|---|
| Middle C | 256Hz |
| D | 288Hz |
| E | 320Hz |
| F | 341Hz |
| G | 384Hz |
| A | 426Hz |
| B | 480Hz |
| C | 512Hz |

**TABLE 2**

| Instrument | Approximate frequency range |
|---|---|
| Piano | 70 → 5200Hz |
| Double bass | 40 → 10 000Hz |
| Violin | 300 → 15 000Hz |
| Timpani | 35 → 5000Hz |

When one note has a frequency double that of another, the two notes sound very different and the one with the higher frequency is higher in pitch. The notes are an *octave* apart.

Does a relationship exist between frequency, wavelength and the speed of sound? In order to answer this question let us do the following calculation:

A soldier walks along a road taking 72 strides a minute. Each of his strides measures 0.5m. What is his speed in metres per second?

The soldier takes 72 strides in 60s
Therefore the number of strides per second $= \frac{72}{60}$ strides

Each stride measures 0.5m
Therefore in 1s he travels a distance $=$ no. of strides/second $\times$ length of each stride

$= \frac{72}{60} \times 0.5m$

Thus the soldier's speed $= \frac{72}{60} \times 0.5m/s = 0.6m/s$

From this we can extract the general formula:
    Speed of man $=$ no. of strides/second $\times$ length of stride
We can relate this argument to sound:
    Frequency of the sound (Hz) $\equiv$ number of strides/second
    Wave-length ($\lambda$) $\equiv$ length of stride(m)

Therefore

speed of sound $=$ frequency $\times$ wave-length

This relationship is useful to scientists and engineers dealing with sound and other wave phenomena.

### 29.10 What determines loudness?

a. In string instruments: the stronger the pluck the louder the note.
b. In wind instruments: the stronger the blow the louder the note.
c. In percussion instruments: the stronger the force the louder the note.

The note is louder when a greater amount of energy is used in its production. How does this energy affect the vibration? In order to answer this question we can do another experiment.

The ruler (as in the previous experiment, page 346) is positioned with a set amount projecting beyond the book. It is set vibrating, first to produce a soft note, then to produce a loud note. The vibrations are observed closely.

It can be seen that when the note is loud the size of the ruler's vibrations (a measure of the energy) are much greater than when the note is soft. The size of the vibration is called *amplitude*.

The amplitude is the distance from the point of rest to the point of maximum displacement.

*QUESTION 9:* What alters in the vibration when a sound dies away?

### 29.11 What determines quality of sound?

Figure 29.15(a) shows the wave form of a pure note. This was produced on the screen of an oscilloscope (see page 334) by the note from a tuning fork received through a microphone.

Figs. 29.14 and 29.15 are photographs specially taken for this book by The Plessey Company Ltd (Plessey Marine).

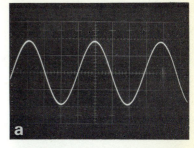

a

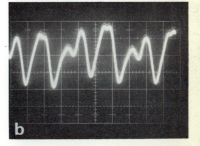

b

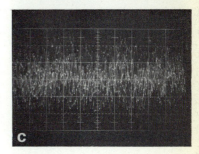

c

347

Figure 29.15(b) shows the wave form of the same note, but this time produced by the note from a piano. Both notes have the same wave-length, but it is clear that the note produced by the musical instrument, the piano, is not pure.

Musical instruments produce one main or *fundamental* note which is "coloured" by weaker notes. These weaker notes, present at the same time, are known as *partials* or *harmonics*. Their frequency is directly related to, in fact a multiple of, the frequency of the fundamental note. The first harmonic has a frequency twice that of the fundamental, and the second harmonic three times that of the fundamental, and so on. For example, middle C of frequency 256 hertz has harmonics of 512 and 768 hertz and so on. It is the varying amounts of harmonics (partials) present that give an instrument its characteristic tone or quality. A violin, for instance, has strong high partials, whereas a cello has strong low partials. Even two instruments of the same type can sound different, because the partials are dependent upon the structure of the particular instrument. A master craftsman can make instruments of outstanding tone, and such instruments are greatly sought after.

## 29.12   The destructive power of sound energy
Figure 29.16 shows a glass shattered not by a blow but by sound energy. The following experiment will help you to understand how this can happen.

A pendulum bob is suspended and set swinging. Every time it reaches the maximum point of its swing it is

**Fig. 29.16**  *A glass shattered by sound energy*

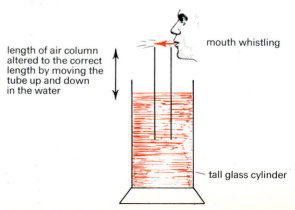

length of air column altered to the correct length by moving the tube up and down in the water

mouth whistling

tall glass cylinder

**Fig. 29.17**  *Resonance in an air column*

348

given a gentle tap in the direction of its next swing. The vibration is seen to increase. The human movement has impressed itself upon the natural movement of the pendulum and made it greater. We carry out a similar action every time we push a child on a swing.

All objects have a *natural frequency* of vibration. Vibration in them can be induced and increased by the presence of another vibration of the same frequency or partials. If you whistle across the top of an air column like that shown in Fig. 29.17, you will (if it is at the correct length) set it vibrating with the same note.

This effect is called *resonance*. Thus the intense resonance set up in a glass by a singer's voice may cause the glass to shatter.

*QUESTION 10:* What use is made of resonance in a string instrument?

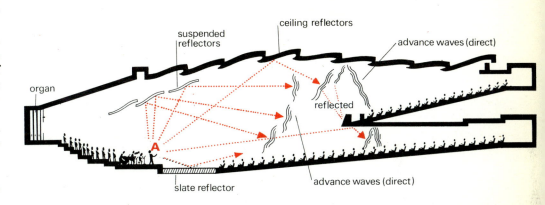

**Fig. 29.18** *Orchestra hall in cross-section*

### 29.13   The concert hall

The sound from an orchestra may be ruined if it plays in surroundings where multiple reflections occur at different surfaces. The reflections reach the ears of the audience after different intervals of time. For instance, in one particular orchestra hall in London it was said that you could hear the concert twice for the price of one performance! Acoustic domes have now cured the fault. In the building of any room, classroom or orchestral hall, where the transmission of sound unhindered by multiple reflections is important, it is very necessary to consider the shape, layout and building materials to be used (see page 354).

Figure 29.18 shows an orchestral hall, in cross-section. Sound in air loses its energy quickly because it spreads out widely from the source in all directions. If then we wish to get sound to a particular point and with a minimum loss of energy, how can we do it?

*QUESTION 11:* You wish to speak to a friend at the other side of the school field. You shout but he does not hear. What can you do with your hands to direct the sound at him?

A cone shape placed around the source of sound appears to channel it in the direction required. This is the principle used in the shaping of a megaphone and the surround of a loudspeaker. But how is it that sound can be directed in this way?

Look at Fig. 29.18 and notice the lines representing sound waves that spread out from point A. Now look at Fig. 29.19 and observe closely the shape of the roof above the concert platform. The architect has designed this roof to direct the sound (which might otherwise be lost) down towards the audience. What does the roof do to the sound waves?

**Fig. 29.19**   *Auditorium of the Royal Festival Hall*

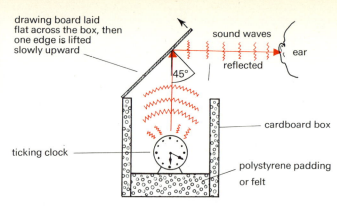

drawing board laid flat across the box, then one edge is lifted slowly upward

sound waves

ear

reflected

45°

cardboard box

ticking clock

polystyrene padding or felt

**Fig. 29.20**   *Experiment to show the reflection of sound*

to the normal as that of the incident wave. In the same way the sound waves from the orchestra are reflected from the roof of the concert hall, down to the auditorium.

Simple methods for the transmitting and direction of sound (e.g. megaphones and speaking tubes) make use of this principle of the reflection of sound waves. This also holds true for devices used to collect sound at a particular point. Notice how nature has put this to good use in the outer ear structure (the pinna) of dogs and cats, for example.

### 29.14   Echoes

Echoes are sound reflections. They may be used (a) to measure the speed of sound (see "Things to do" page 357), (b) to measure distance. In geology, for instance, seismic shooting (detonation and the recording of the resultant sound waves) indicates the depth and position of certain rocks. At sea, the depth of the sea-bed can be found and mapped by using an echo-sounder. This sends out pulses of sound from a loudspeaker and records the time taken ($t$) to receive the echo at a microphone.

Set up the apparatus shown in Fig. 29.20. The sound of the clock is clearly heard when the piece of board is held at an angle (as shown) above the top of the sound-proof box. The board is *reflecting* the sound waves. They obey the laws of reflection in the same way as light, that is, the reflected wave leaves the surface at the same angle

$$\text{Depth of the sea-bed} = \frac{\text{speed of sound in water} \times \text{time }(t)}{2}$$

*QUESTION 12:* What is the depth of the sea-bed when an echo is received in 40 seconds? Speed of sound in water is shown in Fig. 29.8.

By the same procedure the echo-sounder can be used to detect objects in the water. This is now being used by the fishing industry; during the war it was used to detect submarines.

## 29.15   Ultrasonic sound

You have probably heard the expression "as blind as a bat". Bats may have very poor sight but they use a most effective form of echo-sounding to find their way about when in flight. They achieve this by giving out sounds of very high frequency (over 20 000Hz) which are above the range of sounds audible to humans. This high-frequency range is therefore called ultrasonic sound. Because ultrasonic sound has a high penetration it is used in industry for finding faults inside metal castings. The fault reflects the sound, giving a quicker echo than would normally be obtained from a perfect casting.

## 29.16   The human ear

The range of audible frequencies for the human is about 20 to 20 000 hertz (20kHz).

Animals have different ranges. Dogs, for instance, have a much higher scale, hence the use of high-frequency dog whistles which are inaudible to humans.

The structure of the human ear is shown in Fig. 29.21. The sound wave travels down the ear canal setting up vibrations of the ear drum. These vibrations are passed

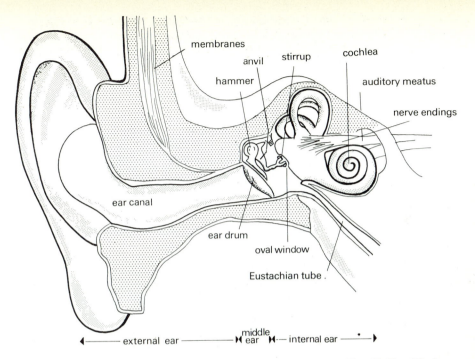

**Fig. 29.21**   *The human ear*

across the middle ear by the three bones, hammer, anvil and stirrup, to the oval window of the inner ear. Within the inner ear they travel through the fluid of the organ known as the cochlea and are detected by sensitive membranes, which in turn stimulate nerve endings. These nervous impulses are then passed along the auditory nerve to the brain, which interprets the sound. In order that the ear drum should vibrate properly and remain intact, the air pressure on either side of it must be equal. This is guaranteed by the presence of the Eustachian tube which leads to the back of the throat.

*QUESTION 13:* Why do colds sometimes cause deafness?

351

## 29.17   Noise

Pressure changes in unpressurized aircraft can be painful to the ear drums. Loud explosions can cause earache or even temporary deafness: our instinctive reaction to them is to put our hands over our ears.

Noise is sound that has a changing frequency (no definite pitch) and irregular vibrations.

That is the scientists' definition of noise. Most of us would say that noise is irritating and disturbing. But it may be more than that. Recent research has tended to show that some noises can do actual physical harm to the body. There has therefore been increasing emphasis recently on the importance of reducing noise in street, factory and home.

## 29.18   Measurement of noise

If we are to control noise levels then obviously we must have a means of measuring the intensity of sound, that is, of measuring sound energy.

In everyday life we tend to compare one sound with another. The ear drum interprets the intensity of the sound by the amplitude of the vibration. We say, for example, that Mr X's car sound twice as loud as that of Mr Y. Scientists use the same method of comparison to obtain the unit of sound level, the *bel*. They have constructed a scale of sound levels expressed in powers of 10.

The sound level of one sound is 1 bel higher than that of another sound if its intensity is 10 times greater.

A *decibel* (db) is one-tenth of a bel, and is approximately the smallest change in sound level to which the human ear is sensitive.

The table on this page gives a list of some everyday sounds and their approximate sound levels expressed in decibels. Figure 29.22 shows *Concorde's* sound level being recorded in decibels during one of its first trial runs.

352

In order to measure sound levels all that is needed is a microphone, amplifier and an ammeter calibrated to read decibels against a standard sound.

**EVERYDAY SOUNDS**

| Sound source | Sound level in decibels (approx.) |
|---|---|
| Sound that is just audible (threshold of hearing) | 0 |
| Ordinary breathing | 10 |
| Whisper | 20 |
| Rustle of a newspaper | 30 |
| Noise in a town at night | 40 |
| Inside a moving train (window shut) | 50 |
| Ordinary conversation | 60 |
| Inside a moving train (window open) | 70 |
| Heavy traffic in city rush hour | 80 |
| Pneumatic drill (near) | 90 |
| Underground train (near) | 100 |
| Thunder or take-off of jet (near) | 110 |
| Noise causing pain (threshold of feeling) | 120 |

The decibel scale does not give us a true picture of sound levels as detected by the human ear. This is because it does not take into account the fact that the human ear is sensitive to differences in frequency as well as to differences in intensity. Another scale of sound level has therefore been worked out based on a unit called the *phon*. A measurement of sound in phons is obtained by comparison against a standard frequency of 1000Hz (1kHz).

## 29.19   Reduction of noise

Noise can be reduced at source, or in the medium that lies between the source and the receiver. Here is an experiment to find out how to reduce noise.

**Fig. 29.22** *Measuring the sound level of* Concorde

**Fig. 29.23** *A broadcasting studio; note the use of sound-absorbing materials*

A sound-absorbing material is useful in house walling, and indeed in house furnishing. Curtains, carpets, pegboard and other sound-absorbing materials are put to full use in broadcasting studios (Fig. 29.23) to prevent undesirable reflection of sound.

*QUESTION 14:* Why does the sound level in a cinema drop when the auditorium is full of people?

## 29.20 Vehicle noise

Motor vehicle noise can now be regulated at source, but this is not easy in the case of aircraft and much research is being done on this matter. Usually the only way of overcoming aircraft noise in the vicinity of an airport is by sound-insulation of the houses and buildings round about. There has been some success recently in reducing the noise made by jet aircraft, but the big problem in the future will be the bangs produced by *supersonic* aircraft. Supersonic means faster than the speed of sound (Mach 1). The bangs occur as aircraft pass through the sound barrier.

A small ringing alarm clock is slowly lowered by means of a string into a deep biscuit tin until it touches the bottom. Changes in the level of noise are noted. This procedure is then repeated several times, but in each case the tin is thickly lined inside with different materials in turn: crumpled paper, wool, latex foam and cork. Noise levels are again noted.

This experiment shows that sound may be reduced by surrounding the source with an adequate layer of a sound-absorbing material.

An ideal sound-absorbing material has a broken rough surface and is soft and spongy.

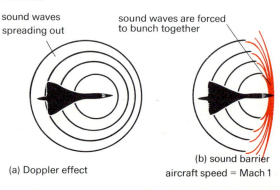

sound waves spreading out

sound waves are forced to bunch together

(a) Doppler effect

(b) sound barrier
aircraft speed = Mach 1

**Fig. 29.24** *Sound waves (a) at aircraft speed below Mach 1, (b) at Mach 1*

354

Look at Fig. 29.24. When an aircraft or car moves, it causes the sound waves in front of it to become bunched together and those behind it are spread apart. Thus, as the object moves towards us our ear receives a higher number of vibrations per second than after it has passed. We hear the change from a high-pitched note to a low one. This is called the *Doppler effect*, after the scientist who first explained it.

When the aircraft reaches the speed of sound, the sound waves are bunched together as a barrier at its nose and we hear one loud noise.

If the aircraft speeds up at this point it pushes through the barrier of waves, creating shock waves which reach the earth as loud bangs, breaking windows and even cracking walls.

*QUESTION 15:* A pilot of a supersonic aircraft would not hear when his engines had cut out at speed above Mach 1. Why?

## 29.21 Communication
Sound is used to convey ideas, feelings and information. A teacher makes great use of the sound of the human voice in order to convey knowledge to his class! The human voice is produced by the vibration of the vocal chords stretched across the trachea (wind pipe). Muscles control their tightness and therefore affect the pitch of the note produced. These sounds are formed into words by the actions of the tongue and the shaping of the mouth and nose cavities. A man's voice has a deeper tone than a woman's because his vocal chords are longer.

It is impossible to transmit sound energy itself over long distances, but it can be converted into electrical energy and then transmitted over cables in the telephone system or by means of radio waves. At the receiving end these energy forms are converted back to sound energy.

Usually a sound disappears very quickly after it is made. But again, by using energy conversions, sound can be "stored" or recorded, on records, magnetic tape or film. These conversions are carried out by equipment that has been developed over the last 100 years and forms the basis of telephones, projectors, record players and tape recorders. These are vital to industrial and leisure activities in our modern age. The action of this equipment is discussed in other chapters.

## 29.22 Gramophone recording
If you examine a gramophone record under a magnifying glass you will see grooves similar to those illustrated in Fig. 29.25. These grooves are made by a small wedge-shaped cutter operated by the varying current which is produced by the conversion of sound energy into electrical impulses. If you can get hold of an old gramophone record and a revolving turntable you only need a pin and a cone of paper to reproduce the sound which was recorded on it (Fig. 29.26).

**Fig. 29.25** *Recording grooves as they appear when magnified*

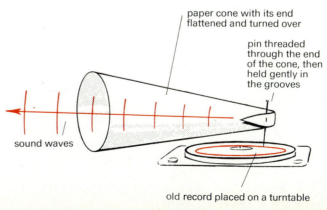

paper cone with its end flattened and turned over

pin threaded through the end of the cone, then held gently in the grooves

sound waves

old record placed on a turntable

**Fig. 29.26** *Construction of a simple gramophone*

The pin is set vibrating as it follows the grooves, and this in turn sets up vibrations in the air in the cone. Thus the sound vibrations that were used originally to cut the grooves are reproduced. The phonograph, developed by Thomas Alva Edison in 1877, worked in this way.

*QUESTION 16:* What form of energy is used to obtain a better amplified reproduction of the sound energy in a modern record player?

In the record-making industry today the grooves are cut on a lacquered disc. Metal is then deposited on this by a process of electrolysis (see Chapter 20), hardened and peeled off to give a metal master disc. This is then used to press out the commercial plastic records on sale in the shops.

### 29.23   Sound on film
Have you ever tried to operate a projector and tape recorder together to match up film and sound for a home movie? It is a very difficult thing to do. This problem of synchronization was tackled in the film industry in the late 1920s when a method of combining film and sound track was developed.

The sound is recorded, as the action and filming take place, on a track that runs down the side of the film. A microphone (similar to the one used in a telephone, see page 197) picks up and converts the varying sounds into varying electric currents. In one method these currents operate a light which exposes in varying amounts the film running past it. When the film is shown the reverse process takes place. A light shines through the sound track and the amount of light getting through depends on the exposure of the film. The light which passes through the film falls on a photo-electric cell. The photo-electric cell converts the varying light intensity into varying electric currents (see page 336) which are amplified and produce sound in a loudspeaker.

356

Another method recently developed and frequently used for home movies is to have a sound track on magnetic tape. The tape is joined to the film to ensure that the picture and sound are synchronized. As in the optical method described above, the sound can be recorded on the magnetic tape as the action is being filmed. The projector for showing the film has a playback head and the magnetic tape attached to the film passes over this as the film runs through the projector.

### 29.24   Recording sound on tape
In the previous section we referred briefly to recording sound on magnetic tape to produce a sound track for a film. The principle of the method is that the sound to be recorded is picked up by a microphone. The microphone turns the sound into electric currents which are amplified before being fed to the record head. The record head is basically an electromagnet and the varying electric currents are converted into varying magnetic patterns on the tape. When the tape is played back, the varying magnetic patterns on the tape induce varying electric currents in the wire round the playback head. The resulting currents are amplified and fed into a loudspeaker. Thus the sound energy that was originally recorded is reproduced exactly.

### 29.25   Recent developments
Many developments have taken place since Bell made his first telephone call and Edison heard his own voice reciting "Mary had a little lamb" from his phonograph. Technicians search constantly to improve sound reproduction. This has resulted in an ever-increasing output of hi-fi and stereo equipment which is constructed to cover the wide range of frequencies produced by musical instruments.

A. Measure the speed of sound by the echo method.

Stand at a distance of 60 to 90m from a high building which will return an audible echo when you bang a metal tray. Note the time between hearing the bang on the tray and its echo. Wait for the same amount of time to pass after the echo has been received, then bang the tray for a second time. Keep on repeating this until you have a sequence of bang-echo-bang-echo occurring at equal intervals of time. Once you can carry out this sequence with skill get a friend to observe the time taken for you to count 0 (first bang), 1, 2, 3, 4 . . . 20 bangs. Measure the perpendicular distance ($x$) between you and the high building.

Between each clap the sound has time to travel $4 \times x$. Therefore the total distance travelled by the sound during 20 bangs $= 4 \times 20 \times x$ m.

$$\text{Speed of sound} = \frac{\text{distance travelled by sound}}{\text{time taken}} \text{ m/s}$$

$$\text{With this method speed (V)} = \frac{4 \times 20 \times x}{\text{time}} \text{ m/s}$$

B. Produce notes of different pitch with a bicycle wheel.

Fasten pieces of flexible card to the two forks of your bicycle wheels so that when the wheels rotate the spokes hit the pieces and set them vibrating. Vary the speed of the wheel to alter the pitch of the note. Do not use stiff card because it will either break your spokes or jam the wheel and throw you over the handlebars.

Frequency of note = the number of times per second the card is hit by the spokes.

Try to produce other notes by similar methods, e.g. holding a card against the rotating wheel of a whisk.

C. Construct a simple telephone.

Obtain two empty fruit tins of equal size and remove their tops neatly. Punch a hole at a central point in the bottom of each can. Thread the ends of a long length of string through each hole. Tie a good strong knot in each end so that when the string is pulled tight between the tins the knots will not pull through the holes. The string must be held tight. Get a friend to use one tin as a mouthpiece while you use the other as an earpiece.

D. Sit down in the garden, or in a room, for a set length of time and list all the sounds you hear. Identify the type of vibration that produces each sound.

E. Find two identical containers with narrow necks. Get a friend to hold the open end of one container near his ear, while you stand next to him and blow across the open end of the other container to produce a note. Your friend should hear the same note from his container. This is because the air in his container resonates in sympathy with your identical vibrating air column.

357

F.   Use two voices to produce a third!

Get a friend to whistle the same note as you are whistling. Then one of you must slightly alter the pitch of your whistle. You should be able to hear a third note! If you alter the pitch only very slightly, the third note will appear to get softer and then louder. The notes are called "beats" and composers put such phenomena to good use in writing music.

G.   Write a project on "Sound waves in nature". Describe how different groups of animals produce and receive the sounds necessary for communication between individuals.

H.   Construct an illustrated chart with the title "Instruments of the Orchestra". For each instrument indicate how the sounds are produced.

I.   Produce an illustrated essay on the recording of sound, dealing also with the historical background.

J.   Do a project on the problem of noise. Include the following topics: measurement of noise, methods of sound-proofing, methods of "silencing" the noise at its source.

1. Explain the following statements:

   a. A boy cannot hear the ticking of a clock when his ears are covered. But if he puts one end of a ruler between his teeth and the other end on the clock he can hear the ticking.

   b. It is possible for motorists who sing in their cars or have their car radios at a certain pitch, to shatter the car's windscreen if there is a slight chip in the glass.

   c. A chimney emits a low-pitched note when the wind is in a certain direction.

   d. Double glazing assists in the sound insulation of a house.

   e. A balloon full of carbon dioxide gas converges sound waves.

2. Briefly explain what is meant by the following terms: (a) frequency of a vibration, (b) quality of musical notes, (c) amplitude of a vibration, (d) an echo, (e) a decibel, (f) resonance, (g) wave-length.

3. With the aid of a simplified diagram, explain the production of the human voice. State how the pitch of the sounds may be altered.

4. State whereabouts the following structures are located in the human ear and the use of each: (a) Eustachian tube, (b) ear drum, (c) hammer, anvil and stirrup, (d) cochlea, (e) auditory nerve.

5. Explain the following statements:

   a. Cats seem to hear noises that we cannot hear.

   b. The starter of a race, who stands at the other end of the track from the time-keepers, uses both a handkerchief and a gun to signal the start.

   c. A megaphone enables a person's voice to be heard over a greater distance.

   d. Supersonic aircraft (flying above the speed of sound) produce in their wake a series of sonic booms.

6. a. List three instruments that make use (i) of vibrating air columns, (ii) of vibrating strings and (iii) of vibrating surfaces.

   b. A guitarist wishes to produce a low-pitched note on his guitar. Explain how he can do this by altering (a) the length of the string, (b) the tension of the string, and (c) the thickness of string.

7. a. How can you tell which organ pipes produce low notes and which produce high notes, simply by looking at them?

   b. How could you use eight milk bottles to produce a musical scale?

   c. A geologist takes his guitar along with him to a research station near the north pole. Explain what adjustments he has to make to his instrument when he arrives.

   d. Why are sounds in an empty room different from sounds in a fully furnished room?

8. Using diagrams, explain why sound from a distant railway is more audible at night than during the day.

9. A bat flying through the air emits a sound and receives its echo in 2s. How far away is the cave wall towards which the bat is flying? (Speed of sound in air = 340 m/s.)

10. Explain why the moon may be described as a "silent planet". Describe a laboratory experiment to confirm your theory.

11. Describe how you would make the ringing of an alarm clock (i) much louder, (ii) much softer.

12. When standing at a distance you can see the spark between the spheres of a Van de Graaff machine before you hear the sound of the discharge. What does this indicate? Name two common occurrences in support of your statement.

13. What is meant by the terms "rarefaction" and "compression", and how do these terms apply to the method by which sound travels through the air? (Give a diagram.)

14. Draw the wave form of a musical note (a) of high pitch, (b) of low pitch, (c) of the same pitch as in (a) but louder.

15. Two identical notes are produced from two sources which are at different distances from a person. The notes are being produced continuously, yet the person hears no sound. Draw a diagram of the wave forms of the two notes showing how it is possible for this to occur.

16. A tuning fork has a frequency of 320Hz. If the speed of sound in air is 340m/s, what is the wave-length of the vibration?

17. Explain why the pitch of a police car siren seems to get higher as the car approaches, then lower after the car has passed by.

18. An echo-sounder on a ship receives an echo 2.5s after sending the signal. At what depth is the ocean bed? (Speed of sound in sea water = 1450m/s.)

19. A man shouts while standing between two facing cliffs. The first two echoes he hears are after 0.5s and after 2s. (a) Where is he standing with respect to the total distance between the cliffs and (b) What is the distance between the two cliffs if the speed of sound in air is 340m/s? Explain why the man will hear more than two echoes.

## CHAPTER 30

# RADIOACTIVITY AND NUCLEAR ENERGY

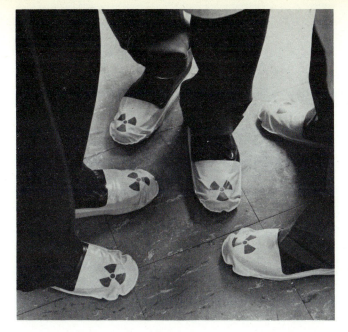

**Fig. 30.1** *Not a fashion show! What is the meaning of the symbol on the overshoes?*

## 30.1 Danger!

The men in Fig. 30.1 appear to be modelling pairs of overshoes! They are not interested in fashion, however, but in protection. In what way do these overshoes protect their wearers? The symbols on the fronts give us a clue: this is an international sign warning of radioactivity. The men are workers in an atomic plant in the United States, and the overshoes protect them against radiation that may be contaminating the floors. An ordinary pair of shoes would not be adequate protection against such radiation.

You will see this symbol on the boxes that contain radioactive sources in your school laboratory. It is also used to mark any area where radioactive materials are in use.

**Fig. 30.2** (a) *Photograph of a damaged section of a human jawbone*

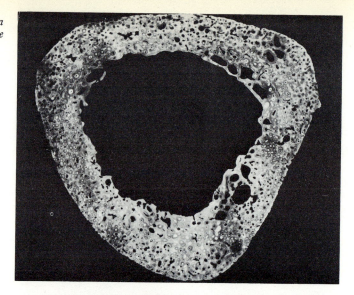

**Fig. 30.2** (b) *Autoradiograph of the same section of jawbone*

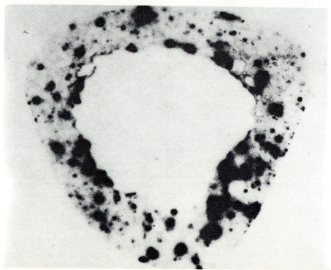

Figure 30.2(a) is a photograph of a damaged section of a human jaw bone. The damage was done in the course of the man's occupation. You might think he was a boxer, but in fact he was a painter of the luminous dials of clocks! In order to keep his brush tip very fine, he had a habit of wetting the bristles with his tongue and the luminous paint he was using got into his mouth. At that time, before their dangers were fully realized, radium salts were used in luminous paints, and the painter's body, particularly his bones, absorbed a dangerous amount of this radioactive radium. This accumulation of radiation caused the damage.

Figure 30.2(b) is an autoradiograph, which has been produced by placing a photographic plate in contact with the bone. When the plate is developed the black areas indicate where it has been exposed by contact with radioactivity. These correspond exactly with the damaged areas shown in Fig. 30.2(a).

A similar exposure of a photographic plate was one of those fortunate "accidents" that so often lead scientists into new fields of study.

## 30.2 The work of Henri Becquerel

In 1895 the French scientist Becquerel was working with fluorescent materials, and in particular with potassium uranyl sulphate which is a salt of uranium. It was only one month after Röntgen's discovery of X-rays and he was trying to find out whether these fluorescent materials gave out X-rays on exposure to light. He knew that X-rays exposed a photographic plate so he wrapped a plate in black paper in order to stop visible light from exposing it, placed his fluorescent material on top of the plate and left them in sunlight. He observed that the fluorescent material glowed, and when he unwrapped the plate and developed it he found that there were a few small spots of exposed emulsion.

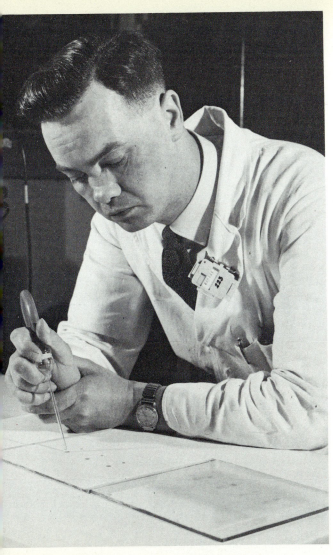

**Fig. 30.3** *This man is working with radioactive materials*

One day there was not sufficient sunlight for him to continue with his experiments so he placed the fluorescent material and photographic plate together in a drawer. A few days later when the sun reappeared Becquerel decided to restart his experiment, but being a very methodical person he first of all developed the wrapped photographic plate he had placed in the drawer. To his amazement the plate was darkened far more than any plate he had used hitherto in these experiments.

Becquerel concluded that uranium salt, without exposure to light, gave off a radiation that was invisible and yet was able to penetrate matter.

From then on other scientists, Marie and Pierre Curie foremost among them, searched for and discovered more substances that gave out a similar radiation. Among these were the elements polonium, radium and thorium. Marie Curie, in 1898, called this radiation *radioactivity*, and believed that it was a property of the atom itself because the radiation remained unaffected by physical and chemical changes.

The amount by which a photographic plate is exposed gives us one method of detecting the level of radioactivity.

*QUESTION 1:* Fig. 30.3 shows a worker in a laboratory using radioactive materials. The man has what looks like an identification badge on the lapel of his coat. This badge is a form of holder. What do you think it contains?

### 30.3 Scintillation counters

Detection methods depend upon the fact that radioactivity is able to excite the atoms of the matter through which it passes. Let us consider once again the luminous paint used for painting the dials of clocks.

The luminosity consists of a series of individual flashes

of light which are termed *scintillations*. These occur so rapidly and at random that under normal circumstances the individual flashes cannot be seen. How are these scintillations produced? Nowadays luminous paint is a mixture of a radioactive material, such as a tritium salt, and a fluorescent material, such as zinc sulphide. The radiation from the radioactive material excites some of the electrons in the molecules of the zinc sulphide. Thus the electrons undergo a change in their energy levels for a fraction of a second. On returning to their normal energy level, the excess energy that they acquired is given out in the form of light. (Safety regulations now ensure that the radiation emitted from luminous paint on clocks and watches does not penetrate the covering glass.)

Fluorescent materials can therefore also be used to detect radioactivity. These are set in the form of screens in instruments called *scintillation counters*.

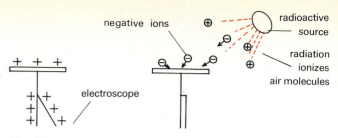

**Fig. 30.4**  *Radiation discharges an electroscope*

charged the negatively charged ions are attracted to the plate of the electroscope, thereby discharging it (Fig. 30.4).

This ionization caused by radiation provides us with other methods for its detection. (In fact, the effect on a photographic plate is the result of the ionization of the chemicals in the emulsion.)

### 30.4  How does radiation affect the air through which it travels?

In Chapter 17 we saw how a flame caused ionization of the air around it, and how the ions formed in this way could be used to carry an electric current through the air between two metal plates.

Let us see whether radiation causes similar ionization.

Charge an electroscope. Hold a lighted match near the disc of the electroscope. Having observed the result recharge the electroscope and bring a radioactive source near its disc.

In both cases the leaves collapse quickly. The radiation from the radioactive source acts in a similar manner to the flame, and ionizes the air through which it travels. The radiation splits the air molecules into negatively and positively charged ions and allows the passage of an electric current in the air. If the electroscope is positively

### 30.5  Wilson cloud chamber

C. R. T. Wilson, an Englishman, was awarded the Nobel prize for his invention in 1911 of this invaluable apparatus for observing the movement of sub-atomic particles.

In the atmosphere, a supersaturated vapour will condense and form a "cloud" of moisture when there are dust particles present to act as nuclei for the water droplets to form on. Charged ions in a cloud chamber act in the same way as dust particles. Cloud chambers are standard equipment in school laboratories (Fig. 30.5), but you can construct one yourself in the way described in the "Things to do" section, page 379.

The radioactive source is placed in the cloud chamber which contains air saturated with water vapour. The hand-pump piston is withdrawn. This reduces the pressure, and the mixture in the chamber expands, cools and reaches the point of supersaturation. For a few seconds the radioactive emission can be detected by the vapour trails (tracks of water droplets — see Fig. 30.5) which form on the ionized particles of air. Permanent records of these tracks may be obtained by taking photographs.

*QUESTION 2:* Do these tracks in the cloud chamber (Fig. 30.5) indicate that radioactivity is an erratic, random process, or an orderly one?

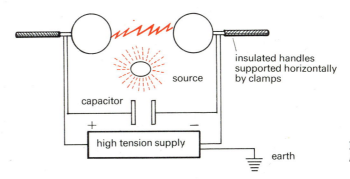

insulated handles supported horizontally by clamps

source

capacitor

+    −

high tension supply

earth

**Fig. 30.6** *Apparatus used to show how a spark counter works*

### 30.6   The spark counter

If a flame is placed beneath the gap between the two highly charged metal balls (Fig. 30.6) a huge spark is seen to pass between the balls. The flame provides ions, which are accelerated in the strong electrical field between the two balls and have sufficient energy to knock off some electrons from other air molecules, thus forming more ions. The build-up of ions continues in this fashion so that a sudden surge of charged particles passes between the two balls, forming a spark. This is what occurs in a spark counter when radiation passes through it as in the following experiment.

The spark counter is connected to a high tension supply (Fig. 30.7) and the voltage between the two electrodes is adjusted until it is at a value just below that when discharge takes place (4000–5000 volts). The radioactive

**Fig. 30.5** *Expansion-type cloud chamber, with tracks visible*

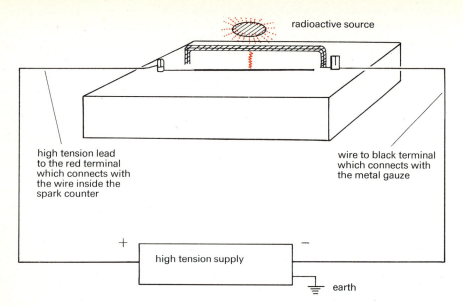

**Fig. 30.7** *The effect of a radioactive source on a spark counter*

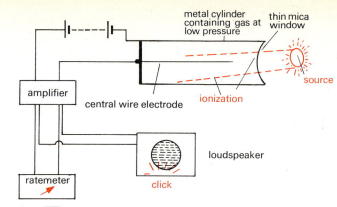

**Fig. 30.8** *The structure of a Geiger-Müller tube with loudspeaker*

source (radium) is held by tweezers* over the gauze electrode of the spark counter. The other electrode is the thin wire, held taut, a short distance below the gauze.

Sparks are seen to jump at random (i.e. the time interval between successive sparks is not constant) between the two electrodes. The passage of the sparks can be heard as well as seen. If the source is moved away from the gauze the sparking ceases. Thus the range of this radiation is limited as it is all absorbed by a few centimetres of air.

*QUESTION 3:* What does the irregularity of the sparks suggest to you about the nature of radioactivity?

* These radioactive sources used in the laboratories are specially prepared and of such a strength as not to be dangerous as long as they are carefully handled according to instructions. They should never be handled with bare hands.

### 30.7  A Geiger-Müller tube

This tube (Fig. 30.8) consists of a thin wire electrode held taut by insulators along the central axis of a cylindrical metal tube in which there is gas at low pressure. The metal tube acts as the second electrode. There is a "window" at one end (a thin membrane, of mica for example) to allow the easier entry of radiation into the tube. The Geiger-Müller tube works on a similar principle to that of the spark counter; but the working voltage (on average about 400V) is such that no spark is produced. The passage of radiation into the tube causes ionization of the gas and a surge of current is created between the two electrodes. These sudden surges of current (pulses) are amplified and operate either (a) a ratemeter, which is a microammeter measuring the rate at which the pulses arrive and indicating this on a scale calibrated in counts per second, or (b) a scaler, which counts the pulses. The pulses can be made audible as clicks through a loud-speaker — the stronger the radiation the more clicks per second. The combination of a G-M tube and either a ratemeter or scaler is often referred to as a G-M counter.

*QUESTION 4:* A piece of cardboard has a map of England drawn on it. Behind the board, in line with a particular town, a radioactive source has been situated. Using a Geiger-Müller counter with a loudspeaker, how would you locate the mystery town?

## 30.8 Experiment to determine the nature of the radiation emitted by a radioactive source

A Geiger-Müller tube with a very thin window is chosen. It is connected to a ratemeter. Three radioactive sources are available, plutonium-239, strontium-90 and cobalt-60.

The tube is clamped in a set position and the ratemeter is switched on. (There will probably be some background radiation which will be audible.) The plutonium source is placed close to the window and at a known distance from it. The ratemeter reading is noted and the clicking rate is heard. The source is gradually moved away from the window and the effect on the pulses is observed. This procedure is repeated using the other two sources.

The plutonium is placed once again in its first position close to the window. A number of materials of different thicknesses are placed in turn between the window and the source. The materials are listed in Table 1 on this page. This procedure is repeated with the other two sources, placed at the same distance from the window. The effect on the pulses is observed in each case.

**TABLE 1**
**PENETRATION OF RADIATION THROUGH VARIOUS MATERIALS**

| Source | Very thin paper | File paper | Perspex 7mm | Aluminium foil | Lead |
|---|---|---|---|---|---|
| Plutonium-239 | Yes | No | No | No | No |
| Strontium-90 | Yes | Yes. There is a negligible decrease in counts per second | No | Counts per second decrease with thickness. None detected when aluminium several mm thick | No |
| Cobalt-60 | Yes | Yes | Yes. There is a negligible decrease in counts per second | Yes. There is a negligible decrease in counts per second with the thickness that stopped the strontium-90 radiation | Decrease in counts per second when the thickness of the lead is increased. However, with every thickness some radiation is still detected |

In the first part of the experiment, radiation is detected from each of the sources. The counts per second decrease as the sources move away from the counter; but far less noticeably in the case of the cobalt source. The effect of the different materials placed in front of each source on the radiation from that source is indicated in Table 1.

The results seem to indicate that the radiation from each source is different, and radiation from the plutonium source appears to be far less penetrating than that from the cobalt source (Fig. 30.9).

**Fig. 30.9**  *The penetrating power of α-, β-, and γ-radiation*

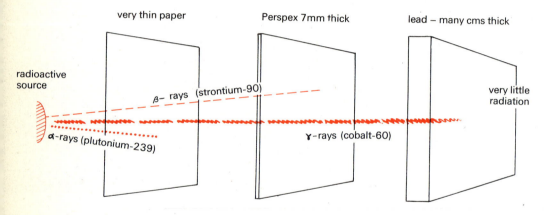

There are three types of radiation, each with its own penetrating power, that may come from a radioactive source. These are:
alpha radiation (α)
beta radiation (β)
gamma radiation (γ)

Some sources give out only one type of radiation, for example plutonium-239 emits only α-radiation, strontium-90 emits only β-radiation and cobalt-60 emits only γ-radiation. Other sources, such as radium, emit all three types of radiation.

368

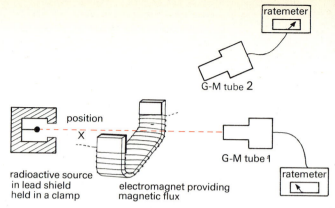

**Fig. 30.10**  *Experiment to show the deflection of β-radiation passing through a magnetic flux*

## 30.9  Experiment to determine whether radiation can be deflected by a magnetic flux (Fig. 30.10)

A radium source is placed in a lead shield which has a small hole in it so that the radiation is kept to a fine beam. The poles of a strong electromagnet are positioned at either side of the radiation beam. A Geiger counter (1) is placed directly in line with the source. The counts per second are recorded. The electromagnet is switched on and another Geiger counter (2) is used to explore systematically the whole area, until its reading of counts per second is at a maximum. Its position is then noted.

A piece of lead (X in diagram), is placed between the source and the two G-M tubes and the effect on the reading of each Geiger counter is noted.

The results show that when the electromagnet is switched on the counts per second recorded by G-M1 decrease; G-M2 records a maximum count when in the position shown in the diagram.

This indicates that the magnetic flux of the electromagnet has deflected part of the radiation emitted by the radium.

When the lead is in position, the reading of G-M2 ceases but that of G-M1 is only slightly reduced.

We conclude that the radiation which has been deflected and stopped by the lead is $\beta$-radiation ($\alpha$-radiation would be unable to travel this distance in air). The radiation which is unaffected by the magnetic flux and the lead is $\gamma$-radiation.

## 30.10 The properties of $\alpha$-, $\beta$- and $\gamma$-radiation

If we apply Fleming's left-hand rule to the deflected $\beta$-radiation we can see that it is deflected in an identical direction to cathode rays subject to the same conditions (page 333). Thus $\beta$-radiation must have the same charge as cathode rays, i.e. a negative charge. $\gamma$-radiation is unaffected by a magnetic flux and does not possess any charge. Does $\alpha$-radiation have any charge?

To help us answer this question we can look at the visible tracks of $\alpha$- and $\beta$-radiation when they are subjected to an intense magnetic field as they travel through a cloud chamber. Such tracks show that $\alpha$-radiation is deflected, but to a lesser degree and in the opposite direction from $\beta$-radiation. $\alpha$-radiation has a greater mass than $\beta$-radiation and a positive charge.

Further experiments have been carried out to measure the ratio of charge to mass ($e/m$) for $\alpha$- and $\beta$-radiation. These have verified that both types consist of streams of fast-moving particles:

$\alpha$-particles are positively charged helium atoms;
$\beta$-particles are fast-moving electrons.
$\gamma$-radiation has been shown to travel at the speed of light and to have a wave-length shorter than that of X-rays.
$\gamma$-rays are electromagnetic waves.

The properties of these three types of radiation are summarized in Table 2 (next page). Figure 30.11 shows $\alpha$- and $\beta$-tracks (in hydrogen). It illustrates the difference in the ionizing power of the particles.

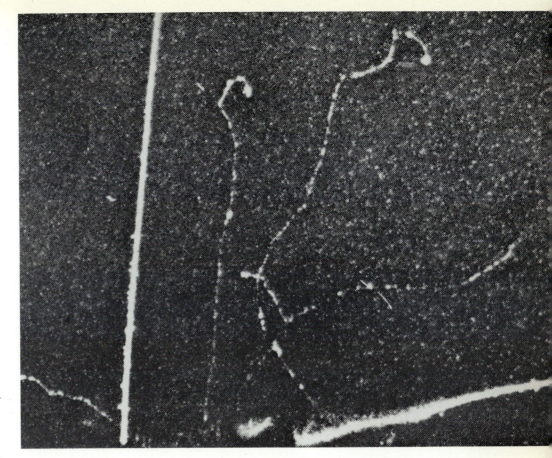

**Fig. 30.11** $\alpha$- and $\beta$-tracks in hydrogen

## 30.11 The strength of radioactive sources

This is determined by the number of emissions (i.e. the number of $\alpha$-, $\beta$- or $\gamma$-rays) given out per second.

The unit is the curie (C). A source strength of 1 curie gives out $3.7 \times 10^{10}$ emissions/second.

The strength of school sources is usually $5\mu$C.

QUESTION 5: How many emissions per second are given by a source strength of (a) one millicurie (mC) and (b) one microcurie (µC)?

**TABLE 2**
**PROPERTIES OF α- AND β-PARTICLES AND γ-RAYS**

|  | α-particles | β-particles | γ-rays |
|---|---|---|---|
| Charge | Positive | Negative | None |
| Penetrating power | 6cm of air Very thin paper | About 500cm of air or about 0.3cm of aluminium | 50 000cm of air or 4cm thickness of lead reduces intensity to one tenth |
| Effect of distance on intensity | Intensity decreases and eventually reaches zero as all is absorbed | Intensity decreases and eventually reaches zero as all is absorbed | The intensity is inversely proportional to the square of the distance from the source—as with all electromagnetic waves |
| Causes fluorescence and exposes a photographic plate | Yes | Yes | Yes |
| Deflected by electric and magnetic fields | Yes (Very small) deflections) | Yes (large deflections) | No |
| Causes ionization | Very strongly | Strongly | Very weakly |
| Detected by spark counter | Yes | No | No |

370

## 30.12 Disintegration theory of radioactivity

Rutherford and Soddy put forward this theory in 1902. It states that radioactive elements are unstable. Their atoms decay naturally, shooting out bits of themselves as α-particles, β-particles and γ-rays, either one kind at a time or more than one kind. In decaying, the radioactive atoms become atoms of different elements. If these too are unstable the process of decay will continue. Eventually a stable form of atom is reached and radiation ceases. For example, radium, after 11 decay processes forming 11 different elements, eventually forms stable atoms of lead. The disintegration theory was a piece of revolutionary thinking in the field of atomic physics because until this time atoms had been thought of as indivisible.

## 30.13 How long does a substance remain radio-active?

No physical or chemical effects can speed up or slow down radioactive decay. Radioactivity is a random process and we cannot determine at any particular time which specific atom is going to disintegrate. But it is possible to discover how long it takes for half the atoms of a certain mass of radioactive material to decay, and this period of time is called the *half-life*.

The half-life of any radioactive material is the time it takes for half the radioactive atoms present at any time to disintegrate.

Radium, for example, is known to have a half-life of 1590 years. Polonium, on the other hand, has a half-life of 140 days.

## 30.14 Radioactivity and the structure of the atom

Scientists have used high-speed particles to investigate the structure of atoms. As a result of the work of Rutherford, Geiger, Marsden, Chadwick, Bohr and others, our ideas on the atom have been changed. It is no longer

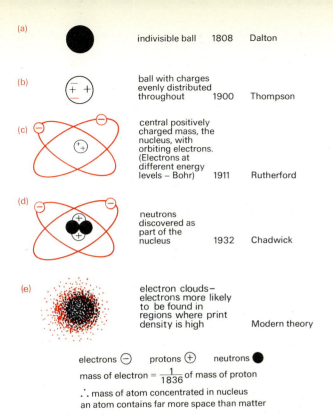

(a) indivisible ball   1808   Dalton

(b) ball with charges evenly distributed throughout   1900   Thompson

(c) central positively charged mass, the nucleus, with orbiting electrons. (Electrons at different energy levels – Bohr)   1911   Rutherford

(d) neutrons discovered as part of the nucleus   1932   Chadwick

(e) electron clouds – electrons more likely to be found in regions where print density is high   Modern theory

electrons $\ominus$   protons $\oplus$   neutrons $\bullet$

mass of electron $= \frac{1}{1836}$ of mass of proton

$\therefore$ mass of atom concentrated in nucleus
an atom contains far more space than matter

**Fig. 30.12** *Development of ideas on the structure of the atom (the helium atom is taken as the example); (d) is the usual method of illustrating atomic structure*

regarded as an indivisible particle, but is thought of as a central nucleus of positive charge surrounded by *electrons* (Fig. 30.12). *Neutrons*, with no electrical charge, and *protons*, with a positive charge, are present in the nucleus. The positive charge of the nucleus is equal to the number of protons that are present.

**TABLE 3**

**ATOMIC NUMBER AND ATOMIC MASSES OF A SELECTION OF ELEMENTS**

| Atomic Number | Symbol | Name | Mass of the most common stable isotope |
|---|---|---|---|
| 1 | H | Hydrogen | 1 |
| 2 | He | Helium | 4 |
| 6 | C | Carbon | 12 |
| 7 | N | Nitrogen | 14 |
| 8 | O | Oxygen | 16 |
| 11 | Na | Sodium | 23 |
| 13 | Al | Aluminium | 27 |
| 15 | P | Phosphorus | 31 |
| 16 | S | Sulphur | 32 |
| 17 | Cl | Chlorine | 35 |
| 20 | Ca | Calcium | 40 |
| 26 | Fe | Iron | 56 |
| 29 | Cu | Copper | 63 |
| 33 | As | Arsenic | 75 |
| 47 | Ag | Silver | 107 |
| 50 | Sn | Tin | 120 |
| 56 | Ba | Barium | 138 |
| 63 | Eu | Europium | 153 |
| 74 | W | Tungsten | 184 |
| 79 | Au | Gold | 197 |
| 80 | Hg | Mercury | 202 |
| 82 | Pb | Lead | 208 |
| 84 | Po | Polonium | — |
| 88 | Ra | Radium | — |
| 90 | Th | Thorium | 232 |
| 92 | U | Uranium | — |
| 94 | Pu | Plutonium | — |
| 103 | Lw | Lawrencium | — |

In a neutral atom the number of protons equals the number of electrons.

The electrons occupy different energy levels. The lower potential (negative) energy levels are occupied by electrons and the higher ones often are not occupied. An electron, if it is given the necessary energy, may move from one energy level to another. If the electrons move down in levels then the difference in energy is emitted as

371

a packet of energy, a *quantum*. The energy is emitted as pulses of electromagnetic waves (photons), for example as light.

The electrons determine the chemical properties of an element because chemical reactions involve their re-arrangement about the nucleus. Therefore it is very helpful to a chemist to have a table of elements based on their *atomic number*, as in Table 3.

Atomic number = number of protons (= number of electrons)
Atomic mass (mass number) = number of protons + number of neutrons

Nuclear physicists use symbols to represent atomic nuclei. For example, a helium nucleus is shown as $^4_2$He. "He" is the chemical formula, the top figure is the mass number and the bottom figure is the atomic number.

### 30.15 Isotopes

Lead that has been formed by radioactive decay has a different atomic mass from other samples of the same element. This difference in atomic mass is explained by the fact that the atoms contain different numbers of neutrons though they have equal numbers of electrons and protons.

Isotopes of an element are atoms with the same atomic number but different atomic masses.

Many elements, including those that are not radio-active, have isotopes.

QUESTION 6: Two isotopes of uranium are uranium-235 ($^{235}_{92}$U) and uranium-238 ($^{238}_{92}$U). How many neutrons does each isotope possess?

### 30.16 Can isotopes be made?

Natural radioactivity is discovered in elements with large unstable atomic masses. It is now possible to make stable

elements radioactive by bombarding them with fast-moving particles, such as high-speed neutrons. In most cases this is achieved by placing the element within the core of a nuclear reactor. The uses of such isotopes are extensive. A few are discussed below.

### 30.17 Some uses of radioactive isotopes

1. *In industry*

a. When oil engineers are about to feed a different grade of oil into a continuous-flow pipeline, they insert a radio-active source into the oil. By using a counter, engineers at the other end can detect when a different grade is about to arrive and they operate a valve to direct the oil into another tank.

b. Test pistons of a car are made radioactive so that their wear is indicated by the level of radioactivity subsequently found in the engine oil.

c. More radiation is absorbed by a thick material than by a thin one. Thus if a radioactive source is placed on one side of a metal sheet, and a counter on the other side, the pulses that are received indicate the thickness of the material. If the metal is made to pass along between the source and the counter, any variations in thickness show up as an increase or decrease in pulse rate. These readings enable the manufacturer to adjust his machinery to give the correct uniform thickness.

d. A similar set-up of source and counter can be used to detect the correct level of material in a container without opening the container. The empty portion of the container does not absorb as much radiation as the full portion.

The great advantage of all of these checking methods is that they can be carried out without interrupting a continuous-flow process.

Figure 30.13 shows the checking of a welded seam in a thick steel belt for a boiler drum. A $\gamma$-source is moved along the inner surface of the weld. On the outer surface

**Fig. 30.13**  *γ-ray photography is used to check a weld*

**Fig. 30.14**  *Radioactive isotopes added to plant nutrients make it possible to check the plant's intake of the nutrient*

there is a film which is exposed by the radiation. The photograph shows up any faults.

### 2. *In agriculture*

a. When isotopes are mixed with another substance they give it a detectable "label". This is the basis of the tracer techniques, by which the passage of various elements may be traced through a plant system. Figure 30.14 shows radioactive phosphorus being injected into the nutrient solution surrounding the roots of barley seedlings. The scientist will later use the Geiger counter to trace the uptake of the phosphorus by the plant.

373

b. Male adults of a species may be sterilized by radiation. By this means reproduction and the survival of the species is rendered impossible. This technique is used in pest control.

c. Radiation kills bacteria, therefore by its use food may be sterilized and stored for longer periods.

3. *In medicine*

a. Sterilization of surgical instruments and hospital clothing.

**Fig. 30.15**(*a*) *Scanner used to detect radioactive elements within body tissues*

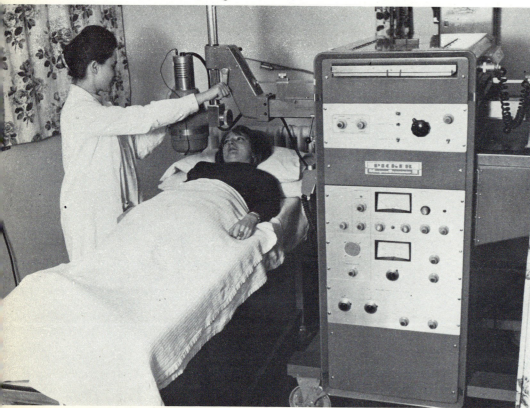

**Fig. 30.15**(*b*) *Scan of a defective thyroid gland*

b. Tracing of elements through the body may be used as a means of diagnosis. In such tracer techniques a radioactive isotope is introduced into the body and its uptake is followed by means of a scanning instrument like the one shown in Fig. 30.15(a). The isotope chosen is an element which is taken into the tissue of the particular organ under study; iodine, for example, is taken up by the thyroid gland. Figure 30.15(b) shows the result of a scan of a thyroid gland. In a normal gland there is an even distribution of the radioactive isotope, but in this case there is a defect in the right lobe of the gland.

c. Some cancerous growths may be controlled or dispersed by means of $\gamma$-radiation, which destorys malignant cells.

4. *In archaeology*

Carbon-14 which is present in the atmosphere is taken up by plants, trees and other living matter. It has a half-life of 5500 years. By detecting the strength of carbon-14 in archaeological specimens the scientist can establish their age to within 100 years.

### 30.18 Dangers of exposure to radiation

The dangers from α-radiation are small because it cannot penetrate the skin. It only becomes a hazard if it is absorbed into the body. But β- and γ-radiation are dangerous, as large doses produce radiation burns, haemorrhaging, changes in the blood and can even affect the nervous system, in which case death is inevitable. Strontium-90, a β-emitter, and caesium-137, a β- and γ-emitter, are contained in the fall-out from atomic explosions. Both these elements have very long half-lives and are absorbed into the body, particularly the bones.

Radioactive sources in the school laboratory must be used with care, and never touched with bare hands. Madame Curie is known to have suffered with radiation burns and to have died from the long-term effects of her beloved radium, which she handled with her bare hands, not realizing the dangers. Papers that she touched when she was writing up reports still possess enough radioactive contamination to expose a photographic plate!

All those who work with radioactive materials today, both in research establishments and in industry, are protected by very strict regulations on procedure, and, where necessary, by the provision of radiation-proof barriers.

### 30.19 Nuclear energy

Figure 30.16 shows the result of a cloud-chamber experiment giving visual proof of the splitting of the atom.

What has happened?

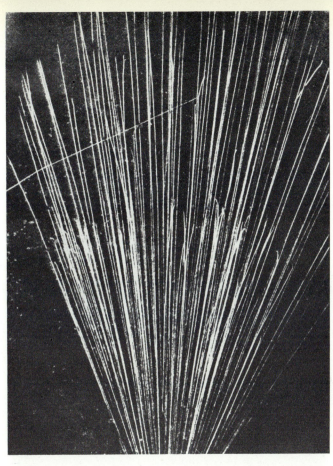

**Fig. 30.16** *Cloud chamber tracks showing the splitting of a nitrogen atom by an α-particle*

The α-particles (seen as straight tracks) are helium atoms with 2 protons. These are travelling through nitrogen gas at high velocity. An α-particle hits a nitrogen nucleus which contains 7 protons. A nucleus containing 9 protons is formed. One proton is then emitted (long track going off to the left) leaving a nucleus of 8 protons. A nucleus of 8 protons is the nucleus of an atom of oxygen (the very short track veering off to the right). The whole process may be represented by the nuclear equation:

$$^{14}_{7}\text{N} \quad + \quad ^{4}_{2}\text{He} \quad \rightarrow \quad ^{1}_{1}\text{H} \quad + \quad ^{17}_{8}\text{O}$$

| nitrogen nucleus | helium nucleus α-particle | hydrogen nucleus proton | oxygen isotope nucleus |

The splitting of the atom was first achieved in 1919, by Rutherford. He achieved what was thought to be impossible. He had split the atom and he had converted the atom of one element into that of another. This latter achievement was what the early alchemists had tried so hard to do — the transmuting of an element.

From then on many other elements were transmuted. It was found that the resulting particles had a greater energy than the bombarding ones. Why was this? Had it something to do with the fact that the total mass of the nuclei formed by the transmutation was always slightly less than the total mass of the reacting nuclei?

What had happened was that mass had indeed been converted into energy, the energy being released as an increase in the kinetic energy of the particles. In these earlier experiments, however, not enough nuclei were split to release any large amount of energy.

QUESTION 7: Bearing in mind that nuclei are positively charged why do you think that the discovery of neutrons in 1932 was so important for the "atom splitters"?

### 30.20 What happens when uranium-235 is bombarded by neutrons?

A uranium-235 atom was first split in 1934. Krypton and barium were the two elements that resulted from this

376

fission. The atomic number of these elements is approximately half the atomic number of uranium-235. Because the uranium-235 nucleus divides approximately into two equal halves, rather like asexual reproduction (e.g. amoeba) the procedure is called the *fission* of uranium (see Fig. 30.17).

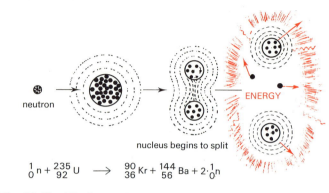

$$^{1}_{0}\text{n} + ^{235}_{92}\text{U} \longrightarrow ^{90}_{36}\text{Kr} + ^{144}_{56}\text{Ba} + 2\cdot^{1}_{0}\text{n}$$

**Fig. 30.17** *The fission of U-235 when bombarded with neutrons*

On fission there is also a loss of mass which reappears as a vast increase in the velocity of the products of fission. These particles moving with great speed bump into and are slowed down by surrounding atoms. Thus, most of their kinetic energy is converted into heat energy. In fact one atom of uranium will release on fission 60 000 000 times as much as energy as that released by one carbon atom when it is burnt (oxidized).

Also on fission there is a release of two or three high-speed neutrons. In order to determine what these do, carry out the experiment illustrated in Fig. 30.18. Set out the matches and then light match A and note what happens.

The flames spread from the first match and eventually set light to all of them by a series of progressions. We have set up a *chain reaction*.

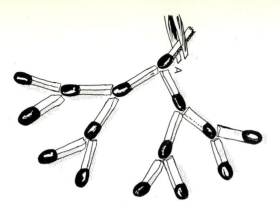

**Fig. 30.18** *Example of a chain reaction*

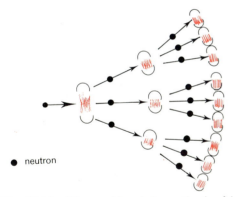

● neutron

**Fig. 30.19** *Diagram of a chain reaction (nuclei only are shown)*

Similarly the neutrons released by fission spread out (like the flames on the matches), and are capable of splitting other uranium atoms. A chain reaction is set up which releases an enormous quantity of energy (Fig. 30.19).

This chain reaction cannot occur naturally because U-235 makes up only 0.7 per cent of uranium deposits. The remaining 99.3 per cent is U-238 which is stable.

Man controls a chain reaction by putting either carbon or beryllium rods in the U-235. These elements, which are used for this purpose in nuclear power stations, are termed *moderators*. Carbon and beryllium slow down the neutrons when these particles collide with their atoms. The first controlled chain reaction was obtained by Fermi on 2 December 1942.

**Fig. 30.20** *Gas-cooled reactor and adjoining power station layout*

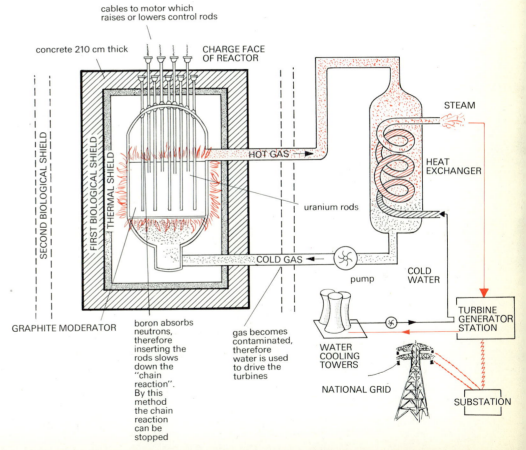

cables to motor which raises or lowers control rods

concrete 210 cm thick

CHARGE FACE OF REACTOR

SECOND BIOLOGICAL SHIELD

FIRST BIOLOGICAL SHIELD

THERMAL SHIELD

HOT GAS

STEAM

HEAT EXCHANGER

uranium rods

COLD GAS

pump

COLD WATER

GRAPHITE MODERATOR

boron absorbs neutrons, therefore inserting the rods slows down the "chain reaction". By this method the chain reaction can be stopped

gas becomes contaminated, therefore water is used to drive the turbines

WATER COOLING TOWERS

NATIONAL GRID

TURBINE GENERATOR STATION

SUBSTATION

**Fig. 30.21** *Reactor charging floor of Calder Hall, Cumberland (first nuclear power station, 17 October 1956)*

### 30.21 Nuclear power stations

Figure 30.20 shows in diagram form a gas-cooled reactor within a nuclear power station. The uranium rods are stacked in channels that have been drilled in the graphite (carbon) moderator. Figure 30.21 shows some uranium rods being transported before being loaded into the top of the reactor (in the background). Carbon dioxide is sent under pressure through the reactor in order to absorb the heat of the reaction. This heat energy may be used to drive a turbine.

Small nuclear reactors are also used to power submarines (see Chapter 11, Fig. 11.8).

### 30.22 Critical mass

In pure uranium-235, every atom is capable of fission; but if the mass is too small most of the fast-moving neutrons escape from the surface before colliding with another atom. As the mass is increased more neutrons are retained and collide with atoms, so starting off a chain reaction. The minimum amount of fissile material required to sustain a chain reaction is called the *critical mass*. If two masses, each slightly less than the critical mass, are placed together, a chain reaction starts; numerous fissions occur in a very short space of time and there is an incredibly vast release of energy. This takes the form of an explosion (an atomic bomb) and a large quantity of harmful $\gamma$-radiation is released on fission.

### 30.23 Nuclear fusion

There is a limit to the energy that can be produced by nuclear fission because the pieces of uranium which are brought together must each be less than the critical mass. However, another reaction exists which enables bombs of unlimited size to be produced. This reaction is the fusing of two heavy hydrogen nuclei. This form of nuclear fusion takes place on the sun, releasing energy, helium and neutrons. However, in order to bring about this fusion of nuclei, the gas must be heated up to, and maintained at, temperatures of millions of degrees. It is only at these high temperatures that the hydrogen nuclei have sufficient velocity to overcome their natural repulsion for one another.

In the case of a hydrogen bomb, the required temperature is achieved by using a uranium bomb as the triggering mechanism.

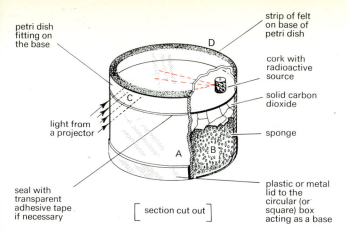

petri dish
fitting on
the base

strip of felt
on base of
petri dish

D

cork with
radioactive
source

solid carbon
dioxide

light from
a projector

C

sponge

A    B

seal with
transparent
adhesive tape
if necessary

[ section cut out ]

plastic or metal
lid to the
circular (or
square) box
acting as a base

**Fig. 30.22** *Construction of a cloud chamber*

A.  Make a cloud chamber (Fig. 30.22).

   You will need to construct two compartments, one for the base and the other for the cloud chamber itself.

a. The base. Select a circular tin A, approximately 10cm in diameter and 7cm deep. It must have a press-on lid. (A small ground-coffee tin with a plastic lid is ideal.) Paint the outside bottom surface of the tin black. Cut plastic foam B to the same depth and diameter of the tin, so that it will fit into the tin exactly.

b. Cloud chamber. Select a transparent box C of plastic or glass which has the same diameter as the base. A biologist's petri-dish or the top of a sandwich box is ideal.

   Cut a ring of felt to the same diameter. Cut out the centre portion so that you are left with a ring which is 1cm in width. Stick this ring D to the top inner surface of the dish. (Note: it is not essential to have circular containers, rectangular ones of similar dimensions will also be suitable.)

c. To use (under the supervision of your teacher). Place solid carbon dioxide (NOT with your bare hands) into the base so that it is several millimetres thick. Press the sponge down on the top of this and **push on the lid. Turn the base over so that the black painted bottom, with the carbon dioxide against its surface, is uppermost.**

   Place a radioactive source in its holder on top of the base. Drop a few drops of methylated spirits onto the felt ring in order to moisten it. Place the top of the cloud chamber on top of the base. Shine the light from a film strip projector through the side of the chamber so that it is well illuminated. As the chamber cools so the movement of the subatomic particles can be observed by their tracks.

B.  Construct 3D models of simple atoms (Fig. 30.23) using polystyrene balls and thick 15A fuse wire. Paint balls in three different colours to represent the different atomic particles. Form the orbits for the electrons from the wire. (In this case we are representing the atoms in a very simplified form with each electron in its own orbit.) These models

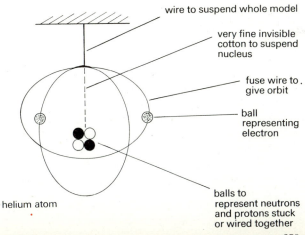

wire to suspend whole model

very fine invisible
cotton to suspend
nucleus

fuse wire to
give orbit

ball
representing
electron

helium atom

balls to
represent neutrons
and protons stuck
or wired together

**Fig. 30.23** *Construction of atomic models*

can then be suspended from the ceiling with their appropriate labels. They make interesting mobiles.

C. Make a large chart showing the elements arranged in order of their atomic weights. Put as many elements on the chart as possible and in each case give their atomic mass and number.

D. Make a *Who's Who* scrapbook of the scientists involved in the discovery of radioactivity and the basic structure of the atom. Put in as many illustrations as possible. The story of Madame Curie is well worth studying in this connection, for it illustrates the devotion of a whole life to discovery for the benefit of mankind.

E. Prepare a number of display charts to illustrate a lecture entitled "Atoms in the Service of Man", dealing with the peaceful uses of radioactivity and nuclear energy. You could include a map of Britain with the nuclear power stations pinpointed by flags, the uses of radioactive isotopes and the epic voyage of *Nautilus*.

F. Get pencil, paper and two packs of cards. Put all the red cards in one pile, and all the black cards in another. The red cards are to represent undecayed atoms, and the black cards will be used to show atoms that have decayed. Take up the pile of fifty-two red cards and deal out four of them. Treat these as atoms that have decayed, putting them on one side and replacing them in the dealing pack with four black cards from the other pile. *Thoroughly* shuffle these cards you hold and then deal out another four. Record under the heading of "turn 1" the number of red cards dealt out in this four, and replace them with black as before. Continue in this way, recording as you go the number of red cards that have to be replaced with black in turns 2, 3, 4 and so on, until nearly all the cards you hold are black.

Repeat the experiment about ten times, recording carefully.

Then take the average of red cards dealt out at each turn. Plot a graph of turn number (representing time) horizontally, against the average number of red cards obtained on that turn. The first dealing of four red cards is to be plotted at nought. From your graph determine the half-life of the sample of cards.

1. Using your library to get the necessary information, describe (a) a particle accelerator, (b) the experiment that Rutherford used to detect the nucleus of an atom, (c) how Chadwick discovered the neutron in 1932.

2. Draw a diagram of the tube of a Geiger-Müller counter and label it. Describe two uses for this instrument.

3. How would you demonstrate that a radioactive source is emitting $\gamma$-rays? Compare the properties of $\gamma$-rays with those of $\alpha$- and $\beta$-particles.

4. (a) Explain the statement that "an atom is mostly space". (b) An atom is electrically neutral. Why is this, in view of the fact that its nucleus contains protons with a positive charge?

5. Explain why Dalton's belief that "all atoms of the same element are alike and indivisible" has had to be modified.

6. There are three isotopes of hydrogen. The simplest form consists of one proton and no neutrons. Draw simple atom models to show the structure of (a) an atom of deuterium (heavy hydrogen) mass 2, (b) an atom of tritium, mass 3. What is heavy water?

7. Draw a diagram of a cloud chamber and explain how it can be used to show the passage of charged particles.

8. Radium has a half-life of 1590 years. How much of 20g will be left in 3180 years?

9. Radium has an atomic mass of 226 and 88 protons in its nucleus. Cobalt has an atomic mass of 59 and 27 protons in its nucleus. Which of these elements would you expect to be radioactive? Explain your answer. Can the other element be made radioactive, and if so how do you think that this would be carried out?

10. A waterworks official suspects that a water main has a slight fracture in it. If he had available a non-poisonous radioactive substance such as sodium carbonate how would he use this to detect the leak?

11. Scientists have a shorthand method for writing the atomic formulae for atoms of elements. They write $^{A}_{Z}E$ where A stands for the mass number, Z represents the atomic number and E is the chemical symbol for the element. Using Table 3 on page 371, write out the atomic formulae in this way for (a) carbon, (b) copper, (c) nitrogen, (d) uranium-235, (e) the three isotopes of hydrogen, (f) sodium, (g) oxygen, (h) thorium.

12. Explain what is meant by a chain reaction? How is a chain reaction controlled in a nuclear reactor? How does the reaction between atoms in the sun's gases differ from that in a uranium nuclear reactor?

13. How many emissions would a radioactive source of $5\mu C$ emit per second?

14. Three radioactive sources give the following results in a series of experiments. What do you deduce about the type of radiation emitted by A, B and C?

| Source | Detected by spark counter | Detected by a thin-walled G-M tube | Detected by thickwalled G-M tube | Deflected by a magnetic field |
|---|---|---|---|---|
| A | Yes | Yes | No | Slightly |
| B | No | Yes | Yes | Yes |
| C | No | Yes | Yes | No |

15. $^{24}_{11}Na$ is a radioactive isotope of sodium. It has a half-life of 14 hours. (a) How many protons and neutrons are there in this atom? (b) How many electrons surround the nucleus? (c) What is meant by the term isotope? (d) What is meant by the term half-life? (e) How much of 100g of the isotope will be left after 42 hours? (f) What time has elapsed when 6.25g of the 100g are left?

radioactive source
in lead block

cloud chamber
magnetic flux directed
downwards into the page

**Fig. 30.24**

16. Figure 30.24 shows a radioactive source situated in a lead block so that only a narrow beam of radiation is emitted. The source emits three types of radiation. A very strong magnetic flux is directed at right angles to the direction of the radiation. The magnetic flux flows downwards into the page. Copy the diagram and draw in the tracks of the three types of radiation as they pass through the magnetic flux.

17. A Geiger-Müller tube is fitted through a cork in the top of a beaker so that the end of the tube is in the beaker. The beaker contains radon gas. The Geiger-Müller tube is connected to a ratemeter and the following readings are obtained:

| Times | 0 | 30 | 60 | 90 | 120 | 150 | 180 | 210 | 246 |
|---|---|---|---|---|---|---|---|---|---|
| Count rate | 100 | 66 | 47 | 32 | 20 | 15 | 10 | 7 | 5 |

Draw a graph of activity (count rate) against time.

How long does it take for the activity to fall from (a) 100–50, (b) 80–40, (c) 60–30? (d) What is the half-life of radon?

18. A sample of a radioactive substance was placed in front of a G-M tube and the table below shows how the rate count ($CR$) varied with time ($T$):

| $T$ | 0 | 20 | 40 | 60 | 80 | 100 |
|---|---|---|---|---|---|---|
| $CR$ | 20 000 | 16 120 | 12 020 | 10 010 | 8100 | 6000 |

What is the half-life of the substance?

19. (a) Outline three ways in which radioactive isotopes are useful to man.
(b) How could you use a source which emits $\alpha$-, $\beta$- and $\gamma$-rays as a pure $\gamma$-source?
(c) An emitted $\alpha$-particle has energy. What form of energy? What happens to the energy as it travels through air?
(d) Why do $\alpha$-particles cause more ionization than $\beta$-particles?
(e) Why must radioactive substances always be handled with tongs?
(f) Why does a charged electroscope discharge when a flame or radioactive substance is brought up to it?

## CHAPTER 31

# MOTION

**Fig. 31.1**  *A speed limit sign*
*(30mph = 48km/h)*

## 31.1  Speed

Every road user must know the meaning of the sign shown in Fig. 31.1. Such signs help our roads to be safe places by limiting the *speed* of vehicles. The speed limits vary because of changes in the type of area through which the road runs.

*QUESTION 1:* In what type of area would you expect to find the speed limit illustrated in Fig. 31.1?

A speed of 30mph means that if a car travelled at this speed for 1 hour it would travel 30 miles.

Speed is defined as the distance travelled in unit time (1 hour, in the example above).

To calculate speed we use the equation,

$$\text{Speed} = \frac{\text{distance travelled}}{\text{time taken}}$$

Two units of speed are metres per second (m/s) and kilometres per hour (km/h). Thus if a car takes 2 hours to travel between two towns that are 96km apart, then the average speed of the car is

$$\frac{\text{distance}}{\text{time}} = \frac{96}{2}\text{km/h} = 48\text{km/h}.$$

**Fig. 31.2** *A comparison of speeds*

You will notice that the word "average" has been used. It would be difficult to find any roads where a driver could drive for long distances without changing his speed in order to corner or to stop at junctions. When a car is travelling at a constant (not changing) speed we say its speed is *uniform*.

Figure 31.2 gives a selection of the maximum speeds achieved by some moving objects. Look at the Figure and you will see the world land speed record. One of the difficulties in setting up such a record is finding an area which gives the vehicle enough distance to reach its

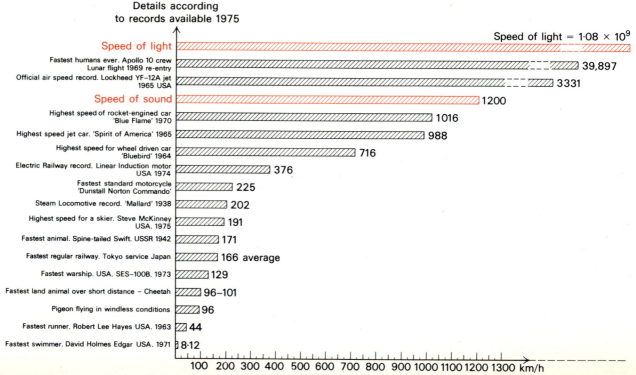

Details according to records available 1975

| | |
|---|---|
| Speed of light | Speed of light = $1.08 \times 10^9$ |
| Fastest humans ever. Apollo 10 crew Lunar flight 1969 re-entry | 39,897 |
| Official air speed record. Lockheed YF–12A jet 1965 USA | 3331 |
| Speed of sound | 1200 |
| Highest speed of rocket-engined car 'Blue Flame' 1970 | 1016 |
| Highest speed jet car. 'Spirit of America' 1965 | 988 |
| Highest speed for wheel driven car 'Bluebird' 1964 | 716 |
| Electric Railway record. Linear Induction motor USA 1974 | 376 |
| Fastest standard motorcycle 'Dunstall Norton Commando' | 225 |
| Steam Locomotive record. 'Mallard' 1938 | 202 |
| Highest speed for a skier. Steve McKinney USA. 1975 | 191 |
| Fastest animal. Spine-tailed Swift. USSR 1942 | 171 |
| Fastest regular railway. Tokyo service Japan | 166 average |
| Fastest warship. USA. SES–100B. 1973 | 129 |
| Fastest land animal over short distance – Cheetah | 96–101 |
| Pigeon flying in windless conditions | 96 |
| Fastest runner. Robert Lee Hayes USA. 1963 | 44 |
| Fastest swimmer. David Holmes Edgar USA. 1971 | 8·12 |

100 200 300 400 500 600 700 800 900 1000 1100 1200 1300 km/h

maximum speed and to keep at this speed whilst travelling in a straight line (Fig. 31.3). So in measuring maximum speeds the body is travelling in a straight line, and we realize that in describing the movement of a body we must not only consider how fast it is going, but also the direction in which it is travelling.

**Fig. 31.3** *Blue Flame Rocket Racer setting up World land speed record (1970). Notice there is a long straight distance*

## 31.2   Velocity

Speed is a general term used in everyday life and it takes no account of direction. The scientist must be more precise and take the direction of motion into account as well as the size of the motion. We use the term *velocity* to mean speed in a given direction. Velocity is therefore a vector quantity (see page 147). It is defined as follows:

Velocity is the distance moved in unit time in a particular direction.

*QUESTION 2:* A driver is driving his car in a northerly direction along the road illustrated in Fig. 31.4. Throughout the journey he makes sure that the needle of his speedometer does not move from 20 mph.

**Fig. 31.4**

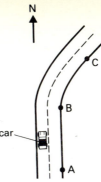

(a) Describe what happens to the speed and velocity of the car along the sections of road (i) A to B (ii) B to C. (b) Would the driver be correct in stating his "average speed" or his "average velocity" for the journey was 20 mph.

If a moving object moves in a constant direction and covers equal distances in equal times it is said to have a *uniform velocity*.

When an object moves in a straight line we can use the term *displacement* instead of distance.

## 31.3   Displacement-time graphs or distance-time graphs

It is often very difficult to look at a mass of figures concerning the motion of a body and from them interpret what is happening to a moving body. However, this task is made a lot easier if we build these figures up into a mathematical "picture", that is, a graph. For example, if we have an object which has moved 10cm during the first second, 20cm after 2 seconds, 30cm after 3 seconds, 40cm after 4 seconds and so on, we can draw a graph of distance (displacement) against time. The result is shown in Fig. 31.5. What we have obtained is a straight line graph. Such a straight line graph indicates that the object is travelling with uniform velocity.

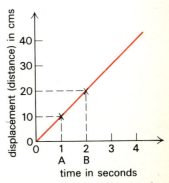

**Fig. 31.5** *Distance-time graph of a body moving with uniform velocity*

385

We can check this mathematically by using the formula:

$$\text{velocity} = \frac{\text{distance}}{\text{time}}$$

for different points on the graph

If the time = 1s (OA on the graph)
Displacement = 10cm.

Therefore velocity $= \dfrac{10}{1} = 10\text{cm/s}$

If the time = 2s (OB on the graph)
Displacement = 20cm

Therefore velocity $= \dfrac{20}{2} = 10\text{cm/s}$

The object is moving with uniform velocity. If the graph is not a straight line then the velocity is not uniform.

*QUESTION 3:* Fig. 31.6 shows some more distance-time graphs. (i) How do the velocities differ for the moving bodies of graphs (a), (b) and (c). The scales are the same. (ii) Describe the movement of the bodies of graphs (d) and (e).

The curve in the graph shown in Fig. 31.6 (e) is the type of distance-time curve that we would obtain for the flight of a ball after it has been thrown up into the air. The ball rises into the air but gradually slows down because of the earth's gravitational attraction (see page 76). Eventually it reaches a point in its flight when it is, for an instant of time, at rest. In other words at this point its velocity is zero. The ball then falls with its velocity increasing because of the pull of the earth.

Throughout this discussion on the flight of the ball we have been considering changes in velocity. We use the word *acceleration* to describe how velocity changes with time.

386

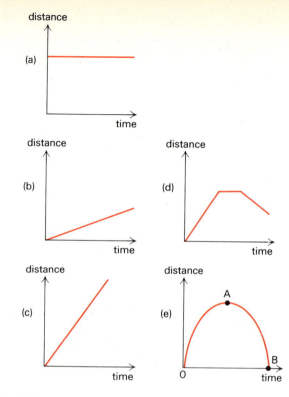

**Fig. 31.6**

### 31.4   Acceleration

When the driver of a car wants to overtake another car, he must accelerate, that is, increase his velocity.

If the velocity of a body is increasing we say that the body is accelerating.

If the velocity of a body is decreasing we say that the body is *decelerating* or *retarding*.

**Acceleration is the change in velocity in unit time.**

$$\text{Acceleration} = \frac{\text{change in velocity}}{\text{time taken to make the change}}$$

The units of acceleration are velocity divided by time. If the velocity is in kilometres per hour and the time is in seconds, then the acceleration is in kilometres per hour per second. If the velocity is in metres per second and the time is in seconds, then the acceleration is in metres per second per second (written m/s²).

Be careful to note that when we talk about acceleration we are always concerned with a *change* in velocity. Thus if a body is moving with uniform velocity it will not be accelerating.

If you look in any newspaper or go to a garage you will find literature which tells you of the cars specifications and performance (see Fig. 31.7.) The acceleration is always given in terms of a change of velocity and the time taken to make the change. By choosing the same change in velocity we can compare the different models of car by comparing the times they take to make this change.

For example, the following details are given by a manufacturer for two different models.

Model A. Acceleration: 0–96km/h in 15.6 seconds.

Model B. Acceleration: 0–96km/h in 13.3 seconds.

If you wanted a sporty car "quick off the mark", which model would you buy?

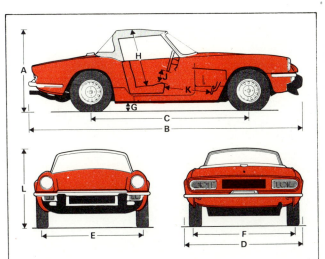

## Dimensions

| | | |
|---|---|---|
| A Overall Height (Hood up) | 3 ft 9¼ in | 1162 mm |
| B Overall Length | 12 ft 4¼ in | 3780 mm |
| C Wheelbase | 6 ft 11 in | 2110 mm |
| D Overall Width | 4 ft 10½ in | 1488 mm |
| E Track (Front) | 4 ft 1 in | 1245 mm |
| F Track (Rear) | 4 ft 2 in | 1270 mm |
| G Ground Clearance (2 up) | 4½ in | 118 mm |
| H Roof to Seat Cushion | 35 in | 890 mm |
| J Steering Wheel to Seat Cushion (Clearance) | 7 in | 178 mm |
| K Seat Squab to Clutch Pedal (Max) | 41½ in | 1055 mm |
| (Min) | 35½ in | 902 mm |
| L Height to Top of Screen | 3 ft 8½ in | 1125 mm |
| Turning circle between kerbs | 24 ft 0 in | 7.3 m |
| Gross vehicle weight, max. | 20½ cwt | 1036 kg. |

**Capacities**

| | | |
|---|---|---|
| Petrol tank | 7¼ galls | 33·0 litres |
| Engine sump | 7 pints | 4·0 litres |
| Gearbox | 1½ pints | 0·9 litres |
| Rear axle | 1 pint | 0·6 litres |
| Cooling system (with heater) | 8 pints | 4·5 litres |
| Heater | 1 pint | 0·57 litres |

**Luggage Compartment**

| | | |
|---|---|---|
| Capacity | 7 cu. ft. | 0·2 cu. m |

## Performance

| Acceleration | Speed range | Time |
|---|---|---|
| Top gear | 30–50 mph | 9·7 secs |
| Top gear | 40–60 mph | 9·4 secs |
| Through gears | 0–50 mph | 8·0 secs |
| Through gears | 0–60 mph | 11·3 secs |

**Maximum Speed**

100 mph, depending on conditions.

*N.B. All performance figures from factory tests.*

**Fig. 31.7** *Details about a Triumph Spitfire 1500 given in maker's leaflet*

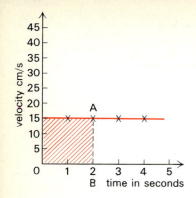

**Fig. 31.8**  *Velocity-time graph*

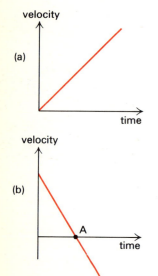

(a)

velocity

time

(b)

velocity

A

time

**Fig. 31.9**

## 31.5  Velocity-time graphs

In the same way as distance can be plotted against time (page 385) so we can plot velocity against time for a moving body. Let us start with a body travelling with a uniform velocity of 15cm/s. The result that we obtain is illustrated in Fig. 31.8. Such a straight line graph which is parallel with the time axis indicates that the body has no acceleration, that is, it is travelling with constant velocity. This graph can also reveal another fact and that is the distance covered by the moving body. We know that velocity = $\dfrac{\text{distance}}{\text{time}}$.

Therefore distance = velocity × time

After 2 seconds the distance covered = 15 × 2 = 30cm
On the graph the velocity can be measured by the line BA, and the time by the line OB

BA = 15 cm/s    OB = 2s

Thus BA × OB = 15 (cm/s) × 2 (s) = 30cm

In calculating the area between the graph line and the time axis we have obtained the distance covered by the moving body. This method can be used no matter what shape the graph takes.

The area under a velocity/time graph is the distance travelled.

*QUESTION 4:* Fig. 31.9 shows two more velocity-time graphs. Can you describe the velocity and acceleration of the body in each case?

Fig. 31.9 (b) is the velocity-time graph of a ball thrown into the air. After reaching its maximum height the ball falls towards the earth and accelerates as it falls. Acceleration has direction as well as magnitude and is therefore a vector quantity (see page 147).

388

## 31.6  Acceleration due to gravity

The acceleration of a ball as it falls towards the earth is caused by the force of attraction of the earth, that is, gravity. In fact we learnt on page 77 that the pull of the earth on a mass of 1kg produces an acceleration of 9.8m/s². In a vacuum different masses fall at the same rate under the action of this force. It is only air resistance under normal circumstances that makes them fall at different speeds.

Fig. 31.10 shows a parachutist falling freely through the air. When he first jumped from the plane his body accelerated under the force of gravity. However, as the velocity of his fall increased, so too did the air resistance. In the end the upward force of air resistance balances the force of gravity downwards on the body. There is no resultant force acting on the body, but it continues to fall with a uniform velocity called the *terminal velocity*.

**Fig. 31.10**  *John Noakes, presenter of* BBC *TV's* Blue Peter, *in free fall over Salisbury Plain. He is the first civilian on record in this country to jump from 8km, about the same height as Everest* (BBC Copyright)

This state of free fall shows us a very important fact. A body moves with uniform velocity when no force acts on it (a special case of uniform velocity is when the velocity is zero). Uniform velocity is the result when a number of opposing forces cancel each other out and the resultant force on the body is zero. Sir Isaac Newton (1642–1727) who studied motion a great deal, recognized this fact approximately 280 years ago and referred to it in his first law of motion. In fact we have already discussed this in Chapter 7 when we dealt with inertia. The law states:

If a body is at rest it will remain at rest, and if it is in motion it will continue to move in a straight line with uniform velocity unless it is acted upon by a force.

*QUESTION 5:* A spacecraft travels in outer space with uniform velocity in a straight line. Why is this and will this motion continue unchanged?

### 31.7 Force, mass and acceleration

From our work so far we know that if we apply a force to a body it changes the speed, that is, the body accelerates. Is there any definite relationship between the force that is applied and the acceleration it produces?

Look at Fig. 31.11. Three boys have made a go-kart out of a box and four wheels. They are on a flat driveway so one way that the boys can have a ride is if the go-kart is pulled by means of a cord attached to the front. Pulling is a force (see page 73). In (a) boy A is pulling his friend B and in (b) boy A and boy C are pulling their friend B. In which case do you think the go-kart will increase speed faster, that is, have the greatest acceleration? We all know that two boys pulling can acclerate the go-kart more than when only one boy is pulling. Such a result can be more formally stated as follows:

the greater the force which is applied to an object the greater the acceleration produced (provided the mass remains constant).

It is easy to believe from everyday experience that a force is required to produce a constant speed and that the speed is proportional to the force. But this is not true. A body does not need a resultant force to maintain a constant speed. To produce an acceleration does need a resultant force.

What happens if the force remains constant and the mass changes? Let us look at Fig. 31.11 and the go-kart again. In (a) boy A is pulling B and in (c) the same boy A is pulling both B and C. Do you think that the go-kart will accelerate better in (c) than (a)? Clearly it will

**Fig. 31.11** *In which direction will the acceleration of the go-kart be greatest?*

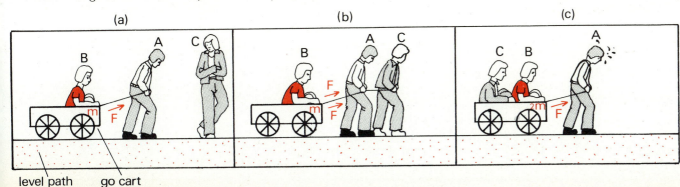

(a)      (b)      (c)

level path      go cart

accelerate faster when only one boy is in the cart. We can state

the greater the mass of the object to which the constant force is applied, the smaller the acceleration which is produced.

Experiments performed in a laboratory with a small trolley (a similar piece of apparatus to a go-kart), or careful measurement using a go-kart (see page 391, Question E) can reveal the exact relationship between force, mass and acceleration which we have so far discovered in general terms. It can be stated as follows:

The acceleration of an object is directly proportional to the force acting upon it, and inversely proportional to the mass of the object.

Summarized in an equation this is

$$F = m \times a$$

where $F$ is the applied force measured in newtons
    $m$ is the mass of the moving body measured in kilograms
    $a$ is the acceleration measured in metres per second per second.

It is this statement which allowed the unit of force, the newton, to be defined on page 77.

A newton is the force that changes the speed of a 1 kilogram mass by 1 metre per second per second.

QUESTION 6: If all three boys wanted to ride together can you suggest what they might do? State what force is being applied to produce the acceleration.

## 31.8 Newton's laws of motion
Throughout this book we have frequently talked about force and motion. On page 74, for example, we discussed how a passenger in a car tends to go on moving forward when the brakes are suddenly applied. This is an example of Newton's first law of motion which states:

If a body is at rest it will remain at rest, and if it is in motion it will continue to move in a straight line with a constant velocity unless it is acted on by a resultant force.

In this chapter we have considered Newton's second law of motion which states:

The acceleration of an object is directly proportional to the force acting on it, and inversely proportional to the mass of the object.

Newton's third law of motion is basically concerned with the fact that forces act in pairs. Every time one body exerts a force on a second body, the second body exerts a force on the first body; these two forces are always equal in magnitude but opposite in direction. One example of this is a boy applying a force to a gate (Fig. 25.20, page 283). Another example is a stone falling towards the earth. The earth attracts the stone and the stone attracts the earth with an equal and opposite force. Newton summarized these effects in his third law which states:

If a force acts on a body then an equal and opposite force must act upon some other body.

A. Make a large chart like in Fig. 31.2 and fill in as many speed records that you can find.

B. Find out about the scale that is used to measure wind speed. What equipment is used to measure wind speed?

C. Take a journey in a lift and make careful note of all the different sensations that you feel when the lift is accelerating, moving at uniform velocity and slowing down. Do this for both up and down journeys. Discuss what you find out with your teacher.

D. Find out how the forces of "g" affect an astronaut on lift-off. How can weightlessness be simulated (produced artificially) during the training of astronauts?

E. Design an experiment to find out about the relationship between force, mass and acceleration. This can be done in a laboratory using, for example, trucks on a railway line, or on a larger scale using a go-kart or bicycle on a tarmac surface. A constant force can be applied by using a rubber band or stretched spring kept at a constant extension. The following is a suggestion for an experiment on a tarmac surface with a bicycle or a go-kart.

(a) Obtain a strong spring which can be attached to the object (bicycle or go-kart) so that when it is stretched a fixed amount it will accelerate the object. The force meter (Fig. 7.11, page 78) would be suitable for this. Attach the spring to the object and practice walking along keeping the extension of the spring constant. You will have to start running after a while. Why? Go on practicing until you can keep the accelerating force constant.

(b) Decide how you can measure the speed of the object. One way would be to get the rider to make a mark on the ground every second. The distance between these marks is the velocity. If you use this method get someone to call out every second so that the rider knows when to make a mark. The acceleration can be calculated by dividing the change in velocity in a given time by the time. You could check that the constant force does produce a constant acceleration by plotting a graph of speed against time. What sort of graph will you get if the acceleration is constant? How can you find the acceleration from the graph? If you use a bicycle which has a speedometer on it all the rider need do is call out the speedometer reading every second. The change of speed divided by the time can then be calculated.

(c) You now need to repeat the experiment using two identical springs stretched by the same amount. We will call this force force 2, i.e. we have doubled the force. Measure the acceleration for force 2. Repeat for force 3 and so on. Check to see if your experiments show that acceleration is proportional to force.

(d) Lastly investigate how the acceleration depends on the mass for a constant force. To do this you will need to add more boys, bricks or tanks of water to the object being accelerated in order to double and treble its mass. If you double the mass what happens to the acceleration?

# THINGS TO WORK OUT

(Where necessary take the acceleration due to gravity as 10m/s².)

1. A speedometer on a go-kart reads 2m/s. (a) If the go-kart is travelling in a straight line what is its velocity? (b) The go-kart goes round a corner with the speedometer still reading 2m/s. Has its velocity changed? What is its speed? (c) How far will it travel in 10s? (d) If it comes to a hill and its speed increases to 4m/s during a time interval of 10s, what is its acceleration?

2. A runner travels 1500m in 4 minutes. What speed did he run (a) in metres/minute, (b) in kilometres/hour?

3. A train is travelling at 100m/s and 5 seconds later it is travelling at 120m/s. (a) What is the increase in speed in 5 seconds? (b) What is the acceleration of the train?

4. A car is travelling at 36km/h. (a) What is its velocity in m/s? (b) How many metres does it travel in 10 seconds?

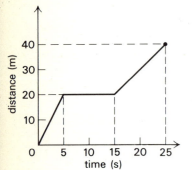

Fig. 31.12

5. The graph (Fig. 31.12) represents the motion of an object moving in a straight line. The vertical axis is the distance travelled in metres and the horizontal axis is the time that has passed. (a) Describe the motion of the object during the 25 seconds. (b) How far did the object travel in the first 5 seconds? How far did it travel in the next 10 seconds? (c) What was the speed of the object during the first 5 seconds? What was it during the next 10 seconds? What is it during the last 10 seconds of its motion? (d) What is the acceleration of the object during the last 10 seconds?

6. Draw a velocity-time graph for a trolley running down a smooth slope, and then coming to rest after running along the flat surface at the foot of the slope.

7. Sketch the graphs showing the following motions: (a) A barge being pulled by a tug at constant velocity. The tug then stops pulling, and the barge slowly comes to rest. (b) A car which accelerates with a constant acceleration for 10s, travels at a constant velocity for the next 5s and then decelerates with a constant deceleration for 5s before coming to rest. (c) The driver of a car starts from rest and goes through all the gears accelerating as fast as he can. He then puts the gear into neutral and travels along a flat road until he comes to rest.

8. You are sitting in a car holding a pendulum (a lump of metal on the end of a string). What would you notice about the motion of the pendulum if (a) the driver drives with his foot on the accelerator so that the needle of the speedometer keeps to the same figure, (b) the driver puts his foot down hard so that the car accelerates, (c) the car turns a bend in the road but at constant speed, (d) the car stops? Explain the observations.

9. This question uses the formula $F = ma$ given on page 390. (a) What force is needed to accelerate a mass of 2kg with an acceleration of 3m/s². (b) What will be the acceleration of a mass of 10kg when acted upon by a force of 20N?

10. A body has a mass of 4kg. If a force of 20N is applied to it (a) what will its acceleration be, (b) how far will it move from its position of rest in 10 seconds and (c) what will its velocity be after 10 seconds? (d) Draw (i) a distance-time graph, (ii) a velocity time graph, illustrating the body's motion.

11. Two men manage to push a car of mass 1000kg at a constant speed along a level road. Three men pushing can accelerate it at 0.2m/s². With what force does each man push?

12. A motor cyclist accelerates at 0.1m/s². (a) If he starts from rest, what is his velocity after (i) 1s (ii) 2s (iii) 3s (iv) 5s? (b) If he is initially moving

at a steady velocity 2m/s before he begins to accelerate what is his velocity after (i) 1s (ii) 2s (iii) 5s? (c) If his velocity after a time $t$ is $v$ and his velocity at time $t = 0$ is $u$, what should be written in the empty box below?

$$v = u + \square\, t$$

13. An object is moving with an initial velocity of $u$ and its acceleration is $a$. Its velocity $v$ after a time $t$ is given by

$$v = u + at$$

Use this equation to find the velocity of a stone when (a) it is dropped over the side of a cliff and has been travelling for 5s, (b) it is thrown vertically upwards with a velocity of 20m/s and has been travelling for 1.5s.

14. Fig. 31.13 shows a car crashing into a wall. Write a brief description of the crash using the words, *safety belt, Newton's first law of motion* and *inertia*.

**Fig. 31.13** (*Reproduced by Courtesy of Director, Transport & Road Research Laboratory, Department of the Environment*)

15. A trolley is on a horizontal table and is being pulled by a constant horizontal force of 5N. The trolley starts from rest and after 1s it has travelled 2m and is travelling at 5m/s. What is (a) the average velocity of the trolley during the first second? (b) the acceleration of the trolley during the first second? (c) the acceleration of the trolley during the following second? (d) the work done by the force when the trolley moves 2m? (e) the energy of the trolley when it has moved 2m? (f) what assumption have you made in answering part (e)?

16. The following table gives the speed of a car starting from rest. A stop watch was started as the car began to accelerate and the reading on the speedometer was recorded for a number of time intervals.
Plot a graph of speed against time.
(a) How long does the car take to reach a speed of 70 km/h? (b) What is the speed of the car after 15s? (c) When is the acceleration zero? (d) When is the acceleration greatest?

| time (s) | 0 | 2.0 | 5.1 | 10.2 | 20.0 | 30.2 | 40.3 | 50.0 | 60.0 | 70.0 | 80.0 |
|---|---|---|---|---|---|---|---|---|---|---|---|
| speed (km/h) | 0 | 30 | 55 | 86 | 116 | 130 | 140 | 145 | 150 | 150 | 150 |

17. A stone is fired vertically upwards using a catapult. It leaves the catapult with a velocity of 50m/s and takes 5 seconds to reach its maximum height.
(a) When the elastic is pulled back what is the name given to the stored energy?
(b) Where did this energy come from?
(c) When the stone reaches its maximum height what is its velocity?
(d) What sort of energy has it got when it reaches its maximum height?
(e) When the stone hits the ground what has happened to this energy?

(f) Gravity causes the stone to decelerate at 10m/s². What will be the stone's velocity 3 seconds after it is fired?

(g) What is the value of the stone's acceleration as it returns to the earth?

(h) How long will it be after the stone is fired before it returns to earth?

(i) Sketch a graph of acceleration against time.

(j) Sketch a graph of velocity against time.

(k) What is the gradient (slope) of your graph of velocity against time?

(l) Sketch graphs of acceleration against time and velocity against time if the stone had been projected upwards with the same velocity but in a situation where the gravitational field was 5N/kg.

18. A falling object accelerates at 10m/s². (a) If it starts from rest, what is its velocity after 4s? (b) What is its average velocity during the 4s? (c) How far does it travel in 4s?

Suppose a body starting from rest accelerates at $a$ m/s² and that it travels a distance $s$ in $t$ seconds. (d) What is its velocity after $t$ seconds? (e) What is its average velocity during the $t$ seconds? (f) How far does it travel in $t$ seconds? (g) Use the equation $s = \frac{1}{2}at^2$ to find how far a body which has an acceleration of 5m/s² travels in 2 seconds.

19. The table below shows the relation between the distance moved by a body starting from rest and the time. It also shows how the speed varies with the time.

| $t$ (s) | Distance (cm) | Speed (cm/s) |
|---|---|---|
| 0.2 | 0.8 | 8.0 |
| 0.4 | 3.2 | 16.0 |
| 0.8 | 12.8 | 32.0 |
| 1.0 | 20.0 | 40.0 |
| 1.4 | 39.2 | 56.0 |

Plot graphs of (a) speed against distance, (b) speed against time.

(i) Is the speed proportional to the distance travelled? Give a reason for your answer. (ii) Is the speed proportional to the time? Give a reason for your answer. (iii) Is the acceleration of the body constant? Give a reason for your answer.

20. The table below shows the time taken by a body to reach a speed of 20m/s for a number of different accelerations. (a) Redraw the table and complete it. (b) Draw a graph of acceleration against time. (c) Draw a graph of acceleration against

$$\frac{1}{time}$$

(d) State the relationship between acceleration and time.

| Acceleration | 1m/s² | 2m/s² | 3m/s² | 4m/s² |
|---|---|---|---|---|
| Time to reach 20m/s | 20s | 10s | | |

21. A bicycle was accelerated by springs attached to it. Each of the springs was identical and extended by the same amount. The time for the bicycle to reach a speed of 15km/hr was measured using first one, then two and finally three springs attached to the bicycle. The results are shown in the table below.

| Number of springs used | 1 | 2 | 3 |
|---|---|---|---|
| Time to reach 15 km/hr (s) | 40 | 19 | 14 |

(a) Draw a graph of time against number of springs.

(b) Draw a graph of $\dfrac{1}{\text{time}}$ against number of springs. (c) State the relationship between time and number of springs. (d) How can you use your graphs to deduce the relationship between force and acceleration. (Hint. Question 20 will help you answer this).

22. Fig. 31.14 shows a graph of speed plotted against time for a model car of mass 5kg. (a) Describe the motion of the car (i) during the first 10s, (ii) during the next 10s, (iii) during the last 10s of its motion. (b) How far did it go (i) between 10s and 30s, (ii) in the first 10s? (c) What is the acceleration during (i) the first 10s, (ii) between 10s and 30s, (iii) during the last 10s of its motion? (d) Use your answers to (c) to sketch a graph of the force acting on the car against the time.

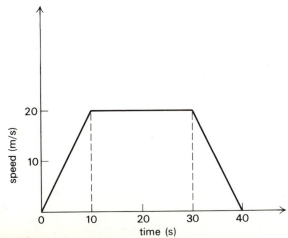

**Fig. 31.14**

**Fig. 31.15**   *The launch of an Apollo spacecraft*

23. Fig. 31.15 is a photograph of the launch of an Apollo rocket. The rocket motors were turned off after about 3 minutes and the rocket "coasted" in

space. Some time later another "burn" was necessary to produce a course correction. This burn lasted for 3 seconds and produced a force of 88 000N. The mass of the rocket was 44 000kg. (a) Explain the meaning of the words "coasted" and "burn". (b) The initial thrust at take-off is many millions of newtons. Why is such a large thrust necessary? (c) What is the acceleration of the spacecraft during the 3 second burn? (d) What was the force acting on each kilogramme of spacecraft during the 3 second burn? (e) What change in velocity occurred during the 3 second burn?

24.  (a) Why do rockets used to put a satellite into orbit round the earth need to be many times larger than the satellite itself?

(b) The energy needed by a rocket to *escape from the Earth's gravitational field* is 62.7 J/kg. Calculation using this figure shows that *the escape velocity from the Earth* is $11.2 \times 10^3$m/s. Explain the meaning of words in italics.

(c) A spacecraft needs an energy of 62.7 J/kg to escape from the earth. When 1kg of petrol is burnt it produces about 50 J/kg. Why is petrol not a good fuel to use in rockets?

(d) The rocket shown in Fig. 31.16 is launched in the direction shown. Draw the path of its motion.

(e) Why doesn't it travel in a straight line?

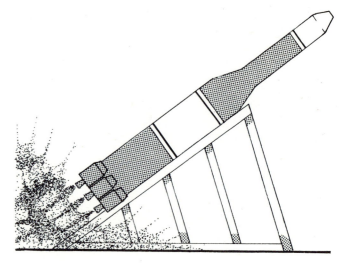

Fig. 31.16

# ENERGY

## 32.1  The energy crisis

In the early 1970's the word energy came far more into general use than ever before. The public were encouraged by all forms of media to learn about energy, its nature, its use and above all its conservation. Notices about the "Save-It" campaign (see Fig. 32.1) were to be found regularly in newspapers and on television. Why was all of this necessary? We were experiencing an "energy crisis". The reason for this was an economic one.

**WHAT ENERGY CRISIS?**

There's plenty of petrol in the pumps. Lots of gas in the pipes. The home fires are burning, and the lamps aren't going out all over Europe.

Funny sort of crisis. Which is probably why we're not doing enough about it.

But although we can't see it or feel it, the energy crisis is costing us a bomb.

In eighteen months, the price of crude oil (which provides almost half the energy we use) has multiplied by five. And all our oil still has to be imported.

The bill we pay is £3,500,000,000 a year Ten million pounds a day. A sum so big, it can't possibly be your problem.

It is, though. That £10 million works out at 20p a day for every man, woman and child in Britain. For a family of four, it's a millstone of £5·60 a week.

You can't shrug it off as a problem for the country to solve. Because the country is nothing more than every man, woman and child in Britain.

Of course, in a few years, North Sea oil will help us pay our way. But we'll still have debts to repay. And North Sea oil won't last forever.

We've simply got to Save It. Not just oil and petrol. But electricity, too, because oil generates a quarter of it. And the less coal and gas we use, the more they're available to take the place of oil.

What's more, we can save it without a lot of fuss and bother. Just with reasonable care.

Turn down a thermostat. Insulate a pipe. Clean out a furnace. Keep your car in tune.

You'll save a few pounds for yourself, and millions for Britain.

**Department of Energy**

**Fig. 32.1**  *One of the press advertisements used in the Department of Energy's "Save It" campaign.* (Crown copyright, reproduced by permission of the Controller, H.M. Stationery Office)

The oil-producing countries realized that their major asset was being used up too quickly and put up the price. This produced shortages all over the world, particularly in the U.K. where most of our energy is obtained from fossil fuels such as oil, coal and natural gas. This rise in price is likely to continue as demand increases. At present the world's consumption of energy is increasing at approximately 5% every year.

However, the world's demand for energy can only be met for as long as the basic resources of energy are available within the earth. As demand increases so does the removal of the resources from the earth. These resources are not infinite, soon they will run out. By the year 2000 AD it is estimated that about 70% of the earth's oil supplies will have been used. Other fossil fuels (such as coal) also have a limited capacity. This is the true energy crisis of the future, for if the present rate of consumption continues the supply of oil and coal will soon run out. It is this crisis that people must become aware of now and we must start to lay plans to (a) cut demand, by using energy more wisely, conserving it and wasting less and (b) find alternative sources of energy.

## 32.2 Reducing our consumption of energy

When we total up our individual energy consumption we must consider all the uses of energy. For example, is our demand for a new car a demand for energy in the same way as switching on our electric light in the home? Both are a demand on energy. Everything that is manufactured is the result of the use of energy within machines. In fact to manufacture a large car one would use up enough energy to run a 1kW electric fire for about 40 000 hours. Thus if we are to cut down on our energy consumption we must consider everything about the way we live.

According to figures given in 1972, each person uses an average of 500MJ of energy per day, and about one third of this energy is used in the home, for such things as heating and cooking. To conserve energy within our homes we are mostly concerned with preventing the loss of heat energy, and in converting our energy supply to the form of heat only when and where we need it.

*QUESTION:* Fig. 32.2 shows a house in cross section. Could you say what action you would take in order to "Save-It" in the places marked A, B, C, D, E and F.

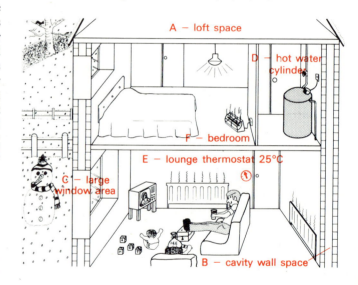

**Fig. 32.2**

Outside the home other savings of energy may have to be made. In transport for instance, think of the number of cars on the roads with only one person in them at any one time. Also a lot of fuel is wasted with cars standing in traffic jams. Could these problems be solved by a cheaper and more flexible public transport system? Could we as individuals adopt different methods of travel within the cities? Figure 32.3 shows a well-known method of transport which may well be used more in the future. Experiments are already being tried out with new methods of transport and two of these are illustrated in Figs. 32.4 and 32.5.

**Fig. 32.3**  *If we don't want to walk we could try two wheels instead of four! A bicycle saves fuel, it is easy to park, it does not pollute the atmosphere and it gives the body exercise*

**Fig. 32.4**  *A 2-seater electric car of the "Witkar" system in Amsterdam. At present there are 35 vehicles and 5 stations where the vehicles are recharged. A subscriber obtains the car by inserting a credit card in a slot at the station. After use the vehicle must be returned to one of the stations*

**Fig. 32.5**  *A 1016kg 4-passenger vehicle suspended by the magnetic forces between an electromagnet and a rail which is made of magnetic material. What do you think are the advantages of such a system? It is estimated that in an urban system these vehicles would be automatic and could run at intervals of 15 seconds or less*

In industry we shall have to consider the efficiency of our manufacturing processes. For example, can the energy loss in the form of heat be reduced? Could the heat released and usually wasted be used elsewhere within the process or within another appliance? Could the waste materials of the process be recycled or used to provide energy? Such solutions are already being tried out (see Figs. 32.6 and 32.7).

Thus the general rule must be to conserve energy and not to throw it away in the form of heat as we so often do.

**Fig. 32.6** *What is different about the "Liverpool Daily Post and Echo" building on Merseyside? This building literally warms itself by making use of the heat given off by machinery, artificial lighting and the bodies of the people in the building. A system of thermal recovery is used which collects and distributes the heat round the buildings. In fact, too much heat is available and the warm air is cooled in "heat exchangers". An air-conditioned environment of 21°C is maintained throughout the year*

## 32.3  Alternative sources of energy

a. *Nuclear fission* (see also pages 375–378)
Britain's first supply of electricity from nuclear power came from Calder Hall in the mid 1950's. More energy is

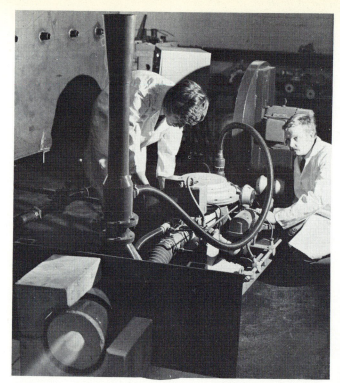

**Fig. 32.7** *A new type of gas burner which recovers much of the heat normally wasted. Such a unit has been used to replace eight burners in a pottery kiln at Stoke-on-Trent, and has cut fuel consumption by 30%*

now supplied by other nuclear power stations, but this is not as much as was first hoped. Generally the reason for this has been the problem of safety, not only in the construction of the reactors but in the danger from the accumulating waste products which are radioactive and must be carefully stored. Also as the rich Uranium sources are used up, the cost of energy supplied in this way may also increase.

## b. *Nuclear fusion*

In one type of fusion reaction the nuclei of deuterium (heavy hydrogen) and tritium are brought together and fused into a single nucleus. At this fusion there is a tremendous release of energy. There are no problems of radiation from the fuel, as the fuel is atoms in sea water and the waste product helium is also not radioactive.

The main problem of producing energy by fusion is that in order to overcome the natural repulsive forces of the nuclei temperatures of 100 million degrees Celsius must be reached. Containing the fuel at this temperature is a problem which scientists have not yet been able to solve and unless there is a sudden breakthrough the possibility of supplying our energy needs by fusion is unlikely to be solved in this century.

## c. *Solar energy*

One page 209 there is a photograph of a solar furnace. In order to provide power on a national scale vast installations of solar panels or solar cells would be necessary. If we could collect all the heat from the sun over 150 miles square of desert in the U.S.A. then there would be enough energy to provide for the whole of that countries needs. For many years solar energy has been used to operate cookers, distillation devices and domestic heating systems in various parts of the world and there are over 400 000 solar energy collectors in use in Japan.

One of the simplest forms of solar collector was built over 12 years ago into the St Georges School, Wallasey, Cheshire. The school was built using heavy thick materials and is well insulated to prevent heat losses. The south side of the school is like an enormous double glazed window. The "greenhouse effect" (pages 221 and 402) is sufficient to maintain a satisfactory temperature within the school even during the coldest winters.

Smaller solar collectors are available for domestic use and can easily be constructed by the individual. The construction of such a panel is shown in Fig. 32.8 and the

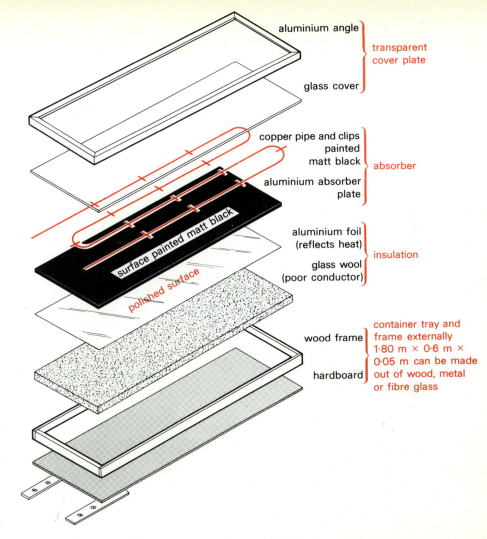

**Fig. 32.8** *Main components of a solar collector panel*

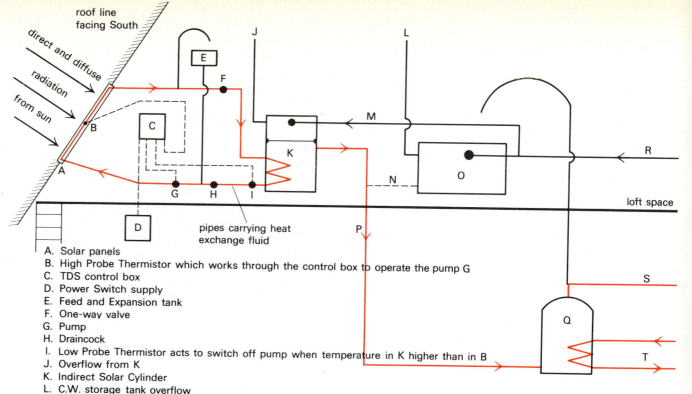

roof line facing South

direct and diffuse

radiation

from sun

pipes carrying heat exchange fluid

loft space

short-wave

solar radiation

direct or diffuse

78% passes through

glass to absorber plate

tray

insulation

black absorber plate

pipes containing fluid in contact with absorber plate

→ long-wave radiation

glass cover usually double glazed to prevent heat loss

**Fig. 32.9** *Short-wave radiation is absorbed by a black plate which then radiates long-wave radiation. This cannot pass back through the glass. Thus the temperature within the solar panel rises ("greenhouse effect"). The fluid within the pipes is heated, and circulates*

A. Solar panels
B. High Probe Thermistor which works through the control box to operate the pump G
C. TDS control box
D. Power Switch supply
E. Feed and Expansion tank
F. One-way valve
G. Pump
H. Draincock
I. Low Probe Thermistor acts to switch off pump when temperature in K higher than in B
J. Overflow from K
K. Indirect Solar Cylinder
L. C.W. storage tank overflow
M. Mains cold feed to Indirect Solar Cylinder
N. Previous cold water feed - disconnected
O. Cold water storage tank
P. Pre-heated water feed to domestic cylinder. Saves on time and fuel
Q. Domestic hot water cylinder
R. Mains feed
S. Hot water outlet
T. Boiler feed and return

**Fig. 32.10** *Diagram of a solar system for household use. Such a system of panels covering 4m² is capable of heating about 130 litres of water to about 60°C on a good summer's day*

principle on which it works is described in Fig. 32.9. A series of such panels is most efficient when used to pre-heat water in the storage tank. Fig. 32.10 shows such a system and Fig. 32.11 is a photograph of a house with solar panels installed in the roof.

**Fig. 32.11**   *The four panels of the system illustrated in Fig. 32.10*

### d. *Geothermal energy*
Geothermal energy is that obtained from heat energy trapped in the rocks beneath the earth's surface. The geothermal power station at Larderello in Italy already provides much of that country's electricity. More research has to be done, but it looks as if this source of energy will also be more suited to local use than on a national scale.

### e. *Energy from the sea*
There is a tidal station at Rance in France (see Fig. 32.12). In this country research has already begun on harnessing the energy of the tides in the Severn estuary.

Research has also begun on harnessing the energy of waves. It is estimated that the total energy needs of the U.K. could be supplied by about 100 "wave generators" situated around the coastlines of Great Britain.

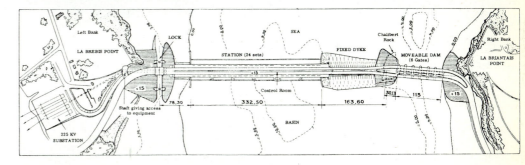

**Fig. 32.12**   *General view and plan of the Rance tidal power station in France, completed in 1967. In 1972 the output of the station was 560 million kilowatt-hours. By opening the gates as the tide rises and closing them at high tide, a pool of area 22km is formed behind the Rance river dam. As the tide lowers the trapped water is allowed to flow out driving 24 electricity-generating turbines*

One of the devices being tested at present is made up of a long line of vanes called "Salter's ducks" (Fig. 32.12b). They are laid parallel to the waves. The waves make the "ducks" tilt up and down on their central axis. Thus the energy of the waves is converted into mechanical energy, and it is this energy which will be converted into other forms, such as electrical energy.

Another device consists of a line of rafts all hinged together (Cockerell's contouring rafts). These are placed at right angles to the waves. Each raft moves about its hinge developing mechanical energy.

Other research is looking into the possibility of driving turbines by using the pressure changes in an air column. These pressure changes would be created by waves swelling in and out of the open bottom of the column.

Some very successful research is being carried out to use the temperature difference between the surface of the sea and the deeper water to produce the energy we need. Power stations using energy from the sea could become a reality by the late 1980s.

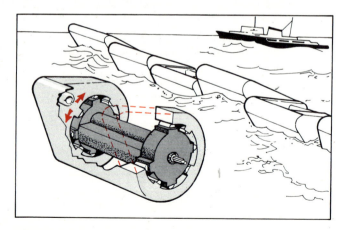

**Fig. 32.12b** *Salter's ducks convert wave energy into mechanical energy*

## f. Wind energy

This source of energy is most suited to supply energy on a small scale to a district or to an individual house. The wind energy is frequently converted into electrical energy. However, wind machines can be used for direct mechanical energy as well as to turn generators. They are certainly coming into use in Britain, where there is a suitable location (Fig. 32.13).

**Fig. 32.13** *A wind generator which has recently been erected at Larkholme County Secondary School, Fleetwood, Lancashire. Research into weather records has shown that this is likely to be a suitable site. An attempt will be made to provide a continuous supply of electricity of limited power which will be put to a number of uses, including some lighting, greenhouse heating and the running of refrigerators.*

## 32.4 What of the future?

No simple solution to our present energy crisis exists. Portable fuels, such as oil and coal, are rapidly running out and no alternatives are readily available, but a car run off a cylinder of hydrogen gas has already been built in Japan. However, on a human time scale, the supply of energy from the sun, from nuclear fusion, from solar energy and from tidal and wave energy appears to be unlimited, but as yet has not been harnessed except on a very small scale. The energy obtainable from these other energy sources is not easily transportable. Even if energy from these sources becomes readily available, how could we fuel our cars and aeroplanes? Much research is going on and the way in which the industrial countries develop in the future may involve considerable changes in our way of life. Certainly we should all be aware of the problem and be prepared to alter our lives as necessary. Reference to Fig. 32.14 shows that the western world is using far more than its fair share of energy. What are the problems concerned with trying to produce a fairer distribution of energy resources? The average person in the western world consumes about 500 MJ per day. What changes would be necessary to our way of life if we had to manage on 100 MJ per day?

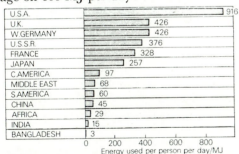

**Fig. 32.14** *Average energy used by each person daily in various countries in 1972. The figures do not include energy from food, or from the burning of wood and animal dung. The global average was 157 MJ per person per day.*

A. Collect newspaper articles, old or new, which discuss the "energy crisis". With your teacher's permission put these on display in your classroom.

B. At home list all the ways by which you could reduce the family's energy consumption. Using your fuel meters (electricity and gas) as a guide see if you can mount a "Save It" campaign within your own house. If you are successful, your parents will be pleased with their fuel bills and they might be generous with the cash they save!

C. Find out about fuel cells. In particular, what kind of fuel do they use and how much pollution results from their use? Do you think they might be used instead of petrol engines in the future? What fuels, other than petrol, are likely to be used to propel motor cars in the next century? What are their advantages and disadvantages compared with petrol?

D. Very large lorries are called "juggernauts". Find out the origin of the word. Will the use of juggernauts save fuel? To answer this question you could begin by finding out how much fuel they use per tonne of load transported and see how this compares with the fuel per tonne for smaller vehicles. Get figures which will help you decide what method of road transport uses the least fuel.

E. Build a small solar collector panel like the one illustrated in Fig. 32.9. Put a thermometer, with its bulb painted black in place of the copper pipe. Put your solar collector in the sunshine and plot a graph of the temperature reading on the thermometer against the time.

1. List some situations you see in everyday life in which energy is wasted. In which of them could the wasted energy be saved and perhaps used to reduce our consumption of energy?

2. Would you rather build your home near to a nuclear power station or down river from a hydro-electric power station built by a large dam? On what grounds did you make your choice? Was it concerned with your emotions? Has emotion been a factor in any of the statements you have heard made about the energy crisis?

3. An average person needs about 11 megajoules of food energy each day. This is about 3kWh. Examine the lighting in your home and estimate the quantity of electrical energy used in a day. Is this more or less than the food energy you need? How much would your food cost each day if you purchased it according to the energy content at 3p per kWh? Is food an expensive form of energy?

4. What do we mean when we talk of an energy crisis? Suggest some ways in which our way of living might change as a result of the crisis.

5. List the common sources of energy which are likely to run out within the next 100 years. In what ways could we be getting our supplies of energy a hundred years from now?

6. If you were staying in a home which was heated by electricity and were asked how the electricity bill could be reduced without those living in the house suffering any discomfort due to lack of heat, what suggestions would you make?

7. The following statements are concerned with the saving of energy. Which of them are true and which of them are false? (a) Electricity can be saved by keeping a deep freeze packed to capacity. (b) Keeping heavy items in the boot of your car will help the tyres to grip better and give you more miles per gallon. (c) More energy is saved by switching off thirty 60W lamps than by switching off a 3kW immersion heater.

8. Study the diagram of the solar collector illustrated in Fig. 32.8 and then answer the following questions. (a) Why are the pipes painted black? (b) Why is there a polished surface underneath the absorber plate? (c) Is the thickness of the glasswool important? Why?

9. Fig. 32.15 shows a sunbather. (a) What kind of energy is falling on her? (b) What is the source of this energy?

**Fig. 32.15**

(c) Does most of this energy reach her by convection, conduction or radiation? (d) If energy is continually falling on her body, why does not her temperature continually increase? (e) Heat energy is being lost from her body. By what process does this loss take place? (f) If she takes a dip in the evening she is likely to be agreeably surprised by the temperature of the water. Why? (g) After her dip she may well be unusually cold. Why? (h) Radiation from the sun falls on the earth at the rate of 0.14W/cm². Estimate how much energy falls on her body (a) each second (b) during 6 hours sunbathing. (You will need to make an intelligent guess at the area of her body.)

10. Scientists in America are investigating the possibilities of setting up highly efficient *solar cells* some 22 000 miles above the earth (Fig. 32.16). The cells would be assembled in space by astronauts flying to the spot in winged Space Shuttles. The electricity generated by the solar cells would be converted into *electromagnetic waves* and beamed back to the earth. At special receiving stations the energy from the electromagnetic waves would be converted into electrical energy and sent over the *grid system* to our factories and homes. (a) Explain the meaning of the words in italics. (b) What would be the best colour for the surface of the solar cells? (c) Would light be a suitable choice of electromagnetic wave to use? Give a reason for your answer. Would microwaves be a better choice? Why? (d) Why will scientists go to all the trouble and expense of setting up solar cells in space when the same cells could be set up on the earth?

11. If energy from the sun falls on the earth at a rate of 0.14W/cm², how much energy falls on a desert 100km × 100km in a day with 10 hours sunshine?

12. Make a list of energy sources which could be used to provide heat in our homes both now and in the future. Put the sources at the top of the list which cause pollution and at the bottom of the list have the ones that do not cause pollution.

13. Reykjavik (Iceland) is heated from hot water which occurs naturally underneath the ground. The temperature of the water at different levels varies from about 95°C to about 135°C. The houses in the city are heated by pumping the hot water which is beneath them through pipes which go through the houses. Sixteen bore holes supply about 8000g of water every minute at an average temperature of about 120°C. (a) What do you think are the main advantages of heating a city in this way rather than using coal or oil? (b) The specific heat capacity of water is 4200J/kg K. How much heat energy is supplied by the sixteen bore holes in 24 hours if the water enters the system at 120°C and leaves it at 30°C? (c) What use do you think the city makes of the waste water which leaves the system at about 30°C?

14. A parabolic reflector of area 1m² focuses the sun's rays onto a metal block placed at the focus of the mirror. The temperature of the block rises by 10K in 3 minutes. (a) If the mass of the metal block is

**Fig. 32.16**

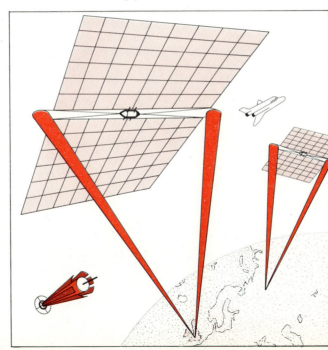

407

0.05kg and its specific heat capacity is 400J/kg K how much heat energy is received by the block in 3 minutes? (b) How much energy is received by the block in 10 hours? (c) If a solar furnace were built in a desert to collect the heat from the sun over an area measuring 200km by 200km, how much energy would be collected in 10 hours? (d) For how long would this energy keep 10 000 electric light bulbs alight, if each bulb is rated at 100W?

15. One manufacturer of double glazing claims that the heat loss through his double glazed window is 0.5J/ m² each second for each degree Celsius temperature difference between the inside and outside surfaces, while for single glazed windows it is 1.5J/ m² each second per degree Celsius difference in temperature. (a) How much energy is lost every second through the windows of a house which has windows of total area 20m² if the difference in temperature between the inside and outside is 10°C? (i) for single glazing (ii) for double glazing. (b) How much would you save on a four weeks' heating bill, by changing from single glazing to double glazing, if the cost of heating is 3p for 4MJ? Assume that there is, on average, a temperature difference of 10°C between the inside and outside temperatures throughout the month.

16. Fig. 32.17 shows someone skating on ice. (a) What sort of energy has she got when skating round the rink? (b) Where does this energy come from? (c) Is energy needed to make the ice? If so, what sort of energy is likely to be used? (d) Is energy needed to maintain the ice once it is made? Why? (e) Is skating easier when the area of the skate in contact with the ice is small or large? Why?

**Fig. 32.17**

# ANSWERS TO QUESTIONS

# Chapter 1

1. (a) The earth's natural source of light is the sun. (b) Here are six examples, but other answers are acceptable: candle, oil lamp, gas light, electric light-filament lamp, paraffin lamp, carbon arc lamp.

2. In order to produce a similar effect you would need a darkened room, dimly lit by light from one source. In front of the source you would need to position the opaque object which is to produce the shadow, for example your own hand. The nearer the object is to the source of light the larger the shadow will be. The shadow will be produced on the wall opposite to the source of light, or on any object in the path of the rays of light.

3. The elliptical orbit of the moon around the earth is not in the same plane as the orbit of the earth around the sun. Thus although the moon comes into the region between the sun and the earth every 28 days it is rarely in direct line with the two, and either passes below or above.

4. An eclipse of the moon occurs when the earth, moon and sun are directly in line, and the earth is between the sun and the moon.

5. Each pinhole will produce a separate image of the object. These could overlap and result in a greater general brightness.

6. The angle at which the waves are reflected by the barrier is the same as the angle at which the incoming waves strike the barrier.

# Chapter 2

1. (a) The light should be incident on the mirror at an angle of 45°. The reflected ray will then leave the mirror at an angle of 45°. Therefore the whole ray bends through 90°. (b) The second mirror should be positioned with its reflecting surface either (i) parallel to the reflecting surface of the first mirror, or (ii) at right angles to the reflecting surface of the first mirror. The light leaving the two-mirror system in (i) travels parallel to, and in the same direction as, the light entering; in (ii) it travels parallel to, but in the opposite direction from, the light entering. (c) Both systems of mirrors are versions of the same instrument, a periscope, which can be used to view objects that are above or below eye-level, or obscured from sight by some obstacle above eye-level.

2. The book surface is a rough reflecting surface. Thus the reflection is diffuse and the rays do not appear to come from a common point. Because of this no image is viewed in the surface. The laws of reflection, however, apply to each individual ray.

# Chapter 3

1. The distortion occurs because the ribs of glass alter the path of the rays of light passing through them.

2. (a) In travelling from air to glass the light is bent towards the normal. (b) In travelling from glass to air the light is bent away from the normal.

3. (a) Diamond reflects $\frac{1}{5}$ of incident light. This is 5 times the amount reflected from glass.

    $\therefore$ glass reflects $\frac{1}{5} \times \frac{1}{5}$ of incident light

    $\therefore$ % of light reflected from glass $= \frac{1}{25} \times 100$

    $\qquad\qquad\qquad\qquad\qquad = 4\%$

    (b) As diamond reflects a greater amount of light (20%) then it must have a higher refractive index.

4. The rays will pass through without undergoing refraction. This is because the rays are travelling along the normal to the surface.

5. The fish-eye lens gives a very much wider-angled view than the standard lens.

**Fig. A1** *Panoramic driving mirror and flat driving mirror: fields of view compared. The curvature of the former enables rays to enter the eye which would not do so if the mirror were flat. Hence the field of view is increased.*

6. Because of refraction of light the fish in the water receives all the light from a 180° semicircular field above the water with-in a cone of apex 98°.

Thus all objects on the horizon are squashed upwards and appear in the air at a strange angle. The view is distorted in the same way as through the fish-eye lens.

The fish sees other objects in the water both directly and as images formed by total internal reflection.

7. The curved surface of the semicircular block allows the light to enter the glass without refraction. This is because it is incident upon the surface at an angle of 90°. Thus the angle at which the incident ray hits the glass-to-air surface can be easily and accurately controlled.

8. A silvered surface tends to flake off, particularly under damp conditions. No such loss of the reflecting surface can occur when prisms are used as reflecting surfaces.

## Chapter 4

1. (a) The images are not identical. When objects are viewed in the surface that curves outwards, the image is always the right way up and smaller than the object. Most of the surroundings to the object are also visible. When the same object is viewed in the surface that curves inwards, the image may be upside down or the right way up. Not so much of the surroundings is visible.

(b) When the object gets nearer to the surface that curves outwards, the image remains the right way up and smaller than the object. The image size increases slightly. When the object gets very

near to the surface that curves inwards the image suddenly becomes the right way up and is enlarged (i.e. magnified).

2. (a) The boy is measuring the focal length of the concave mirror, because parallel rays from a distant object are brought to a point on the focal plane of the mirror. (b) A convex mirror diverges rays of light therefore one cannot form a real image of the distant object on the screen.

3. The centre of curvature is twice as far from the centre of the mirror as the focus is.

focal length × 2 = radius of curvature

4. On switching over to light source B, the motorist would cause the beam from the headlight to become "dipped". This is because the reflector associated with the source B is positioned at such an angle that it reflects all the light downwards. [This type of headlamp has the following advantages over other types: (a) the intensity of the light is greater, (b) there is never any heat damage, (c) there are no moving parts (compared to lamps where a shield covers part of the bulb).]

5. A convex mirror gives a much wider range of view of the area in front of it. This is because the shaping of its surface allows it to collect rays of light from areas at the side. The diminished image means that the field of view is larger.

6. (X) Shaving mirror. Dentist's mirror used inside the mouth. (Y) Car driving mirror. It provides the motorist with a wide-angled view to the rear of the car. See Fig. A1.

7. The word "eye" refers to the lens of the camera.

8. You would obtain the value of the focal length of the convex lens. The parallel rays of light from the distant object are brought to a point on the focal plane of the lens.

9. (a) The stronger lens is the one with the short focal length. (b) If the lens is stronger then it must be re-

412

fracting the light to a greater extent. In order to do this the lens surfaces must be sharply curved. Thus a strong convex lens will be a very "fat" lens and a strong concave lens will be a very "thin" lens (see "Things to do" section).

10. When a convex lens produces a virtual erect and enlarged image, then the lens is termed a magnifying glass.

## Chapter 5

1. Diamond has a high reflectivity, and it is cut to produce internal reflection. However, not only is the light reflected but it is dispersed into colours. These flashes of colour are called the fire of the diamond. The light coming from a good diamond will form a complete spectrum when it falls on a white screen. Cut-glass and chandeliers are attractive because they, too, disperse the light in the same way as a cut diamond.

2. There is no atmosphere to scatter the light in outer space.

3. The distances $x$ and $y$ must be equal to the focal lengths of each lens. The first lens produces parallel rays which enter the prism. The second lens converges the parallel colours of the same wave-length to a common focus. Thus there is no merging of the colours on the screen and the spectrum is said to be pure.

4. The second prism must be placed in an upside-down position by the side of the first prism in order to re-combine the colours.

5. Cyan allows blue and green light to pass through. Yellow allows red and green light to pass through. Thus the colour that emerges from the filters will be green.

6. (a) Red spots on a black ball, (b) yellow spots on a green ball, (c) greenish spots on a greenish ball.

7. The photographer must use the yellow filter. This will absorb and therefore reduce the quantity of light falling on the film from the blue sky. This portion of the film will therefore not be greatly exposed and will appear black in the final print. The filter will allow light through from the clouds, thus this portion of the film will be exposed, and will appear white in the final print.

8. The machine is a paint-mixing machine and is used to produce exact colour shades required by customers. It does this by using standard colours and mixing them in varying quantities.

9. The dark lines in the absorption spectrum correspond to the lines in the sodium line spectrum. The sodium vapour has absorbed these wave-lengths of light from the white light. They correspond to the wave-lengths of light emitted by sodium when it is energized in the form of a vapour.

10. Harmless electromagnetic waves: radio waves, light waves. Dangerous electromagnetic waves: cosmic rays, gamma rays.

## Chapter 6

1. The retina is the light-sensitive screen of the eye. The yellow spot is directly in line with the lens of the eye, and it consists of a mass of nerve endings which make it the most sensitive area of the retina. The blind spot has no nerve endings because at this spot the optic nerve enters the retina.

2. The images on the film would appear to move at a faster speed than normal, resulting in unnatural, jerky and rapid movement.

3. A convex lens is used to correct the eye defect of long sight, as indicated in the figure.

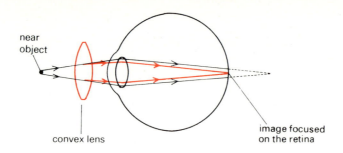

near
object

convex lens

image focused
on the retina

**Fig. A2** *Correction of long sight*

4. The ciliary muscles have been affected by the onset of old age. They are not supple enough to contract or relax as they should, therefore the power of accommodation is impaired.

5. (a) The slide should be placed on the platform upside down. (b) The paper should be moved across the right-hand face of the slide. This is because the lens inverts the image.

6. The image is inverted. This does not present a great problem to astronomers as they tend to record data in the form of photographs.

7. The concave mirror must be placed at a distance from the light source equal to the mirror's radius of curvature. In this way, light which would otherwise be wasted is directed forwards to illuminate the slide.

8. (a) The film must be positioned in the projector gate upside down and the wrong way round laterally. (b) The projector would have to be moved away from the screen. (c) The fluff would be located in the bottom right-hand corner of the gate of the projector.

## Chapter 7

1. (b) A stone being whirled around on the end of a piece of string; a tennis ball being returned over the net. (c) A piece of plasticine or rubber being squeezed into a shape; a rubber balloon increasing in size as air is forced into it.

2. (a) No. The force of friction would then be opposing any attempt to set the buckets in motion. (b) The force of gravity acts vertically downwards and is not opposing the horizontal force that sets the buckets in motion. (c) Yes. The mass or inertia of a body is the same whatever the force of gravity. You cannot set a space rocket in motion simply by giving it a push, even if it is in outer space!

3. The extension of the spring is proportional to the force. A straight-line graph through the origin always tell us that the two quantities plotted are proportional to each other.

4. Yes, it would be possible. The scale on a spring balance is linear (that is, the scale markings are evenly spaced). This means that if the force is doubled the extension is doubled, therefore the extension is seen to be proportional to the force.

5. It measures weight because the scale reading depends on the force needed to extend the rubber band. For a given mass in the pan the scale reading would be less on the moon than on the earth, because the force of gravity on the moon is less than the force of gravity on the earth.

## Chapter 8

1. $$\text{Density} = \frac{\text{mass}}{\text{volume}} = \frac{1.78 \times 10^4 \,(\text{kg})}{2 \,(\text{m}^3)} = 8900 \text{kg/m}^3$$
If $1\text{m}^3$ has a mass of 8900kg
$0.5\text{m}^3$ has a mass of $8900 \times 0.5 = 4950\text{kg/m}^3$

2. Relative density $= \dfrac{\text{density of lead}}{\text{density of water}} = \dfrac{11\ 300}{1000}$

$$= 11.3$$

## Chapter 9

1. The smaller the area of the object in contact with the plasticine the greater is the depth of penetration.

2. (a) The force on the book is equal to the weight of the cube and is therefore 0.72N. (b) 0.72N acts on an area of 4cm² and hence the pressure or force acting on 1 cm² is determined as follows:

$$\frac{\text{force in newtons}}{\text{area in square centimetres}} = \frac{0.72}{4} = 0.18\text{N/cm}^2$$

3. If the liquid had a greater density the weight of liquid supported by the bottom of the tank would be greater. This force acts on the same area and hence the pressure is greater.

4. Pressure increases with depth. The greater the height of the tank above the tap the greater is the water pressure at that tap. Therefore the water pressure at the ground-floor taps is the greater.

5. (a) The atmospheric pressure pushing down on the surface of the mercury in the bowl. (b) The height of the mercury column would be less because atmospheric pressure is less at the top of a mountain.

6. (a) Because atmospheric pressure decreases with increase of altitude, the barometer readings can therefore be converted to height readings. (b) An altimeter does not show height above ground. If the pointer is set on zero at sea-level, then the altimeter will read height above sea-level. A pilot normally sets his altimeter to zero before take-off and the meter then tells him his height above the aerodrome from which he took off. Before landing at another airport he radios that airport for the atmospheric pressure there, and resets his instrument accordingly.

7. In the absence of X, water would only issue from the pump as the piston was moved down. With the chamber X present, water enters X during the downstroke and the air in X is compressed. On the next upward stroke the air in X expands, thus maintaining the flow of water.

## Chapter 10

1. The particle would move about in a random manner as shown in the inset to Fig. 10.1. The situation is similar to that of the tennis ball being hit by the marbles. The ball moves first in one direction and then in another, depending on the direction in which a marble is travelling when it hits the ball.

2. Millions of molecules are colliding with the table-tennis ball every second, causing almost equal forces on all sides. In any short interval of time there may be a few more molecules hitting one side or another, but the ball is so massive in comparison with the forces involved that no visible movement occurs.

3. The colour would gradually spread until both gas-jars were filled with a coloured vapour.

4. The temperature will fall. It is the faster moving molecules which escape and hence the average speed of the remaining molecules is lower. A lower average speed means a lower temperature. (We will discuss this cooling of an evaporating liquid again in Chapter 23.)

5. The volume will halve because the pressure has been doubled.

6. The fact that the surface area is as small as possible

indicates that the molecules are attracting one another.

7. The detergent lowers the cohesive forces (i.e. the surface tension is reduced), hence the skin effect is reduced and the razor blade or needle sinks.

## Chapter 11

1. We have seen (page 85) that pressure acts in all directions and increases with depth. The pressure, and hence the total force, is greater on the bottom face than on the top face, and it is this difference in force on the bottom and top faces which causes the upthrust.

2. The upthrust is equal to the weight of liquid displaced.

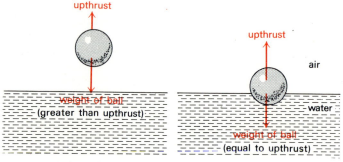

Fig. A3

3. When the ball is under the water the upthrust is greater than the pull of gravity and there is an overall (or resultant) force upwards on the ball; hence it rises (see Fig. A3). When it is in the air the weight of the ball acting downwards is greater than the upthrust due to the displaced air, and the ball falls (See Fig. A3.)

4. It will sink further in a liquid of low density. It will always sink until the weight of liquid displaced is equal to its own weight. Since the weight of liquid displaced is always the same it must displace more liquid of low density.

416

5. Since the Dead Sea contains a large amount of dissolved salt (you can see deposits of it in the photograph), its density is much greater than ordinary sea water. A considerable volume of a bather's body is still above the surface when she is displacing a weight of water equal to her own weight.

6. Weight of copper sulphate displaced = 0.24N
   Mass of copper sulphate displaced
   $$= \frac{0.24}{9.8}\text{kg} = 0.0245\text{kg} = 24.5\text{g}$$

   But this mass of copper sulphate is displaced by 20.4cm³ of stone, and therefore has a volume of 20.4cm³.
   Density of copper sulphate solution
   $$= \frac{24.5}{20.4} = 1.2\text{g/cm}^3$$

## Chapter 12

1. The professional footballer needs about twice as much energy as a typist.

2. (a) No. The force is not moving once the tissues have been compressed. You could just as easily keep him squealing by using some sort of clamp, screwing it up and leaving it in position (work would only be done while the device was being screwed up). (b) Yes. You need to apply a force to the brick to lift it, and the force moves, so work is done.

3. (a) Yes, twice as much work. This is because he does exactly the same amount of work in lifting the 1N onto the first shelf as he does in lifting it from the first shelf to the second, since the distance in each case is the same. (b) Yes, ten times as much work.

4. (a) $4 \times 2 = 8$J     (b) $2 \times 2 = 4$J

5. (a) Potential energy. The strap has energy because of its state (it is stretched and work has been done

in stretching it). (b) Kinetic energy. The ball has energy because of its motion.

6. If the work is measured in joules, and the time in seconds, then the power is in J/s.

7. Work done in 1 second $=$ force $\times$ distance
$$= 500(\text{N}) \times 1(\text{m})$$
$$= 500\text{Nm} = 500\text{J}$$

8. You can easily do the experiment, and you will find that the answer to (a) is "Yes". The force of starting friction is greater than the force of sliding friction. The answer to (c) is "No".

## Chapter 13

1. Spanner (b). The greatest turning effect is produced by the spanner with the longest handle. A simple way to verify this is to try to open a door by pushing it at different distances from the hinge. A handle on the part of the door nearest to the hinge would not work.

2. If the force is doubled the distance is halved, or the product of force multiplied by the distance is a constant.

3. (a) The only force turning the beam clockwise is the 1N and this is acting at 9cm from the fulcrum.
Moment $= 1(\text{N}) \times 9(\text{cm})$
$$= 1 (\text{N}) \times 0.09(\text{m}) = 0.09\text{Nm}$$
(b) $0.2(\text{N}) \times 20(\text{cm}) = 0.2(\text{N}) \times 0.2(\text{m})$
$$= 0.04\text{Nm}$$
(c) $0.5(\text{N}) \times 10(\text{cm}) = 0.5(\text{N}) \times 0.1(\text{m})$
$$= 0.05\text{Nm}$$
Sum of (b) and (c) $= 0.04 + 0.05 = 0.09\text{Nm}$
This is equal to the clockwise moment.

4. Taking moments about the fulcrum:
$400 \times 0.5 = $ effort $\times 1$   $\therefore$ effort $= 200\text{N}$

5. For maximum stability the centre of gravity should be as low as possible and the area of the base as large as possible. When the matchbox was opened the centre of gravity was raised and it became less stable.

## Chapter 14

1. When the effort moves down 10cm the load rises 1cm and hence the $VR$ is $\frac{10}{1} = 10$. Can you see why it is called velocity ratio? (Hint: if the effort moves at 10cm/s how fast does the load move?)

2. (a) $\dfrac{\text{load}}{\text{effort}} = \dfrac{10}{1} = 10$. (b) 10N move through 1cm (0.01m), therefore work done $= 10 \times 0.01 = 0.1\text{J}$. (c) An effort of 1N moves through 10cm (0.1m), therefore work done $= 1 \times 0.1 = 0.1\text{J}$.

3. Efficiency $= \dfrac{\text{work done on load}}{\text{work done by effort}}$
$$= \frac{10 \times 0.01}{2 \times 0.1} = \frac{0.1}{0.2} = \frac{1}{2} \times 100 = 50\%$$
The work got out of a machine can never be greater than the work put into it. Hence the efficiency can never be greater than 1.

4. The bottom pulley acts like the one in Fig. 14.3. The top pulley simply changes the direction of the force. The effort is therefore 1 divided by 2, i.e. 0.5N.

5. The effort, in addition to raising the load, has to raise the bottom pulley and also overcome friction.

6. We imagine the string to remain fixed and the bottom pulley and load to rise 1m. This leaves a total of 2m of slack (1m on each side of the pulley). The effort must therefore move 2m to take up the slack.

7. (a) $VR = \dfrac{R}{r} = \dfrac{10}{2} = 5$. (b) Less than 5 because of friction. If there were no friction the $MA$ would be 5 because $\dfrac{MA}{VR} = 1$ for the perfect machine.

8. The effort moves $d$ and the load moves $h$, therefore
$$VR = \frac{d}{h}.$$

9. $VR = 2$ or $\frac{1}{2}$. When the big wheel goes round once the little wheel completes two revolutions. Can you see why there are two answers?

10. Lever and screw (inclined plane).

## Chapter 15

1. (a) In 1 second you would swim 3m downstream and the current would take you 4m. You would therefore be travelling at 7m/s. (b) You would swim upstream 3m every second but the stream would carry you back 4m, and you would be moving backwards at 1m/s. (c) This problem is more difficult. All we can say at the moment is that your velocity will be between 1m/s and 7m/s. We shall now investigate this problem further.

2. Suppose two forces of 3 units and 4 units are used to pull a toboggan along. If both forces are applied by parallel ropes so that both forces act in the same direction, the total force is 7 units. If the ropes are tied to opposite ends of the toboggan and pulled in opposite directions then the net force (we call this the resultant force) acting on the toboggan is 1 unit (if the two forces were equal the toboggan would not move). We shall now describe an experiment to see how we can add forces when they do not act in the same straight line.

3. Rope (a) is the more likely to break. Fig. A4 shows why. In each case the weight of the man is represented by the line of length $F$ units drawn downwards. The tensions acting on each half of the rope must have a resultant equal to the weight of the man. The magnitude of $T$ in (b) to give a resultant of $F$ is much less than in (a). (If you are crossing a ravine in this way and there are crocodiles in the water below, use as long a rope as possible!)

418

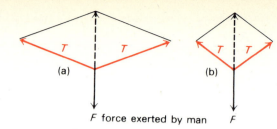

(a)　　　　　　(b)

F force exerted by man　F

**Fig. A4** *The tension in the rope (a) is greater than the tension in the rope (b)*

## Chapter 16

1. Bring each end up to one end of a suspended magnet. If attraction occurs in both cases the bar is unmagnetized, but if one end repels the suspended magnet then the bar is magnetized. Attraction is not a test for magnetism because all magnets are attracted by unmagnetized magnetic material. *Repulsion is the test for magnetism.*

2. Bring each end in turn up to the north-seeking pole of a suspended magnet. The end of the clockspring that repels the magnet is the north pole.

3. (a) As stroking continues so all the elementary magnets become aligned as in Fig. 16.6(a). A point is naturally reached when no further alignment can take place, and the magnet cannot become stronger. The magnet is said to be *saturated*. (b) Heating destroys magnetism. The extra energy of vibration of the heated atoms puts the elementary magnets out of alignment. (c) The north pole of the bar magnet attracts the south poles of the elementary magnets in the nail, and the elementary magnets become partially and temporarily aligned (the bar magnet is hardly likely to be strong enough for the alignment to be complete). (d) If the bar were simply rubbed backwards and forwards with a magnet, then as soon as the elementary magnets started to become

aligned in one direction they would be pulled back in the other direction. (e) The elementary magnets lose their alignment.

4. At A the magnet's field is much stronger than the earth's field, and the compass needle will set parallel to the magnet with its north pole pointing south (compare Fig. 16.12). At B it is very nearly out of the influence of the magnet, and will set along the earth's field with its north pole pointing north.

5. The south-seeking pole (south pole) will dip downwards in the southern hemisphere.

# Chapter 17

1. The atoms of the strip contain a positively charged nucleus surrounded by negatively charged electrons. When the strip is rubbed, some of the electrons that are furthest away from the nucleus (and therefore not strongly attracted by the nucleus) pass onto the duster. The duster is now negatively charged. The strip, on the other hand, has lost electrons and this means that some of the protons in the nuclei are no longer neutralized by negative charge, therefore the strip now has a positive charge.

2. (a) X repels the acetate strip and therefore has the same kind of charge as the acetate strip, i.e. it is positively charged. (b) You should be surprised! No material like Y has ever been discovered. If a material like Y existed it would mean that there also existed a third kind of charge which was neither positive nor negative but something different.

3. Free electrons on the cap are repelled down to the leaf and rod. The leaf and rod are now negatively charged and repel each other.

4. The electroscope is negatively charged. When the comb is brought up to the cap the greater divergence of the leaf must mean that more negative charge has been *repelled* from the cap to the leaf and rod. Hence the comb must be negatively charged.

5. Since there is an attractive force, the charge on the top of the paper must be unlike the charge on the comb. The charge on the top of the paper is positive.

# Chapter 18

1. When the terminals are connected a current of electrons will flow round the circuit. All the electrons on the negative terminal are repelling each other, and the electrons are attracted by the positive charge on the positive terminal. When the circuit is complete the electrons will be repelled away from the negative terminal and move towards the positive terminal.

2. The current through each ammeter will be 0.5A. It is a common mistake to suppose that because the electrons leave the negative terminal and flow to the positive terminal, that the current through the ammeter connected to the positive terminal is less then the current through the ammeter connected to the negative terminal. The current entering a bulb is the same as the current leaving it.

3. Yes. The volume of water passing every point in the water circuit must be the same. If more water flowed into a radiator than flowed out of it, then the volume of water in the radiator would be continually increasing and the radiator would eventually burst.

4. (a) 3C, (b) 4C.

5. Each bulb is connected across the battery, and since the bulbs are identical the same current must flow in each. We know that the same current

leaves a bulb as enters it so $I_1 = I_2 = I_3 = I_4 = 0.2A$. $I_5 = 0.4A$. It may help if you imagine the wires as roads that are filled with cars travelling bumper to bumper. Suppose four cars leave the negative terminal of the battery every second. At the first junction two cars turn left every second and two go straight on. Along the bulb parts of the road two cars are passing every point per second. At the far junction four cars every second arrive at the junction so four cars per second return to the battery. In this analogy, the number of cars per second corresponds to the number of coulombs per second (C/s).

6. Each battery will give a coulomb 3 joules of energy, and hence two batteries will give a coulomb 6 joules of energy. The voltmeter will read 6V.

7. Using equation 1 we have:

$$\text{Resistance} = \frac{10V}{2A} = 5\Omega$$

8. The wire and the copper sulphate. These are the only conductors showing graphs with a straight line through the origin.

9. (a) $1kW = 1000W = 1000J/s$, (b) $1000J/s$ for $(60 \times 60)s$ is $(1000 \times 60 \times 60)J = 3\,600\,000J$. A joule is a very small unit of energy.

10. Total power consumption $= \dfrac{5 \times 60}{1000}kW$

Energy consumed $= \dfrac{5 \times 60}{1000} \times 10kWh = 3kWh$

Cost $= 3 \times 3 = 9p$

## Chapter 19

1. Imagine your eye at E looking along the wire and the current flowing away from you. The lines of flux go round in the direction of the hands of a

clock. (a) The north pole underneath will therefore move towards you, the reader, out of the page. (b) If the north pole were above the wire it would move away from you, into the page.

2. Reference to Chapter 16, for example to Fig. 16.12, shows that the lines of force leave the north pole of the bar magnet and go to the south pole. In Fig. 19.4 the lines of flux leave the end A. The end A is therefore a north pole, and the north pole of a compass will be repelled by it.

3. Look back at Fig. 16.7. When a magnetic material is in a magnetic field the domains set along the lines of flux. When the current is flowing in the solenoid there is a strong magnetic field and the domains set along the lines of flux. The substance will become magnetized. In Fig. 19.4 the end nearest to A would be a north pole.

4. As the magnet is withdrawn the domains will be urged first one way and then the other by a force which is gradually growing weaker. When the magnet is some way from the solenoid it will be demagnetized. This is a very good way to demagnetize a watch that has become magnetized.

## Chapter 20

1. The copper goes to the *negative* cathode. Since unlike charges attract, the copper travelling to the cathode is positively charged.

2. (a) The positive hydrogen ions will travel towards the negative cathode. (b) Hydrogen ions arriving at the cathode gain an electron and become hydrogen atoms. Two hydrogen atoms join together to form a hydrogen molecule. The chemical equation showing the change may be written thus:

$$H^+ + H^+ + 2 \text{ electrons} = H_2$$

This is what takes place at the cathode during the electrolysis of water. At the anode hydroxyl ions ($OH^-$) arrive and the reaction is:

$$4OH^- - (4 \text{ electrons}) = 2H_2O + O_2$$

# Chapter 21

1.  (a) The easiest way to pass the parcel is for A to hand it to the nearest person to him, this person then hands it on to the next person, and so on until the parcel reaches B. This is an illustration of heat transfer by conduction. (b) A moves along the carriage and hands the parcel to B. This is an illustration of heat transfer by convection. (c) A can throw the parcel over the roped-off area to B. In this case the parcel is travelling in a different manner, which can be compared to heat transfer by radiation. The illustration would be a better one if A were a magician, and he could make the parcel disappear from his hands and reappear in the hands of B, the movement of the parcel being invisible. Heat transmitted as radiation becomes heat energy when it is absorbed by any object in its path.

2.  (a) Conduction (heat is conducted from the plate to the base of saucepan, through which it is conducted to the contents). (b) The red-hot element gives off radiant heat which is reflected by the polished back plate. (c) Heat escapes in the form of convection currents. (d) Hot air is convected up and away from the radiator. This often leads to the blackening of a wall behind a radiator, because dust carried by the air is deposited on the wall.

3.  A metal surface feels much colder than a wood surface. This is because metal is a good conductor of heat and wood is a bad conductor of heat.

4.  Copper is a good conductor, and ensures that the rods are heated evenly by the heat source.

5.  (a) Yes. Worn inside out, the fur provides a lining of trapped air which is a poor conductor of heat, therefore the heat of the body is not lost. A fur-lined coat is warmer than a fur coat, though perhaps not as glamorous. (b) Snow contains many pockets of air, which in fact insulate the plants from the cold.

6.  (a) The fire would go out because it would not be able to draw in currents of air containing oxygen. (b) The people in the room would become drowsy, because the process of breathing uses up oxygen and produces carbon dioxide. No fresh supplies of oxygen would enter the room and the amount of carbon dioxide would increase. This situation could become dangerous, because without a draught poisonous fumes from the fire could enter the room instead of leaving by the chimney or flue. This is why ventilation ducts are essential for all forms of heating systems.

7.  (a) Hot water rises. To obtain the correct circulation of hot water in the system, the hot cylinder must be higher than the boiler. (b) The water that has been heated rises to the top of the boiler. It would be inefficient to cool this down again by drawing in cold water at the top of the cylinder. Cold water must enter the bottom of the boiler where it is heated directly, and the correct circulation within the boiler is maintained.

8.  (a) The surfaces of petrol tanks in hot countries are polished shiny surfaces. They reflect heat so that little heat is absorbed by the petrol — a necessary precaution if it is not to ignite. (b) Fallen leaves are dark surfaces, thus they absorb heat and the snow underneath is melted because of the rise in temperature. The white snow tends not to absorb much heat, and the melting is therefore slower in exposed areas.

9.  Radiators give out their heat mainly by convection.

Thus their colour does not affect their heat transmission very much and many people prefer light colours to match their walls and woodwork.

10. Clouds prevent the loss of radiant heat from the earth's surface. Thus when they are not present there is a rapid loss of heat and a large drop in temperature, and frosts are likely to occur.

11. (a) Glass is a bad conductor, therefore the glass walls prevent heat conduction; (b) the vacuum prevents any conduction or convection of heat; (c) the inside silver surface is a bad radiator. Radiant heat is reflected back from the outside silver surface.

12. The statement is correct in saying that "it keeps the heat in" because the material prevents a flow of heat across it. However, saying that "it keeps the cold out" suggests that cold can travel like heat. This is not so. Heat is a form of energy able to be transmitted. Cold is an absence of heat energy. In the same way, darkness is an absence of light energy — darkness does not come, it is light which goes.

## Chapter 22

1. (a) The bridge would buckle out of shape on a hot day if it were fixed at both ends. There would be no room for expansion to take place. (b) The weights on the cables would move (i) downwards on a hot day, taking up the extra length of the cables formed by expansion, (ii) upwards on a cold day as the cables contract. Thus, by the movement of the weights, the correct tension of the cables is maintained throughout the changes in temperature. (Correct tension is essential for the running of the locomotives. This can now be regulated by the use of a gas hydraulic tensioning unit.)

2. The shelves inside a cooker fit only very loosely, thus there is room for them to expand and yet still be easily removed from a hot oven.

3. The cast iron pin should be put through a hole on the other side of end A. Then when the bar expands the pin is forced against the end of the apparatus and eventually snaps.

4. (a) Milk bottles are made from thick glass. Glass is a bad conductor of heat. The inner surface of the bottle expands when heated by the boiling water, but the outer surface remains the same size. The stresses that are set up cause cracking. (b) This is again caused by unequal expansion. Paper is a bad conductor of heat, therefore the surface nearer the heat expands more than the inner surface, and the cover curls.

5. The aluminium expanded by the greater amount, and because of this it forms the outside of the curve. When the bottle cap was put in hot water the metal expanded more than the glass and so the cap came off easily.

6. When the bimetallic strip is heated the brass expands more than the iron. Thus the strip curves over to the left (as viewed in the diagram) and the ratchet also moves in this direction. The movement of the ratchet rotates the cog wheel and the pointer in a clockwise direction. The pointer indicates the rise in temperature. Such thermometers may be seen on top of the radiators of cars made in the 1930s, and on mantelpieces in many homes.

7. (a) The flask, being in direct contact with the heat, expands before the water. Thus the level of the water in the flask drops. (b) The experiment indicates that the expansion of liquids is greater than that of solids for the same temperature rise. If the solid (i.e. the flask) had expanded more than the liquid, the level of the liquid in the flask would have fallen.

8. (a) The level of the liquid in the thermometer rises, but the index is left behind. (b) The level of the liquid in the thermometer falls, dragging the index with it, so that the position of the index records the minimum temperature. The top of the index marks the minimum temperature.

9. (a) About 1500°C (accurately this is 1535°C). (b) Solder begins to set at 220°C, but it remains "pasty" until it finally sets at 180°C.

10. Underneath the ice there is water. This indicates that ice forms on the top of a pond and gradually increases in thickness downwards. At first, however, the layer of ice is very thin.

11. (a) If the micrometer screw is kept in contact for too long, it too will be heated and undergo expansion. Thus it will not give accurate readings. (b) When the reading is constant this indicates that the metal rod has undergone its maximum expansion for this rise in temperature. It will expand no more.

12. Lowest temperature on the moon $= -150°C$
Highest temperature on the moon $= 100°C$
Therefore rise in temperature $= 250°C$
Coefficient of expansion for steel $= 0.000\,012/°C$
1m of steel heated through 1°C (1K) expands by 0.000 012m. Therefore 150m of steel heated through 250°C expands by $0.000\,012 \times 150 \times 250 = 0.450$m or 45.0cm.

## Chapter 23

1. (a) The ice changes to water when it melts in the warmer orange squash. (b) The steam changes to water when it cools and condenses on the cooler surface of the dish.

2. (a) The horizontal portion of the graph represents the period when the naphthalene is undergoing no change in temperature, although it is still being supplied with heat. Thus the napthalene must be changing state at this point, i.e. melting. (b) The melting point of napthalene is 80°C (353K).

3. (a) When ice melts it needs to obtain its latent heat of fusion (the heat needed to change it from solid to liquid). It does this by removing heat from its surroundings, and the drink around the ice is cooled. (b) When steam condenses it gives up its latent heat of vaporization (the heat needed to change it from water to steam) to the surroundings. Thus the dish and its contents, on which the steam condenses, are heated.

4. When boiling water comes into contact with the skin it cools to body temperature and gives up heat that causes burning. When steam comes into contact with the skin, it condenses and gives up its latent heat of vaporization; it then cools to body temperature, thereby delivering two massive quantities of heat to the skin and causing a much more serious burn. The calculation of actual quantities of heat delivered is considered in Chapter 24.

5. "Spirit" (a form of alcohol) is a volatile liquid and therefore it evaporates at a quicker rate than water when under the same conditions. Thus a spirit-based pen provides a faster drying ink than a water-based one.

6. A warm, dry and windy day (c) is the ideal drying day.

7. Because the man has completely surrounded the refrigerator he prevents the circulation of the cooling air around the condensing coils. The condensing unit is therefore unable to transfer its heat away and the liquefaction of the gas is impaired. The danger is that the refrigerator may overheat.

8. (a) A fog is formed of droplets of condensed water vapour which have formed around particles of dust and dirt in the air. (b) When there are people

423

in the car, the air becomes saturated with water vapour from their breathing. The windows are cold because of the lower temperature outside the car, and the water vapour condenses on these colder surfaces.

9. The boys have pressed the snowflakes together between their hands. They have subjected the snow to pressure.

10. Copper is a good conductor of heat, string is a bad conductor of heat. Thus the string cannot transmit the latent heat of fusion needed to melt the ice beneath it.

11. The pond should have sloping sides. When water freezes, it expands. If the sides of the pond are straight then all the pressure of the ice caused by this expansion is directed outwards onto the sides. Thus the sides will eventually crack. If the sides are sloping the ice can "slide" up them as it expands and does not exert so much direct pressure.

12. At high altitudes the air pressure is lower than at sea-level and water boils at a temperature below 100°C (373K). Food, therefore, on the whole cooks much more slowly, and some foods cannot be cooked properly at temperatures below 100°C (373K).

## Chapter 24

1. A lagging of felt or any similar material acts as an insulator. Thus the loss of heat energy from the container to the surroundings is reduced. As heat loss is not taken into account in the calculation, trying to eliminate it will naturally make the result more accurate.

2. The specific heat capacity of water is 4200J/kgK
   ,, ,, ,, ,, paraffin is 2100J/kgK
   ,, ,, ,, ,, mercury is 140J/kgK
   If identical masses of these substances are given the same quantity of heat then the temperature rises that each goes through will be directly related in the same way as are their specific heat capacities. Thus 1kg of paraffin needs 2100J to raise its temperature through 1K, but for the same temperature rise 1kg of water needs 4200J. Therefore, given the same heat energy:

   the temperature rise for paraffin will be double that of water,

   the temperature rise for mercury will be 30 times that of water.

3. (b) 12 600J   (c)   16 800J
   (e) 12 600J   (f)   16 800J
   (h) 50 400J   (i)   210 000J

4. (a)  Mass of copper = 2kg

   Specific heat capacity for copper = 380J/kg K

   Temperature rise = 308 − 283K = 25K (25°C)

   Quantity of heat energy $= (mcT)_{copper}$
   $$= 2 \times 380 \times 25J$$
   $$= 19\ 000J$$

   (b)  Mass of water = 1kg

   Specific heat capacity of water = 4200J/kg K

   Initial temperature = 283K (10°C)

   Quantity of heat supplied by heater per second
   $$= 2100J$$

   Quantity of heat supplied by heater in 1 minute
   $$= 210 \times 60J = 126\ 000J$$

   Heat supplied by heater = heat absorbed by water
   $$= (mcT)_{water}$$

   $126\ 000 \qquad\qquad = 1 \times 4200 \times (T - 283)$
   $$= 4200T - (4200 \times 283)$$
   $126\ 000 + 1\ 188\ 600 = 4200T$
   $$T = \frac{1\ 314\ 600}{4200}K$$
   $$= 313K\ (40°C)$$

   Final temperature will be 313K (40°C)

5.  (a)  Mass of turpentine = 0.5kg

Rise in temperature = 15K (15°C)

Heat supplied by heater = $1.35 \times 10^4$J

Heat supplied by heater = heat absorbed by turpentine

$$1.35 \times 10^4 = 0.5 \times c \times 15 = (mcT)_{\text{turpentine}}$$

Therefore $c = \dfrac{1.35 \times 10^4}{0.5 \times 15}$

$= 1800$J/kg K

(b)  Mass of cylinder = 1kg

Rise in temperature = 40K

Power of heater = 60W = 60J/s

Time = 9 minutes 20 seconds = 560s

Electrical energy = heat energy supplied by heater = $60 \times 560$J

Heat energy supplied by heater = heat energy absorbed by cylinder

$VIt = (mcT)_{\text{cylinder}}$

$60 \times 560 = 1 \times c \times 40$

$$c = \frac{60 \times 560}{40} \text{ J/kg K}$$

$= 840$J/kg K

The specific heat capacity of the material making up the cylinder is 840J/kg K. This value is the specific heat capacity of aluminium.

6.  Water has a very high specific heat capacity, therefore as well as requiring a large quantity of heat to raise its temperature by one degree, it will also give out a similar quantity of heat on cooling through one degree. Clearly, it will give out a larger quantity of heat than substances with smaller specific heat capacities when both are cooled through the same temperature range. Thus, a warming pan full of water would give off more heat than one containing an equal mass of cinders at the same temperature, and would heat the bed more efficiently.

7.  The thaw would take place very slowly because on a dull day there would not be a great quantity of heat available to melt the snow.

8.  When 1g of boiling water comes into contact with the skin it cools down to body temperature.

For each kelvin that the 1 gramme of water cools through it gives out $\frac{4200}{1000}$J of heat. Thus on cooling from 373K (100°C) to 309.8K (36.8°C) it gives out $4.2 \times 63.2$J of heat (309.8K is the temperature of the body). When 1g of steam comes into contact with the skin it condenses to water at 373K (100°C) and gives out 2260J of heat. Further to this, it cools from 373K (100°C) to 309.8K (36.8°C) and gives out $4.2 \times 63.2$J of heat.

Thus 1g of steam gives out more heat (2260J) to the skin than 1g of boiling water.

## Chapter 25

1.  Group B will probably complete the instructions first. This is because it is less easy to create order from chaos than the reverse procedure.

2.  The crankshaft converts the up-and-down motion of the piston into a rotary motion.

3.  Four cylinders are used because they give a smoother motion and more power than a single cylinder.

4.  In a 2-stroke engine (a) the ports are opened and closed by the piston, (b) the gas is compressed and forced into the cylinder above the piston (in a 4-stroke engine the atmospheric pressure is sufficient).

5.  There is no sparking plug in a diesel engine. The ignition in this engine is achieved by compression.

6.  (a) There are fewer moving parts, (b) the assembly is therefore easier and quicker, (c) there is less wear on the moving parts, (d) the engine is more compact.

7. The gate will open inwards into the field as the boy jumps off. It is the force backwards resulting from the boy's movement forward which causes this. (See also Chapter 31, page 390.)

8. A rocket carries its own supply of oxygen in the form of liquid oxygen contained in fuel tanks.


## Chapter 26

1. In (a) the right-hand end of coil A has electricity flowing through it in an anti-clockwise direction, thus it becomes a temporary north pole. The left-hand end of coil B has electricity flowing through it in a clockwise direction. This end therefore becomes a temporary south pole and attracts the north pole of coil A.

   In (b) the two facing ends of coils A and B now both have electricity flowing through them in an anti-clockwise direction. Thus both ends become temporary north poles which repel one another.

2. (a) Oersted was the name of the scientist. (b) His experiment revealed that there is a magnetic flux around a conductor carrying an electric current.

3. (a) Because this instrument measures milliamperes it is called a milliammeter.
   (b) 10 divisions is equivalent to 1     milliampere
       1 division is         ,,        0.1        ,,
       2 divisions is       ,,        0.2       ,,
   (c) If a.c. is supplied to a moving coil ammeter then the instrument gives no deflection.

   An alternating current is one in which the electrons flow first in one direction and then the other. (See Chapter 27.)

   With such a current flowing through the galvanometer the direction of the magnetic flux of the coil is changing continually. Thus the coil may be simply thought of as "not knowing which way to turn" and the resultant force is zero. (If the current is alternating very slowly the coil will follow the oscillations.)

4. (a) No matter in what direction the current flows through the repulsion-type ammeter, the two iron rods still become induced magnets with like poles together, and a force of repulsion exists between them. In the attraction-type ammeter the coil still acts as a magnet thereby drawing the soft iron rod towards it. (b) An alternating current is one in which the current reverses many times per second. We have explained that the meters give a deflection no matter in what direction the current is flowing. Thus moving-iron meters can be used to measure alternating current.


## Chapter 27

1. A galvanometer with a centre zero would be the correct instrument to use in detecting the electric current in coil B.

2. If the wire is moved along the line AB, this is horizontal and parallel to the magnetic flux between the two poles. Thus the conductor does not cut the magnetic flux, therefore no e.m.f. is induced within the conductor.

3. The deflection of the galvanometer needle will be to the left. The current will flow in such a direction as to produce a temporary south pole at the end of the coil opposing the movement of the magnet's south pole towards it.

4. In the horizontal position the coil is cutting the magnetic flux at the greatest rate.

5. When the dynamo is speeded up the frequency of the a.c. will be increased.

6. The deflection of the needle will be to the right. The induced current flows in this direction in the secondary coil in order to produce a temporary north pole at the end of the coil facing the north pole of the primary coil. Thus the current obeys

Lenz's law in attempting to oppose the change that induces it.

## Chapter 28

1. A current flows in the red circuit when the metal plate is positive and the filament negative. This would occur if negative charge given off by the filament were attracted by positive charge on the metal plate (unlike charges attract). When the metal plate is negative no current flows, so no positive charge is being attracted to it from the filament. Statement (b) is therefore the correct one.

2. (a) The emission from the heated filament travels in straight lines. (b) The fact that it is deflected by a magnetic field shows that the emission carries charge.

3. Upwards.

4. It takes no current. A moving-coil voltmeter takes a current and hence the p.d. being measured is reduced.

5. Electrons are emitted from the zinc when ultra-violet light is shone onto it. When the electroscope is negatively charged the electrons emitted are repelled away, but when the electroscope is positively charged the emitted electrons are immediately attracted back to the zinc.

## Chapter 29

1. It is difficult because you have to move the page with your fingers and every movement makes a sound.

2. (a) Close a door, (b) rub hands together, (c) pluck a rubber band, (d) blow a whistle.
   The words are: bang, rub, pluck, blow.

3. The vibrations cease as the hands absorb the energy of movement. The sound therefore also dies away.

4. The particles vibrate up and down, that is at right angles to the direction of travel of the wave.

5. It rests on the ribs of the chest and the sound is conducted through these. The stethoscope picks up vibrations and concentrates them.

6. On the moon there is no atmosphere, therefore no sounds can be transmitted. The astronaut would be unable to hear any sounds behind him.

7. It is easier to hear sounds at night because the layers of air immediately next to the rapidly cooling ground are colder than the air above them. Sounds travelling from cold to warmer layers of air are refracted at each layer and bent generally downwards, therefore sounds that would normally escape upwards and not be heard are directed downwards to the ground and heard very clearly.

8. Your four instruments should be listed here:
   Vibrating strings (string instruments): violin, double bass, cello.
   Vibrating air column (wind instruments): tuba, French horn, trumpet, trombone, clarinet, bassoon, flute, oboe.

9. The amplitude gets less as the energy and sound die away.

10. A sounding box full of air serves to amplify the sound produced by the vibrating string, because vibrations are set up in this air.

11. The natural reaction is to cup your hands around your mouth and to shout through them in the direction required.

12. Speed of sound in water $= 1450$m/s
    Time for sound to travel to sea-bed and back $= 40$s

    $$\text{Depth of sea bed} = \frac{\text{speed of sound} \times \text{time}}{2}$$
    $$,, \quad ,, \quad ,, \quad ,, = \frac{1450 \times 40}{2}$$
    $$= 29\,000\text{m}$$

13. Colds cause a build-up of fluid in the Eustachian tube, thus the pressure on either side of the ear

427

drum cannot be equalized. This makes the ear drum less sensitive and therefore hardness of hearing results. A sensation of popping of the ears may also be felt at these times.

14. People act as sound absorbers, therefore the sound-energy level falls.
15. The sound from the engines never catches up with the aircraft cabin because the aircraft is travelling faster than its own sound waves.
16. Electrical energy.

## Chapter 30

1. The holder contains a small piece of film wrapped in black paper. When this is developed it indicates the amount of radioactivity to which the wearer has been exposed. Thus a check is kept on the safety of laboratory workers.
2. The vapour trails show that the radioactivity may be emitted at any time, in any direction. Thus radioactivity is a random process and not a regular one.
3. The irregularity of the sparks reinforces our belief that radioactivity is a random process.
4. The Geiger-Müller tube must be directed towards the board and moved up and down so that it scans the whole area of the map. A greater number of clicks per second will be heard when the Geiger counter nears the position of the radioactive source. The exact position of the source is located at that point where the number of clicks is greatest.
5. (a) 1 curie gives out $3.7 \times 10^{10}$ emissions per second

   1 millicurie (mC) = one-thousandth of a curie $(10^{-3})$

   Thus 1 millicurie gives out $3.7 \times 10^{7}$ emissions per second.

(b) 1 microcurie ($\mu$C) = one-millionth of a curie $(10^{-6})$

Thus 1 microcurie gives out $3.7 \times 10^{4}$ emissions per second.

6. Uranium has 92 protons.

   Mass number of uranium-235 = 235 = number of protons + number of neutrons

   Number of neutrons in uranium-235 = 235 − 92
   $$= 143$$

   Mass number of uranium-238 = 238

   Number of neutrons in uranium-238 = 238 − 92
   $$= 146$$

7. Until this date, $\alpha$-particles, or protons, had been used to split the nucleus of the atom. But as the nucleus of an atom is positively charged, it tended to repel the positively charged particles. To overcome this the particles had to be accelerated to extremely high velocities. A neutron, however, has no electrical charge, thus the nucleus of the atom does not repel it. Therefore there is a greater chance of "scoring hits" with neutrons.

## Chapter 31

1. 30mph (48km/h) appears in built-up, residential areas.
2. (a) (i) From A to B the speed of the car is constant, so is the velocity as its direction does not change.
   (ii) B to C the speed of the car is constant but its direction is changing and so its velocity is changing.

   (b) As direction changes during the journey, the driver should say his average speed is 20mph. Indeed his speed remained constant at 20mph but his velocity changed, because his direction changed.
3. (i) In (a) the body is stationary (the distance does not change as time passes). In (b) and (c) the bodies are moving with uniform velocities. The

different slopes of the graph lines indicate that the velocities are different (graph scales are the same). The greater the incline of the graph line, the greater is the velocity.

(ii) In (d) The body starts moving forward with uniform velocity. It then stops and stays still before moving backwards with uniform velocity.

In (e) The curve of the line tells us that the velocity is not uniform. The body's velocity is decreasing between the points O and A. At A the velocity is zero. Between the points A and B the velocity is increasing.

4. In (a) The body is moving with a velocity which is changing by equal amounts in equal time intervals. Thus its acceleration is constant.

In (b) The body's velocity is decreasing regularly, thus its retardation is constant. At A it has no velocity, from then on it accelerates uniformly. Notice that after the point A the velocity has become negative (i.e. it has changed direction).

5. In outer space the gravitational attraction of the earth is very small. Thus until the spacecraft comes under the influence of another gravitational field, or another force, it will continue moving in a straight line and with uniform velocity.

6. A change from the driveway is what is required! The boys should find a hill (but well away from traffic). The force applied to the trolley under these conditions is a reduced gravitational force. Such a force gets larger as the slope of the hill gets steeper. See Fig. A5.

**Fig. A5**

**Chapter 32**

1. A. Line the loft with insulation.
   B. Fill the cavity with suitable insulation, foam or small polystyrene balls.
   C. Double glaze the windows.
   D. Lag the hot water cylinder with an insulating jacket.
   E. Turn the thermostat down!
   F. Switch off lights and fires not being used.

**Photographs** (Fig. nos)

**ACKNOWLEDGMENTS**

Ambrotel Films London  *1.4*
W. Ashhurst Esq  *19.5*
Argonne National Laboratory USA  *30.2*
Associated Press Ltd  *31.3*
Avon Rubber Co. Ltd.  *12.2*
Beken of Cowes  *13.21*
Peter Benison Ltd  *32.15*
Bernsen's International Press Service Ltd  *32.4*
Bradville Ltd  *A1*
British Aircraft Corporation Ltd  *20.7, 21.28, 25.24*
British Broadcasting Corporation  *6.9, 6.10, 29.1, 29.11, 29.23, 31.10*
British Gas Corporation  *32.7*
British Insulated Callender Cables Ltd  *18.5*
BMC Ltd  *12.9(b)*
British Railways  *22.8, 22.11, 25.25*
Central Press Photos Ltd  *17.14, p. 329*
R. V. Coleman and G. G. Scott, American Institute of Physics  *16.10*
Colombia Pictures Corporation Ltd  *2.10, 4.1*
Council of Industrial Design  *13.22*
Crompton Leyland Electricars Ltd  *p. 293*
G. Cussons Ltd  *25.3*
Dupre Vermiculite Ltd  *21.10*
Electric Power Storage Ltd  *20.6*
English Electric-AEI Machines Ltd (Small Industrial Motors Div.)  *26.1, 27.26*
Expandite Ltd  *22.7*
The Fairey Co. Ltd  *12.11*
Editor of *Flight Deck*  *12.5*
Ford Motor Co. Ltd  *25.21*
French Embassy  *32.12*
Greater London Council  *29.19*
Griffin & George Ltd  *7.11, 27.10(a)*
Guide Dogs for the Blind Association  *1.1*

Thos & Jas Harrison Ltd  *14.6*
H.M. Stationery Office  *32.1*
Michael Holford, London  *5.4*
Iliffe Specialist Publications Ltd  *3.6*
Irwin & Partners Ltd  *24.7*
Israel Government Tourist Service  *11.7*
Jaguar Cars Ltd  *12.9(a)*
Japanese Cameras Ltd  *6.8, 6.11, 6.12, 6.15*
Prof. B. V. Jayawant & Research Team, Wolfson, Magnetic Suspension Project, University of Sussex  *32.5*
Keystone Press Ltd  *17.13*
King's Lynn Glass Ltd, CoID photograph  *13.22*
Lancashire County Council  *32.13*
Leybold Heraeus GMBH & Co.  *10.9*
Liverpool Daily Post & Echo  *32.6*
London Express News & Feature Services  *29.22*
McCann–Erickson Advertising Ltd  *p 71*
Marconi-Elliott Microelectronics Ltd  *18.1(b)*
G. Maunsell & Partners  *22.2*
Metro Goldwyn Meyer  *p 1*
Minnesota Mining & Manufacturing (3M) Co Ltd  *2.4*
Erwin W. Müller  *10.8*
John Murray Ltd  *19.5*
NASA  *31.13*
James Neill & Co. (Sheffield) Ltd  *16.1*
W. B. Nicolson Ltd  *30.5*
Northrop Corporation USA  *7.5*
Norwegian National Tourist Office  *14.11*
Reyrolle Parsons Ltd  *25.27, 27.14*
Philips Electrical Ltd  *5.13*
Photain Controls Ltd  *28.12*
Pilkington Bros Ltd  *3.1, 3.2, 3.4, 21.1(a)*
The Plessey Co Ltd (Plessey Marine)  *29.14, 29.15*
Pye Group Ltd  *18.1*
*Radio Times* Hulton Picture Library  *20.3(b)*
Raleigh Industries Ltd  *32.3*
Rank Organization (Bush Murphy Div.)  *4.18*

*Fig. 30.11* was first reproduced in *Radiation from Radioactive Substances*, Rutherford, Chadwick & Ellis (Cambridge University Press, 1930).

**Diagrams**

# INDEX